Wiley-IS&T Series in Imaging Science and Technology

Panoramic Imaging

Wiley-IS&T Series in Imaging Science and Technology

Series Editor:
Michael A. Kriss

Consultant Editors:
Anthony C. Lowe
Lindsay W. MacDonald
Yoichi Miyake

Reproduction of Colour (6th Edition)
R. W. G. Hunt

Color Appearance Models (2nd Edition)
Mark D. Fairchild

Colorimetry: Fundamentals and Applications
Noboru Ohta and Alan R. Robertson

Color Constancy
Marc Ebner

Color Gamut Mapping
Ján Morovič

Panoramic Imaging: Sensor-Line Cameras and Laser Range-Finders
Fay Huang, Reinhard Klette and Karsten Scheibe

Published in Association with the Society for Imaging
Science and Technology

Panoramic Imaging

Sensor-Line Cameras and Laser Range-Finders

Fay Huang

Institute of Computer Science and Information Engineering, National Ilan University, Taiwan

Reinhard Klette

Department of Computer Science, The University of Auckland, New Zealand

Karsten Scheibe

German Aerospace Center (DLR), Germany

A John Wiley and Sons, Ltd, Publication

Registered office
John Wiley & Sons Ltd, The Atrium, Southern Gate, Chichester, West Sussex,
PO19 8SQ, United Kingdom

For details of our global editorial offices, for customer services and for information
about how to apply for permission to reuse the copyright material in this book
please see our website at www.wiley.com.

Library of Congress Cataloging-in-Publication Data

Huang, Fay.
   Panoramic imaging : sensor-line cameras and laser range-finders / Fay Huang,
   Reinhard Klette, Karsten Scheibe.
      p.   cm. — (The Wiley-IS&T series in imaging science and technology)
   Includes bibliographical references and index.
   ISBN 978-0-470-06065-0 (cloth)
   1. Photography, Panoramic.   2. Remote sensing.   3. Three-dimensional imaging.
   I. Klette, Reinhard.   II. Scheibe, Karsten.   III. Title.
   TR661.H83 2008
   778.3′6—dc22

                                                                    2008036181

A catalogue record for this book is available from the British Library.

ISBN 978-0-470-06065-0

Set in 10/12pt Times by Integra Software Services Pvt. Ltd. Pondicherry, India
Printed in Great Britain by CPI Antony Rowe, Chippenham, Wiltshire

Contents

4 Epipolar Geometry — **81**
4.1 General Epipolar Curve Equation — 81
4.2 Constrained Poses of Cameras — 89
   4.2.1 Leveled Panoramic Pair — 89
   4.2.2 Co-axial Panoramic Pair — 95
   4.2.3 Symmetric Panoramic Pair — 96
4.3 Exercises — 97
4.4 Further Reading — 98

5 Sensor Calibration — **99**
5.1 Basics — 99
   5.1.1 Camera Calibration — 99
   5.1.2 Extrinsic and Intrinsic Parameters — 101
   5.1.3 Registration and Calibration — 102
5.2 Preprocesses for a Rotating Sensor-Line Camera — 103
   5.2.1 Precalibration — 103
   5.2.2 Correction of Color Shift — 106
   5.2.3 Radiometric Corrections — 110
   5.2.4 Geometric Corrections — 110
   5.2.5 Correction of Mechanical Vibrations — 112
5.3 A Least-Square Error Optimization Calibration Procedure — 113
   5.3.1 Collinearity Equations — 113
   5.3.2 Difference between Planar Capturing Surface and Panoramic Cylinder — 115
   5.3.3 Parameters and Objective Functions — 116
   5.3.4 General Error Criterion — 117
   5.3.5 Discussion — 119
5.4 Geometric Dependencies of R and ω — 120
   5.4.1 Three Methods — 120
   5.4.2 Focal Length and Central Row — 121
   5.4.3 Point-based Approach — 124
   5.4.4 Image Correspondence Approach — 125
   5.4.5 Parallel-line-based Approach — 129
   5.4.6 Experimental Results — 135
5.5 Error Components in LRF Data — 138
   5.5.1 LRF Used in Experiments — 138
   5.5.2 Error Measurement — 138
5.6 Exercises — 143
5.7 Further Reading — 144

6 Spatial Sampling — **145**
6.1 Stereo Panoramas — 145
6.2 Sampling Structure — 145
   6.2.1 Outward Case — 146
   6.2.2 Inward Case — 149

Preface

I find panoramic vision fascinating. Imagine, if we humans had panoramic vision, we would not have words for in front and back (only relative to something else). As a result, we wouldn't use metaphors based on them, like "this is a backward society" or "this is the frontier of science". Perhaps, then, we would be thinking differently. . . . Change eyes, change thinking. . . (Yiannis Aloimonos)

This book is about accurate panoramic imaging in the context of close-range photogrammetry, computer vision, computer graphics, or multimedia imaging, for applications where (very) high-resolution data need to be recorded for documenting or visualizing three-dimensional (3D) scenes or objects. For example, large-screen virtual reality systems often use panoramic images for various types of virtual tours.

One of the deplorable consequences of computers being used for image processing is that many grossly distorted pictures are provided on the Internet, and some have even been published in books and journals. Many landscape photographs get ignorantly stretched horizontally, to produce fake panoramic views. This book addresses high-quality panoramic imaging rather than panoramic imaging as already popular (at various levels of quality) for non-professional digital cameras, typically using some "stitching software" for combining a set of images into a larger image.

Sensor-line cameras have been designed since the 1980*s*, originally for "push-broom" viewing from an aircraft or satellite. (The book briefly addresses recent applications of airborne sensor-line cameras in Chapter 10.) A sensor-line camera has basically just a line of CCD or CMOS pixel sensors, and records images line by line. Such a camera may appear "somehow" simpler with respect to its hardware design than a sensor-matrix camera, where images are captured by a rectangular array of pixel sensors. However, having only one image line at a time defines particular difficulties for creating a "consistent" two-dimensional (2D) image.

Panoramic images may be captured by a sensor-line camera through rotation. In the case of a free-hand rotation, subsequent geometric rectification of recorded image lines (e.g., on a cylindrical surface) would be *really* difficult. In this book we basically assume that rotation is accurately performed by using a computer-controlled turntable. During such a rotation, a panoramic image is captured on a cylindrical surface (in contrast to a planar image when using a sensor-matrix camera).

The geometry of this cylindrical surface comes with new challenges with respect to sensor calibration, stereo viewing, stereo analysis, and so forth. (Note that stereo analysis or stereo visualization of panoramic images becomes more and more essential for various imaging applications.)

A line of pixel sensors consists of "many" (say, 10,000) sensor elements. During the rotation, we take, say, more than 100,000 individual shots, line by line. All these line images define a panoramic image of 1,000 megapixels (i.e., of at least several gigabytes) or more on a cylindrical surface. Additional efforts (when using a rotating sensor-line camera) thus pay already off with benefits regarding image resolution.

The architecture of the rotating sensor-line camera is also a perfect match with the architecture of a (rotating) laser range-finder, which becomes more and more popular for generating 3D object or scene models. A (rotating) laser range-finder produces a dense "cloud of points", representing 3D locations of visible surface points in a given scene. These "clouds of points" need to be processed and triangulated. Color images taken by a rotating sensor-line camera can then be used for rendering of triangulated surfaces. This is of value in many applications such as (environmental) surveillance or for architectural photogrammetry.

This book illustrates that a multi-sensor approach (i.e., combining laser range-finder and rotating sensor-line camera) allows a digitization or 3D modeling of indoor or outdoor scenes, illustrates important progress, but also indicates areas of future research. Typically, cameras or laser range-finders have to be placed at various *attitudes* or *poses*, each defined by a location (say, in an assumed world coordinate system), an up-vector, and a direction of an initial viewing axis (so as to give a unique specification of a sensor's attitude; obviously, viewing directions change during rotation). The book discusses the calibration of sensor attitudes, and the fusion of data recorded at multiple viewpoints.

The authors have used rotating sensor-line cameras for nearly ten years; experience with laser range-finders is a little more recent. Working with (rotating) sensor-line cameras has proved to be more difficult in general than using a sensor-matrix camera. But the two stated benefits (high resolution of recorded image data, and matching sensor architectures) are worth the extra effort.

This book reports on theoretical issues and also on practical experience with rotating sensor-line cameras, and laser range-finders, with regard to image correspondence, stereo sampling, image acquisition, camera design, and so forth. One common question is how to choose suitable camera parameters to obtain an optimized pair of stereo viewable panoramic images. We show that the decision may be based on simple distance estimates in a given 3D scene.

Sensor calibration is an important subject in this book. A least-squares minimization-based approach is proposed for the rotating sensor-line camera, and two more options are briefly discussed, suggesting areas of future research.

The book may be used as a textbook for graduate courses in computer science, engineering, information technology, or multimedia imaging. Each chapter contains exercises, which are supported by an accompanying website (see also links on authors' home pages).

Chapter 1 provides a historic review of panoramic imaging and used sensor technology, also briefly discussing those areas which are not covered in this book. Chapter 2 discusses used sensors (rotating line cameras and range-finders). Chapter 3 is basically about the mathematical fundamentals used throughout the book, with a focus on spatial positioning of sensors. Chapter 4 provides epipolar curve equations, which will prove to be useful not only in stereo applications, but also in sensor calibration, which is the subject of Chapter 5. The potential distribution of stereo samples is discussed in Chapter 6, and formulas for optimization of sample distributions are provided in Chapter 7. A more refined analysis of stereo geometry and parameters in relation to given 3D scenes is contained in Chapter 8. Chapters 9 and 10 discuss how sampled depth values can be visualized by means of triangulated surfaces, also using captured panoramas for surface texturing. Chapter 10 also discusses some applications towards 3D rendering of aerial sceneries, such as city scenes captured from an airplane.

The authors acknowledge comments by, collaboration with or support from various students and colleagues at The University of Auckland and DLR (Berlin and Munich) over the past seven years, and we also thank friends and family for very important patience and help. We also thank Illustrated Architecture (Berlin) for their collaboration.

Fay Huang, *Reinhard Klette*, and *Karsten Scheibe*
Yi-lan, Auckland and Berlin

Series Preface

My first introduction to panoramic imaging was my junior high school (in the 1950's) and senior high school class pictures. The class members, segregated by height, were lined up on four rows of bleacher seats that extend over 40 feet. We were asked to hold still for about 30 seconds as the "strange" camera did its "scan" of the bleacher bound class. Some students, of course, ran from one end of the formation to the other to get into the picture twice. The students in the center of the picture were in sharp focus and those at the ends looked a little out of focus and slightly distorted. None of this bothered me at the time for I knew just the rudimentary aspects of optics, cameras and film. I had no idea if the camera had a curve film plane to go with a rotating lens and slit or was just a large format camera with a wide-angle lens. As I became more interested in photography and purchased a good 35mm SLR, I would attempt to take panoramic images by using a good tripod, some form of level indicator and a variety of lenses. I, like so many other amateur photographers, would then try to piece the images together to form a useable print, never really satisfied with the results due to changes in exposure, color balance and translational (vertical) variations. All of this was before the "age of digital images" where one could scan negatives or prints into digital files and use some sort of software to align them properly, make the exposure and color balance uniform and then crop to get the best looking aspect ratio for the final print. While there were many specialized cameras developed to take panoramic images, very few of them were used by the causal picture taker. The Advantix™ cameras provided a "panoramic look" by using a wide-angle lens (28 mm) and a framing aperture that gave the look of a panoramic image. Digital still and video cameras, introduced in the mid-90's and now clearly dominant in most fields of imaging, when combined with the "cheap" computer power and ample image oriented application software now makes it possible to create panoramic images at home. These historical antidotes provide a reference point for what is now possible with modern panoramic cameras.

While consumer imaging will always drive the digital camera markets, there are many scientific, industrial, cartographic, robotic, and military uses of panoramic imaging, often combined with range finding equipment to form three-dimensional representations of the original scene. These new panoramic cameras provide researchers and end-users new tools to study and evaluate real-world scenes to a degree not possible a decade ago. The sixth offering of the Wiley-IS&T Series in Imaging Science and Technology by Dr. Fay Huang, Dr. Reinhard Klette, and Dr. Karsten Scheibe, *Panoramic Imaging: Rotating Line Cameras and Laser Range Finders*, provides the imaging community with a detailed primer on how to understand the nature of panoramic imaging, a guide on how to build cameras using rotating, linear arrays, and how to combine the imaging data with depth information from laser rangefinders to create accurate three-dimensional images. The reader will appreciate the clarity and detail of the complex mathematical formulations of the geometric projections required to reconstruct an accurate, three-dimensional panoramic image. Every aspect of understanding the nature of the scene and its representation by the panoramic camera is covered in great detail. This text is a must for any researcher in computer vision, robotic imaging, remote sensing, image understanding, image synthesis and related topics.

Dr. Klette, who has a Ph.D. in Mathematics and a Dr. Sc. degree in Computer Science, has been involved in imaging research for over 35 years with over 280 refereed publications and numerous presentations. His current research interests focus on Image and Video Technology, Digital Geometry and Topology, Computer Vision, Panoramic Imaging, and Biomedical Image Analysis. He was the head of an interdisciplinary research project 'Intelligent Microscopy' at Jena University and Carl-Zeiss Company (1979–1984), the head of an AI research unit at the Academy of Sciences, Berlin (1984–1990), Professor for Computer Vision at the Technical University of Berlin (1990–1996), and currently a Professor for Information Technology at The University of Auckland (since 1996). He also served as the Director of the Center for Image Technology and Robotics (1996–2005), and Deputy Head of Department (1996–2002, 2005). Dr. Klette is an active participant and editor in many professional scientific and engineering societies. He has mentored 17 Ph.D. students and 99 M.S. students over his distinguished career.

Dr. Fay Huang received her Ph.D. in Computer Science from The University of Auckland, New Zealand in 2002 and is currently an Assistant Professor at the Institute of Computer Science and Information Engineering, National Ilan University in Taiwan, Republic of China. Her research interests are in Computer Vision, Computer Graphics, Virtual Reality and Computer Art. Dr. Huang has over 40 publications (journal and conference papers, book chapters, and technical reports, for example at IMA Minneapolis) in her relatively short, but active career.

Dr. Karsten Scheibe received his Ph.D. in Computer Science from Georg-August-Universität Göttingen in 2006. He is currently working at the German Aerospace Center (DLR), Optical Information Systems, Information Processing for Optical Systems. Dr. Scheibe research focuses on image and signal processing and modeling and simulation of imaging systems. Dr. Scheibe has published 30 papers in the area of computer graphics and robotic vision. He contributes at DLR to major satellite imaging, 3D visualization and aerial 3D imaging projects.

Michael A. Kriss
Formerly of the Eastman Kodak Research Laboratories
and the University of Rochester

Website and Exercises

The websites http://www.mi.auckland.ac.nz/books/HKS/ and http://www.wiley.com/go/ hkspanoramic accompany this book. Each chapter in the book ends with a few exercises. The website may be found useful when solving those exercises.

Some of the exercises are labeled as being possible lab projects (say, assignments in a course). There is example source code available on the site. (Just download the zip file numbered as the corresponding exercise.) Example source code needs typically some visualization or window management, and therefore it uses some common platform libraries (e.g., Microsoft Foundation Class, DirectX, OpenGL, Windows GDI, etc.). Those have to be adapted to the used platform. In principle, the source code for most of the exercises is edited and compiled with Microsoft Visual C++ .NET, without using a special platform library.

Solutions to exercises, which are simple console programs, should be compilable with your own favorite developer environment. Readers who are using Microsoft Visual C++ .NET just load the project file with the extension *.vcproj. If you are using a different developer environment make sure that you load all *.cpp and *.h files as sources into your project. Additionally, link external libraries (e.g., glut.lib) as documented in your developer environment. Solutions to exercises, which need those external libraries, also contain a separate readme file where these external libraries are listed. Make sure that you are downloading the compatible library for your platform (e.g., glut.lib needs to be available for the majority of platforms).

Sometimes, source code requires more detailed documentation, and this is not discussed in this book. In such cases, the zip also contains a pdf file detailing some aspects of the source code (e.g., drawings for the geometric understanding of the actual implementation).

All the sources provided are for educational purposes only, and the authors and publisher cannot be made responsible for any outcome using those sources.

List of Symbols

Image notations

$E_\mathcal{P}$	panoramic image (two-dimensional image)
$E_{\mathcal{P}_i}$	the ith panoramic image, for integer i, when multiple panoramic images are considered at the same time
$E_{\mathcal{P}_L}, E_{\mathcal{P}_R}$	the left or the right panoramic image in the stereo image case
M	planar matrix image width in pixels
W	panoramic image width in pixels
H	image height in pixels
(x_o, y_o)	image center (measurements, in real numbers)
(i_c, j_c)	image center (pixel position, in integers)

Parameters of rotating sensor-line camera

τ	CCD size or pixel size (in micrometers, assuming square CCD cells)
f	camera's focal length (in millimeters)
υ	camera's vertical field of view
R	off-axis distance (in meters)
ω	principal angle (i.e., the camera's viewing angle)
γ	angular unit (i.e., angular distance between adjacent projection centers)

Stereo analysis

d	image disparity in pixels
θ	angular image disparity

A scene's range of interest (RoI)

D_1	radius of the RoI's inner cylindrical frontier (in meters)
D_2	radius of the RoI's outer cylindrical frontier (in meters)
V_1	height of the RoI's inner cylindrical frontier (in meters)
H_1	distance between the camera's optical center and the RoI's inner cylindrical frontier with respect to the camera's viewing direction (in meters)

Coordinate systems

W	origin of world coordinate system
O	origin of sensor coordinate system
C	origin of camera coordinate system
(X, Y, Z)	coordinates in a 3D Euclidean coordinate system
(x, y)	coordinates in a 2D Euclidean image coordinate system
(ϑ, φ)	coordinates in a 2D non-Euclidean spherical image coordinate system
(φ, L)	coordinates in a 2D non-Euclidean cylindrical image coordinate system
(i, j)	coordinates of an image pixel

Geometric transforms

R	rotation matrix
t	translation vector

Geometric objects

$\mathcal{H}$	a circle defined by possible positions of a camera's optical center in 3D space
$\mathcal{L}$	projection ray or line in 3D
$\wp$	plane in 3D
ν	image vector

Screen parameter

H_S	the height of the displaying screen in pixels

General mathematical symbols

A, B, C	real numbers
M, N	integers
α, β	angles
i, j, k, m, n	indices
p, q	2D points or vectors
P, Q	3D points
M, R	matrices
f, g	functions
f'	derivative of a unary function
$\mathcal{A}, \mathcal{B}, \mathcal{S}$	sets

1

Introduction

This chapter provides a general introduction to panoramic imaging, mostly at an informal level. Panoramas have an interesting history in arts and multimedia imaging. Developments and possible applications of panoramic imaging are briefly sketched in a historic context. The chapter also discusses the question of accuracy, and introduces rotating sensor-line cameras and laser range-finders.

1.1 Panoramas

A panorama is defined by a wide field of view. Obviously, a single panoramic image thus contains more information or features than a "normal" image. This has potential for understanding the geometry of three-dimensional (3D) scenes, or for estimating the locations of panoramic sensors within a 3D scene.

1.1.1 Accurate Panoramic Imaging

Panoramic images are already part of our daily lives. They may be generated with relatively inexpensive tools, and basically by anyone with a digital camera after spending a few minutes reading the manual.

However, the accuracy which is required makes a difference: if panoramic images or 3D models, derived from panoramas, have to satisfy high-quality demands (as in close-range photogrammetry, wide-screen visualization, or in many computer vision applications, such as in industrial inspection or accurate object modeling), then the geometry of the panoramic sensor needs to be understood, sensors have to be calibrated, and image capturing has to follow strict rules. This is the starting point for this book, which is about *accurate* panoramic imaging, where accuracy requirements might be described in terms of fractions of a millimeter, even for large objects such as an important building.

Panoramic Imaging: Sensor-Line Cameras and Laser Range-Finders   F. Huang, R. Klette, and K. Scheibe
© 2008 John Wiley & Sons, Ltd

The authors predict that accurate panoramic imaging will become available for a wider community in the near future, for example by means of tools for correcting perspective or radial distortions, or perhaps automated incremental generation of 3D models, and mapping of images onto those models for further refinement.

1.1.2 Importance of Panoramas

Potentially, panoramas provide improved support for the visualization or inference of 3D world features; but the tradeoff is that a larger amount of data is acquired (in comparison to conventional approaches with more restricted fields of view). However, as faster computers and larger bandwidth became available, applications and studies using panoramas became widely accessible. We mention a few aspects which highlight the importance of panoramas:

Immersion. The importance of a panoramic image rests, obviously, in its large field of view. The impression of immersion can be fully achieved by a display method (e.g., a facility known as "The Cave") which allows an equally wide field of view, possibly enhanced by 3D stereo viewing.

Figure 1.1 shows an important panoramic image, documenting the impact of the first weapon of mass destruction (a small nuclear weapon, unleashed from the US B-29 bomber *Enola Gay*). The bomb detonated 600 m above the center of Hiroshima, a city of about 400,000 people, on 6 August 1945 at 8.15 a.m. local time. An actual *panorama* (not shown here), formed from those 16 photographs, is on display at the Hiroshima Peace Memorial Museum. Images like these provide shocking evidence of the disasters at Hiroshima and Nagasaki. In order to understand the importance of a worldwide ban on weapons of mass destruction we must not be allowed to forget events like these, and wide-angle views may help to initiate wide-angle thinking.

Realism. A photorealistic quality of a "virtual world" can be achieved by a "dense" (with respect to locations of viewpoints) set of panoramas within a scene of interest. As with other virtual reality systems, stereo visualization can also be implemented with free navigation in a virtual panoramic world. A virtual reality world of synthetic scenes, generated by computer graphics approaches only, can hardly achieve a comparable photorealistic quality for complex 3D scenes.

Simplification. For example, stereo reconstruction of 3D scenes can be based on 360° stereo panoramic images; this bypasses a complicated (and possibly erroneous) process of merging multiple depth-maps in traditional multi-pinhole-image approaches.

Localization. A robot or aircraft equipped with panoramic vision can recognize a location by comparing an incoming panorama with memorized maps or images, and this strategy improves robustness because of more information, provided by a larger visual field.

Compression. A panorama composed from an image sequence of a video camera may be seen as a compact version of the original video data, without redundancy.

Before embarking on a more technical discussion of panoramas, we provide a brief but informative historic overview, starting with panoramic paintings, continuing with panoramic photographs, and adding a few comments on panoramic digital imaging in general.

Figure 1.1   Six of 16 photographs taken by Shigeo Hayashi in 1945 from the rooftop watchtower of the Hiroshima Chamber of Commerce and Industry. Courtesy of Hiroshima Peace Memorial Museum.

1.2 Panoramic Paintings

The word "panorama" was introduced by the Irish painter Robert Barker (1739–1806) when describing his 1787 paintings of Edinburgh. However, panoramic paintings date back to earlier times. Panoramic presentations in Asia, such as Chinese scrolls, may be cited as historic origins.

1.2.1 Chinese Scrolls

The name "panorama" is typically not used in art history publications on Chinese landscape paintings. A typical format of a Chinese landscape painting is either a *handscroll* or a *hanging scroll*.

The handscroll – so called because it is unrolled by hand horizontally – is a unique way to view the landscape one section at a time (from one end to the other), as if touring through the picture. Vertical hanging scrolls are a dominant form of wall decoration in Chinese domestic architecture. They are easily rolled up and stored, and are changed frequently or viewed only on special occasions.

Landscape paintings are a popular Chinese tradition, showing poetry in nature. For over a thousand years, since the T'ang dynasty (618–907), Chinese landscape paintings have been understood as both an intimate expression of nature and a way of conveying profound emotions. They play an important role in expressing Chinese spirituality and philosophical values. Literally translated, the Chinese characters for landscape mean "mountain and water". Almost all Chinese landscape paintings depict mountains or water in some abstract style. The landscape is a composite of many elements of nature that invites the viewer to a spiritual journey of the imagination.

Landscape painting reached a (first) period of prosperity during the Five Dynasties (906–960) and the Song (960–1279) era. Chinese paintings, especially in the horizontal handscroll format, are read like a traditional written text. They begin on the right and come to a conclusion at the left-hand end. Like a conventional narrative, these landscapes have a beginning, middle, and end. Painted around the year 1000, Fan Kuan's *Travelers among Streams and Mountains* (2,064 × 1,035 mm) is one of the most famous hanging scroll landscapes in Chinese art history; see Plate 1. It shows tiny figures on a road which passes through a landscape dominated by a huge, looming mountain peak.

Later landscape paintings during the Ming (1368–1644) and Ch'ing (1644–1911) dynasties begin to introduce more people into the scene. In comparison to the "mountains and water" style, there is also a return to greater realism. Typically, the landscapes become intricately detailed visions, almost as if they are painted from the air, from a bird's-eye view of the world. In the Ch'ing dynasty, a landscape painting may also include a painting of a city. One of the best-known, long handscroll format paintings of the Ch'ing dynasty is *Along the River during the Ch'ing-ming Festival* (356 × 11,528 mm); see Plate 2.

This work is based on an original painting by Chang Tse-tuan (painted in early 12th century). It beautifully illustrates life and customs during the Sung dynasty at the capital of Pien (K'ai-feng). This theme, popular in the Northern Sung dynasty (960–1126), has been copied often throughout the years. The one shown in Plate 2 is one of the most famous, made by court painters of the imperial painting academy in the time of the Ch'ien-lung Emperor, who reigned 1736–1795. Street entertainment, commerce, folk customs of the Ming and Ch'ing dynasties, daily household life, and architecture as shown reflect very much the life and appearance of the period when it was painted. The lively activities include a theatrical performance, monkey show, acrobatics, and a martial arts ring to lend a festive air to the scenery.

These Chinese paintings can certainly be classified as *panoramas*, meaning wide-angle views of real-world scenes. Today, "panorama" is considered to be an English word, and there are different ways to translate it into Chinese. The format of a handscroll painting is most relevant to "cylindrical panoramas" as discussed in this book, although they are very often presented in a "linear" (i.e., translational) multi-perspective approach, while the ones to be discussed in this book typically have projection centers along a circular path.

1.2.2 European Panoramic Paintings

Figure 1.2 shows (in gray levels only) one of the early panoramic paintings by Robert Barker and his son, Henry Aston Barker (1774–1856). The first exhibition of a panoramic painting by Robert Barker was in 1787, with the city of Edinburgh painted around the (inner) wall of a rotunda. It had to be viewed from the center of the room to give an illusion of reality.

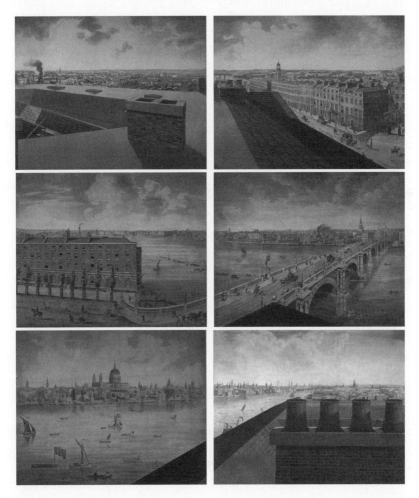

Figure 1.2 Six plates forming a 360° panorama, *London from the Roof of the Albion Mills*, by Robert Barker and his son, Henry Aston Barker, London, 1791. The figure shows gray-level copies of the originally 1792 color aquatints by Frederick Birnie, a panoramic print series; the size of each of these six prints is 425 mm × 540 mm, with an overall length of 3,250 mm. Courtesy of the University of Exeter Library, Bill Douglas Centre.

Barker obtained a patent for these panoramic paintings. His exhibitions of panoramas were very influential: panoramas or *dioramas* (a related way of displaying wide-angle scenery, often also by decorating the foreground with various, sometimes moving, objects, often also with audio) became an early form of mass entertainment in the larger cities of Europe or North America, before the advent of the cinema.

However, panoramic paintings were popular in Europe much earlier than the late 18th century. The Bayeux Tapestry (which is actually embroidery, not tapestry; see Plate 3) might be one of the earliest examples. It tells a story, similar to Chinese scrolls, and was probably commissioned in the 1070*s* by Bishop Odo of Bayeux, half-brother of William the Conqueror, and (again, probably) embroidered in Kent, England.

At an exhibition in London (around 1870) a panoramic view of the entire Earth could be seen from a platform at the centre of a large hollow sphere, which was painted on the inside to represent the Earth's outer surface.

Panoramic pictures or dioramas developed into "multimedia shows" (which are still shown today in Gettysburg, Uljanowsk, Bad Frankenhausen, and other historic sites). Wide-angle paintings of military battles, historic scenes, or landscapes became popular in a few countries such as United States, Russia, France, UK, and Germany.

Paintings of very large dimensions are popular in modern art. Typically, today they no longer aim to provide photo-realistic presentations of extensive scenes.

1.3 Panoramic or Wide-Angle Photographs

The first photograph was taken by Nicéphore Niepce in 1826 near Chalon-sur Saône, France (this photograph still survives, in recognizable form). Soon after, photographers began to capture wide-angle photos of landscapes (e.g., Edouard Baldus in France, and Maxime Du Camp during his travels in Egypt). General-purpose panoramic photography developed around 1900 into wide-angle architectural photography or aerial photography, contributing to the development of *photogrammetry*.

1.3.1 Historic Panoramic Cameras

Panoramic cameras have been built since the second half of the 19th century. For example, in England in 1860 Thomas Sutton designed a wide-angle (120°) camera whose "lens" was a hollow glass sphere filled with water; this camera was then built by Paul Eduard Liesegang in Germany. Albrecht Meydenbauer (1834–1921), a German architect, designed a camera in 1867 which used the first wide-angle (105°) optical lens.

As an alternative to a single wide-angle shot, cameras could also take a continuous shot during a full 360° rotation. For example, the sophisticated "Cyclographe" panoramic cameras of Jules Damoizeau, built between 1890 and 1894, rotated by means of a spring mechanism as the film was fed past the shutter at the same speed, but in the opposite direction. A camera with a pivoting lens, called a *périphote*, was built in 1901 by Lumière in Lyon. During exposure, the lens rotated 360° (see Figure 1.3); no film transport was needed for this model.

Of course, cameras could also be used for multiple shots in different directions, during a 360° rotation of the camera on a tripod. However, these separate images could not be used for creating a seamless panorama, until the occurrence of digital stitching techniques (see Section 1.4.2).

Figure 1.3 A panoramic camera (*périphote*), built in 1901, with a lens rotating 360° (manufacturer: Lumière Frères, Paris). Reproduced by permission of George Eastman House, Rochester, NY (see Plate 20).

1.3.2 Photogrammetry

Engineers of many disciplines were also quick to realize the potential use of photography as a tool for both civilian and military applications. This was true for areas such as architecture, and more significantly in cartography.[1] In particular, some pioneering photographers attempted to apply their art to the science of measuring physical spaces at an accuracy corresponding to their latest tools. Of course, photography is in general more concerned with creating images that meet aesthetic and commercial demands than with capturing accurate images required for scientific measurements.

Close-range photogrammetry, as a method for recording and monitoring architecture, originated in the work of Albrecht Meydenbauer (see previous subsection). He was the first to use the term "photogrammetry", and he pioneered the photogrammetric recording of Islamic architecture in the Middle East in the 1870s.

Aerial photography or *wide-range photogrammetry* was pioneered by Aimé Laussedat, using balloon photography or aerial photography supported by a string of kites (see Figure 1.4).

---

[1] Plane surveying (with the *dioptra*, an instrument used for measuring angles since about 300 BC) was developed in Hellenistic culture, and then revived in the Netherlands in the 16th century. The distance between one pair of points is measured, and then angle measurements are used to calculate a *triangulation network*. In the 1930s, the British Colonial Survey made a geodetic survey of Nigeria (which is very flat) by running a tape measure over that colony. Only a minor role was played by angle measurements (mostly for altitude estimation), and a network of triangles was calculated by *trilateration* (not by triangulation). Consequently, by 1939 Nigeria was the most accurately mapped country on Earth. (Computer vision techniques, which use projected light patterns, calculate 3D points by triangulation.)

Figure 1.4   George Lawrence, *San Francisco in Ruins* (1906), a panorama taken with a camera on a string of kites (top); magnified detail (bottom). These photos are modified from a version of the panorama freely available from the Library of Congress (Washington, DC).

Photogrammetry became the field of engineering concerned with the task of documenting and measuring scenes based on photographs. Photogrammetry is basically a measurement technology in which the 3D coordinates of points of an object are determined by measurements made in two or more photographic images taken from different positions or with different viewing directions.

The development of photogrammetry can be subdivided into four periods. Each period is characterized by technological or methodological innovations which made photogrammetry incrementally more flexible and more effective. *Graphic* or *plane table photogrammetry* (*c*.1850–1900) combined terrestrial photographs and topographic maps; early experiments with aerial photographs demonstrated the potential of this new discipline. *Analog photogrammetry* (*c*.1900–1960) emerged with the advent of stereoscopy (i.e., the use of corresponding points in binocular images for identifying 3D positions by triangulation) and of airplanes. The first cameras specially designed for aerial imaging were built around 1920.

Analytical photogrammetry (since around 1960, and still active) is characterized by a "rigorously correct least squares solution, the simultaneous solution of any number of photographs, and a complete study of error propagation" (Doyle, 1964), which was possible with computer technology. Captured photos are still chemically processed, but are today scanned afterwards into digital representations.

Digital photogrammetry began around 2000, but has yet to achieve wide acceptance in the photogrammetric industry; if and when it does so it will edge out analog photography, establishing a completely digital processing chain from image capture to image analysis.

This book contains material related to digital photogrammetry. The digital camera used and characterized by a rotating sensor line, is basically a "side-product" of camera design for digital photogrammetry.

1.4 Digital Panoramas

Script, printing, photography, and computer technologies have all had a defining impact on the progress of human mankind. Digital panoramas are a fine example of the combination of photography and computing technology.

1.4.1 Image Mosaicing

Image mosaicing is often just a single component of a rather complex project. Unlike the stitching approach (see below), an image mosaicing method assumes motion-uncontrolled image sequences (see Figure 1.5(b)), or just sets of images (e.g., taken by different satellites on different days, at different elevations, and with various resolutions above the same location).

Mosaicing is simplified if the non-rotational motion of the camera can be controlled (to some extent). For example, a geostationary satellite, which is constantly looking down at a fixed angle, provides ideal conditions for mosaicing. Figure 1.6 illustrates that manual translation of a camera ("manually controlled motion") was sufficient in this case to produce "fairly accurate" mosaics for these two panoramic views of a mummy. In total, eight normal-sized rectangular digital photos were combined to create this mummy mosaic (with less projective distortion compared to a photo taken at greater distance with the same camera, and of resolution roughly equal to 4,000 × 8,000 pixels, which was beyond the capabilities of non-professional digital cameras in 2001, when the mosaic was created).

The problem of accurate mosaicing appears in photogrammetry, computer vision, image processing, and computer graphics. Application areas of mosaicing include change detection,

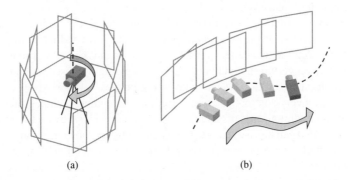

(a)   (b)

Figure 1.5   (a) Motion-controlled and (b) motion-uncontrolled scanning.

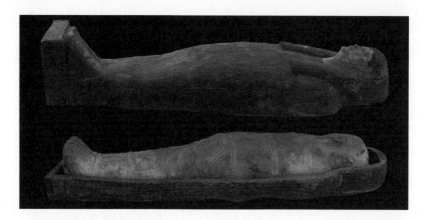

Figure 1.6 Mummy in the Auckland Museum: the figure illustrates that "careful" manual translational camera movement allows "fairly accurate" mosaicing: while walking along the mummy, four digital images were taken; these were manually aligned and photometrically unified (i.e., eliminating "seams" between images, thus forming a high-resolution projection of the mummy into one image plane).

video compression and indexing, photo composition and editing. One of the oldest applications, dating back to the early 20th century, is the construction of large aerial views (maps).

1.4.2 Panoramas by Stitching

The simplest way of generating a single digital panorama is by "stitching" together a set of overlapping digital images. The process is simplified if those images (i) are taken by rotating a camera (see Figure 1.5(a)), (ii) show "overlapping" scenery with identifiable corresponding "landmarks", (iii) are all taken within a short interval of time, and (iv) without much motion or changes (e.g., of illumination) in the recorded scenes. An alternative to rotation is a simple camera motion defined by translation (e.g., like a satellite which is constantly looking down at a fixed angle).

Stitching maps adjacent images into a panorama by registering them at first, and then unifying image values (by merging and blending) so that no visually disturbing "seams" would indicate where the stitching took place; see Plate 4.

Figure 1.7 illustrates stitching in a horizontal direction, and Figure 1.8 in a vertical direction. In the latter case, projective distortions become more noticeable (e.g., for buildings, and also for the tree on the left). Another challenge is stitching for dynamic scenes. To produce Figure 1.9, images had to be taken very fast, and manual post-processing was used to eliminate "ghosts" or blurry transitions at "seams" between contributing images.

Stitching is today a very common technique for home, commercial, and professional panoramas. Some customer digital cameras already come with special software for image stitching. The panoramas created can be "played" (e.g., visualized by rotation and zooming) with audio, or combined into multi-view panorama applications, where links among viewpoints may be shown in panoramas.

Stitched panoramas can be used for 3D scene animations. Apple's QuickTimeVR system was one of the first systems to suggest that a traditional CAD-like modeling and rendering process

Figure 1.7 Canadian Niagara Falls. All three panoramas show basically the same scene but were captured at different locations. Each panorama consists of more than four images, and was stitched using a program which came in 2000 with a Canon PowerShot digital camera.

Figure 1.8 Automatic vertical stitching: tree in the Australian rain forest at Natural Bridge (left), Montserrat monastery in Spain, near Barcelona (centre left), and Arco Naturale on Capri (centre right, and right; projective distortions of 3D shapes are less apparent in this case).

may be skipped for visualizing 3D scenes. The "multi-node version" (i.e., multiple panoramas taken at different locations, also called "nodes") of this system also uses "environment maps" to illustrate relative locations of all the nodes and the user's current position. The user can "hop" from one node to another, and can turn around 360° for different views at each location.

Figure 1.9 A Haka competition in Auckland: manual post-processing required after automated stitching.

A multi-node panorama, possibly with an environmental map, is a possible implementation of a virtual reality (VR) system. To contrast this with VR systems solely using computer graphics for generating synthetic worlds, it is called *image-based VR*. A particular trend of research on image-based VR systems requires a ("rough") 3D model of the scene, and panoramic images are used (similarly to texture maps in computer graphics) for texturing the 3D scene. The 3D scene model, in such an image-based approach, is constructed based on features shown in the images. (This book will discuss an integrated *image- and range-scan-based approach* for combining a detailed 3D scene model, calculated by using a laser range-finder, with high-resolution panoramic images for texturing, which potentially enables us to create accurate documentations or a high-quality VR system.)

Stitched panoramas, using non-professional digital cameras, are of smaller size or resolution than those addressed in this book. Also, stitching does not ensure geometric accuracy of visualizations, but the methods described in this book potentially do.

1.4.3 Catadioptric Panoramas

A catadioptric camera system enables us to record a full "half sphere image" in one shot. The word *catadioptric* means pertaining to or involving both the reflection and the refraction of light. A catadioptric camera system is engineered as a combination of a quadric mirror and a conventional sensor-matrix camera; see Figure 1.10(a). Catadioptric camera systems provide real-time and highly portable imaging capabilities at affordable cost.

There are only two possible combinations which satisfy a "single projection center constraint": one is a hyperboloidal mirror used in conjunction with a sensor-matrix camera, and the other is the (more theoretical) configuration of a paraboloidal mirror with an (assumed) orthographic projection camera. Both *catadioptric sensors* allow that all the reflected projection rays intersect at a single point, and hence possess a simple computational model which supports various applications. Both sensor models are illustrated in Figure 1.11.

The major drawbacks of the catadioptric approach include low resolution near the center of an image, non-uniform spatial sampling, inefficient usage of images (i.e., there is a self-occluded or mirror-occluded area in each captured image), severe distortions and image blurring due to aberrations caused by coma, astigmatism, field curvature, and chromatic aberration. These drawbacks suggest that catadioptric panoramas are not suitable for those recognition or inspection types of applications where high accuracy or high image resolution is required (as in close-range photogrammetry).

(a)   (b)

(c)

Figure 1.10   (a) a commercial catadioptric camera system. The shape of the mirror has been emphasized by an added black curve for clarity. (b) an image taken by this panoramic sensor. (c) a panorama produced by rectifying the above image (courtesy of N. Ohnishi and A. Torii).

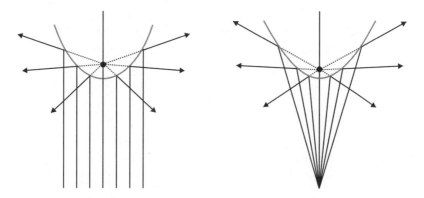

Figure 1.11   Catadioptric panoramas: Parabolic mirror with orthographic projection (left), and hyperboloidal mirror with perspective projection (right).

However, they have various application areas, and proved to be very efficient in particular for robot navigation. They are also used for capturing wide-angle views of dynamic scenes (i.e., in those applications where the accuracy drawbacks listed above can be neglected).

This book does not discuss catadioptric panoramas further.

1.4.4 Stereo Panoramic Imaging

Stereo viewing of photographs became very popular in the 1870s. However, no stereo viewing of panoramic photographs is known for the pre-digital age.

The book will discuss how a stereo pair of panoramic images may be obtained and optimized with respect to a given 3D scene (for best effects in stereo viewing). Such a "binocular" panorama approach enables us to create *parallaxes* (i.e., shifts of projected surface points, defined by their positions in the "left" and "right" panorama), as known from "normal" pairs of stereo images. These parallaxes may also be used for stereo analysis, not only for stereo visualization.

To create a stereo panorama visualization is not straightforward, because panoramic and "binocular stereo" image acquisition setups are intrinsically incompatible. However, the stereo effects can still be achieved due to the fact that human stereo perception has tolerance towards parallax errors.

Anaglyphs define a common stereo visualization method which requires a pair of red-green (or red-cyan, for RGB) glasses to perceive depth. An *anaglyphic panorama player* is illustrated in Plate 5. It shows an interface of a 1998 interactive player of stereo panoramas: the environmental map is on the left, and the bottom line of the dialog box says that the globe is available for interactive manipulation (i.e., rotation). The player came with auto-spin options, manipulable objects embedded into the stereo panorama, and the environmental map.

Simple (non-interactive) animations of 3D objects within a panoramic scene are waving of leaves, random passage of cars through a scene, or special effects such as illumination changes (e.g., a sunset).

1.5 Striving for Accuracy

Many applications in computer vision, computer graphics (e.g., image-based rendering) and photogrammetry demand spatial (geometric) or color accuracy, very high resolution, or minimized radial distortion, which excludes the use of simple stitching or mosaicing.

For example, the quality of close-range photogrammetry (e.g., for static architectural scenes) depends on image resolution to support very accurate representations of scene geometry. Computer animations in a professional context often specify needs for high resolution, photometric correctness, or geometric accuracy.

1.5.1 A General Perspective on Panoramic Sensors

Informally speaking, a panoramic sensor sends rays into the 3D world. A ray emerges at a *center of projection* and collects information (such as color, range, or intensity) about the first surface point it hits in 3D space (possibly also involving recursion as known from raytracing in computer graphics).

In an abstract geometric sense, measured data are mapped onto a *capturing surface*, which may be understood as being a plane in the case of a "normal camera" (note: a CCD[2] or CMOS[3] matrix of sensor elements basically defines a photosensitive rectangle), or a sphere,

---

[2] Charge-coupled device.
[3] Complementary metal oxide semiconductor.

Figure 1.12 A high-accuracy panoramic image (view towards Rangitoto Island, Bastion Point, Auckland, 2002). Each column of this image consists of 10,200 color pixels; the size of the panorama is about 4 GB (see Plate 21).

cube, cylinder, and so forth in the case of a "non-standard panoramic sensor" (and this defines *spherical*, *cubic*, or *cylindrical panoramas*).

Extreme cases of panoramas are defined by a full $360° \times 360°$ *spherical view* or *cubic view*, a full $360°$ *cylindrical view*, or a *translational planar view* (described above as being image mosaics).

A panorama can also be defined by a subrange of one of the examples listed, as long as it still allows a wider view compared to normal photographs. We may still call it a cylindrical or spherical panorama. The image in Figure 1.12 is a cylindrical panorama, and this will become clearer later in the book. With respect to the capturing surface, catadioptric panoramas may be called *hyperboloidal panoramas*.

A *camera* typically collects color values (i.e., surface reflectance values depending on given lighting) which form an *image*.

A *range-finder* records distances (at accurately recorded horizontal and vertical angular increments) between the center of projection and surface points which generates a *range-scan*. The particular case of returning intensity only (rather than color) forms a *gray-level image*. A range-finder typically enables us to obtain such a gray-level image in addition to its range-scan (but not a color image); a pair consisting of a range-scan and a gray-level image (see Figure 1.13) is basically geometrically aligned because sensed by the same sensor at the same viewpoint.

1.5.2 Rotating Sensor-Line Cameras

A "normal digital camera" combines a (CCD or CMOS) sensor matrix with some optics, all packed nicely into a box which also containing various electronic components. Now imagine that the matrix, consisting of $M \times N$ sensor elements (each recording a single pixel), degenerates in such a way that there is only one column of sensor elements (say, $N = 1$; for example, similar to those used in a flatbed scanner). The benefit is that sensor technology enables us to produce such a *sensor line* for very large values of M, say M greater than 10,000, but producing sensor matrices of $10,000 \times 10,000$ elements is still a challenge today.

A digital camera, with the sensor matrix "shrunk" into a single sensor line, may now be placed on a tripod and rotated, taking many images "column by column" during such a rotation. This defines a *rotating sensor-line camera*, a panoramic sensor which may record $360°$ panoramic images within a time frame needed for taking many shots during one full rotation. Such a sensor is not only more economic (compared to the use of a, say, $10,000 \times 10,000$ sensor-matrix camera), it also comes with several benefits for recording panoramic images, which will be discussed in this book.

(a)

(b)

Figure 1.13 Two panoramic scans showing the office of King Ludwig in Neuschwanstein castle in Germany: (a) a gray-level image; (b) range data visualized by means of a gray-level picture.

In the 1990*s*, theoretical studies by various authors pointed out that the use of a rotating sensor-line camera, where panoramas are shot line by line, each line with its own projective center, enables us to control conditions for improved stereo analysis or stereo viewing. Basically, this was the start of a new category of digital panoramas, defined by super-high resolution and geometric accuracy (but "not good" for movement in the scene).

Sensor-line cameras were designed for digital aerial imaging (using a *push-broom technique*) in the early 1980s. Three or more such sensor lines, optics, and a frame grabber together define a *sensor-line camera* for use in today's airborne image sensors (e.g., for creating "3D maps" of cities or interesting landscapes). (Having multiple sensor lines in one camera is of benefit for stereo panoramic imaging.)

These sensor-line cameras were positioned and rotated on tripods, thus generating panoramas. For example, the *Wide Angle Airborne Camera* (WAAC) of DLR[4] Berlin-Adlershof was

---

[4] The German Air-and-Space Institute (DLR = Deutsche Luft- und Raumfahrt) has various institutes in Germany, with its head institute near Munich.

Figure 1.14 A panoramic gray-level image captured at Friedrichshafen (Germany) in 1995, the first use of a rotating sensor-line camera. Courtesy of DLR Berlin-Adlershof.

Figure 1.15 Another early (1995) panorama using a rotating sensor-line camera shows the Gendarmenmarkt in Berlin. Courtesy of DLR Berlin-Adlershof.

used in 1995 for taking a panoramic image (a view from the roof of Dornier in Germany). Figure 1.14 shows the image taken at this first use of a rotating sensor-line camera. This panoramic image is of very high resolution, with each of its columns consisting of more than 5,000 pixels. Another panoramic image from 1995, also taken by the rotating WAAC, shows a square in Berlin; see Figure 1.15.

These studies and experiments led in 2000 at DLR Berlin to the design of a commercial rotating line camera which had about 10,000 color pixels in its rotating line sensor. The image shown in Plate 6 was taken with this camera during the "Robot Vision 2001" workshop in Auckland. People actually managed to be in this photo up to seven times. This amusing "multi-presence" effect was possible due to the use of a rotating sensor-line camera; for this particular panorama, the total capture time was about 5 minutes.

Panoramic images of very high resolution are today required in various applications, such as inspections of pipelines (as an example of industry applications in general), scanning of house facades for 3D city maps, or accurate indoor and outdoor documentation of selected architectural sites. In the latter two cases, panoramas are used for texture mapping, and the 3D geometry of the object surfaces is typically acquired by a laser range-finder (on a moving platform, or at multiple locations).

1.5.3 Laser Range-Finder

A *laser range-finder* (LRF) or *laser scanner* determines distances to opaque objects; the technique is also known as laser imaging detection and ranging (LIDAR). Applications of laser scanning are, for example, in the building industry, geology, seismology, remote sensing, atmospheric physics and cultural heritage.

An LRF determines the distance to an object or surface using laser pulses (similar to radar technology, which uses radio waves instead of light). Each individual pulse is directed along one scan ray, and the returned range value identifies one point in 3D space. Figure 1.16 illustrates the use of such a device: for this model of an LRF, the laser pulses radiate through the rectangular window, and the small circular window covers a sensor-matrix camera for capturing (relatively low-resolution) color images, sufficient for identifying locations. To the right the figure shows such a color image and a visualization of a range-scan in image form.

A range-scan actually defines a "cloud" of points in 3D space, which represents visible surfaces by those (possibly noisy) discrete points. Figure 1.17 shows one view of the office of King Ludwig in Neuschwanstein castle in Bavaria: the millions of measured points on surfaces in this office are located in 3D space, and "textured" with the corresponding gray level measured for this point by the range-finder.

These isolated points need to be mapped into meshed (e.g., triangulated) surfaces, and the surfaces may be "smoothly" rendered using gray levels recorded by the range-finder. However, color is typically required, and color panoramic images may be used for proper rendering.

The scan geometry of range-finders and rotating sensor-line cameras is very similar, and this supports accurate rendering of 3D surfaces, generated from range-scans, using color panoramic images, recorded with a rotating sensor-line camera. Figure 1.18 illustrates a view into such

Figure 1.16   Use of a laser range-finder in 2005 for generating 3D models of buildings on Tamaki campus, The University of Auckland.

Figure 1.17 "Cloud of points" illustrating the same range-scan as shown in Figure 1.13(b). A partially visible (dark) disk "behind" the lower corner shows the position of the scanner (the turntable of the scanner occluded the points below the scanner).

Figure 1.18 A view into the 3D reconstructed (using multiple laser range-scans) throne room of Neuschwanstein castle, with color texture generated by multiple, very-high resolution panoramic scans. Spatial resolution on surfaces is of the order of 10 mm, and texture resolution is about 1 mm (see Plate 22).

a very high-resolution 3D model, calculated in 2004, textured by panoramic image data. The book explains in a few chapters the operations involved in obtaining results like this.

1.6 Exercises

1.1. What kind of data are available when using a laser range-finder which scans 3D scenes incrementally (rotating horizontally around its axis, and scanning at each of these positions also vertically within a constant range of angles)?

1.2. [Possible lab project] Implement a program which enables anaglyph images to be generated. Instead of simply using the R (i.e., red) channel of the left image and the GB (i.e., green and blue) channels of the right image for creating the anaglyph RGB image, also allow the R channel to be shifted relative to the GB channel during 3D viewing for optimum 3D perception.

When taking stereo pairs of images (with a common digital camera), make sure that both optical axes are about parallel, and shift the camera just orthogonally to the direction of the optical axes. Corresponding to the distance between human eyes, the shift distance should be about 50–60 mm. However, also carry out experiments with objects (to be visualized in 3D) at various distances, and try different shift distances.

Report on subjective evaluations of your program by different viewers (e.g., which relative channel shift was selected by which viewer).

1.3. [Possible lab project] Put a common digital camera on a tripod and record a series of images by rotating the camera: capture each image such that the recorded scene overlaps to some degree with that recorded by the previous image. Now use a commercial or freely available stitching program (note: such software is often included in the camera package, or available via the web support provided for the camera) to map the images into a 360° panorama. When applying the software, for the projection type select either "cylindrical" or "panorama".

1.4. [Possible lab project] Record a series of images by using the camera on a tripod but now do not level the tripod. The rotation axis should tilted by some significant degree to the surface you want to scan. When taking the pictures go for some unique geometries (i.e., straight lines on buildings, windows or simple geometries). For this experiment it is not important to have a full 360° scan.

1.7 Further Reading

Gibson (1950) – see, for example, the experiments described on pages 183–187 – explains the importance of wide-angle perception for human visual perception of the world. For edited volumes on panoramas in the context of machine vision, see Benosman and Kang (2001), Daniilides and Papanikolopoulos (2004) and Daniilides and Klette (2006). The basic notions specified for this book are unified from the diversity of notions used in those edited volumes.

For Chinese scrolls, see Soper (1941), Priest (1950) and Lee and Fong (1955). These sources do not provide an answer to the question "how and why the longish landscape painting was developed originally", which would be interesting to know (maybe because it was just the best way of illustrating the beauty of "mountain and water"). For the Bayeux Tapestry, see *http://www.bayeuxtapestry.org.uk*. James Wyld's (1812–1887) inverted globe on Leicester Square in London (1851–1862) was an important example of wide-angle viewing; for a large clear image of this globe see Williams (1978, pp. 128–129).[5] See also the introductory chapter in (Benosman and Kang, 2001) about the history of panoramic imaging, especially in Europe.

For the history of photography, see Willsberger (1977) and Gernsheim (1982). Willsberger (1977) also lists and illustrates some early examples of panoramic or stereo cameras. Koetzle (1995) contains detailed stories about early photographs. See also Davis and Stanbury (1985) for early photography in Australia, also containing stereo image pairs and panoramic images. For a forum with current discussions about (3D crosseye) stereo panoramas, see *http://www.panoguide.com/forums/galleries/4519/*.

Standard references about photogrammetry are, for example, Albertz and Kreiling (1989), Luhmann (2003), Kraus (1997) and Gruen and Huang (2001). Luhmann et al. (2006) is a recent book on photogrammetry focusing on close-range photogrammetry and its techniques and applications. For the briefly listed periods of photogrammetry, see Albertz and Wiedemann (1996). See Schwidefsky (1971) and Burtch (2007) on the work of Albrecht Meydenbauer; on Aimé Laussedat, see Burtch (2007). Historic photographs may also be used for some 3D modeling; for example, Ronald F. Keam, in his historical studies of the Rotorua geothermal region (especially the volcanic eruption of Tarawera), has done much photogrammetric analysis by microscopic examination of original negatives of 19th-century photographs; see Keam (1988), which is a magnificently illustrated book.

For early papers on image mosaicing in computer vision, see Mann and Picard (1994), Anandan et al. (1994), Kumar et al. (1995) and Irani et al. (1995); applications of image mosaicing are described in Burt and Adelson (1983), Anandan et al. (1994), Irani et al. (1995), Sawhney and Ayer (1996) and Lee et al. (1997). Recently, image mosaicing also addresses arrays of stereo cameras (Zhu et al., 2007) or multiple fisheye cameras (Liu et al., 2008), the latter for creating a bird's-eye view.

For the beginning of stitching-based panoramic imaging, see Chen (1995); see also Ayache and Faverjon (1987) and Davis (1998). Error analysis for this kind of panoramic image generation has been characterized and discussed in, for example, Kang and Weiss (1997) and Wei et al. (1998). A comparative analysis of various stitching techniques was given in Chen and Klette (1999). A recent tutorial on stitching is Szeliski (2006).

For catoptrics, the science of reflecting surfaces (mirrors), see Hecht and Zajac (1974). This field was established by Euclid in his book of the same name. For catadioptric panoramic imaging, see Zheng and Tsuji (1992) and Baker and Nayar (1999). Applications include robot navigation, teleoperation, and 3D scene reconstructions.

Multi-viewpoint panoramic imaging has been widely accepted for applications such as virtual travel, real estate, or architectural walk-throughs; see Irani et al. (1995), Chen (1995) and McMillan and Bishop (1995) for its beginnings. Multi-viewpoint applications can be supported by environment maps. Polyhedral environment maps had been proposed by Greene (1986) to render either a far-away backdrop or reflections of specular objects closer to the viewer. Dynamic panoramas (Rav-Acha et al., 2005) are a recent development for allowing flexible (also in time!) panoramic viewing based on captured video.

Johansson and Börjesson (1989) describe experiments for studying wide-angle space perception, using (synthetic) patterns and a spherical image projection model. Wide-angle stereo image analysis or visualization will be addressed in later chapters of this book, with more references. For a start, Huang and Hung (1997) designed a first system for creating stereo panoramas. Wei et al. (1998) introduced a new stereo panorama visualization approach, based on a general optimization of color filters for full-color anaglyph stereoscopic viewing. The incorporation of stereo object movies into a visualization of stereo

---

[5] This book starts with a provocative preface on the role of measurements in science, citing an 1883 lecture by Lord Kelvin (Kelvin, 1889) as follows: "When you can measure what you are speaking about, and express it in numbers, you know something about it; but when you cannot measure it, when you cannot express it in numbers, your knowledge is of a meagre and unsatisfactory kind; it may be the beginning of knowledge, but you have scarcely, in your thoughts, advanced to the stage of science, whatever the matter may be."

panoramas was also discussed in Hung et al. (2002). For omnidirectional stereo imaging (S. Peleg and students), see, for example, *http://www.ben-ezra.org/omnistereo/omni.html* and references and links on that page. See also Suto (2008) for software for generating stereo panoramas (and examples of those, with auto-spin and at good resolution).

A rotating sensor-line camera was originally (in the computer vision literature) an abstraction of the approach by Peleg and Herman (1997) where panoramas are created by selecting columns in subsequent video frames (i.e., also allowing a rotating video camera). At the same time, those already existed in hardware; see, for example, Reulke and Scheele (1998) for the WAOSS/HRSC camera (the WAAC, mentioned in Section 1.5.2, was a predecessor of this camera).

For laser-range finders in general, and their accuracy in particular, see Boehler et al. (2003) and Sgrenzaroli (2005). The principle of a laser range-finder using triangulation is explained, for example, in Klette et al. (1998). See *http://citr.auckland.ac.nz/~rklette/talks/05_China.pdf* for some illustrations of the approach of combining data recorded by rotating sensor-line cameras and range-finders.

Panoramic or 3D viewing may benefit from current developments in screen technology (e.g., for visualizing 3D images or video). See, for example, Blundell (2007).

2

Cameras and Sensors

This chapter starts by recalling a camera model sometimes referred to in computer vision and photogrammetry: the pinhole camera. It also discusses its ideal mathematical and approximate implementation by means of a sensor-matrix camera. We recall a few notions from optics. Panoramic sensors are basically defined by the use of "panoramic mirrors", or controlled motion of such a sensor-matrix camera, which may "shrink" into a sensor-line camera. We conclude with a brief discussion of laser range-finders as an alternative option for a panoramic sensor.

2.1 Camera Models

We consider cameras and sensors at varying levels of abstraction. For example, when using a camera or sensor, we have to take into account as many parameters as possible, to ensure perfect modeling of the image recording situation, and thus high-quality imaging. However, when talking about basic camera or sensor design, we may be more abstract. We start the chapter with such an abstract notion.

2.1.1 Capturing Surface and Central Point

A panoramic camera or sensor captures data about a 3D scene along projection rays. (We neglect for the moment the fact that these rays are actually refracted by some optical system.) In some abstract sense, to be specified for each sensor, these rays map 3D data onto some surface, which can be modeled by a simple geometric shape. For example, a sensor matrix and a "normal optical system" may be understood as defining a light-sensitive *rectangle*. A fisheye lens may capture "a full half-sphere of directions" into a sensor array, and we may consider either a spherical surface patch or a rectangle as being the capturing surface of this camera (with proper specification of corresponding projection equations). For a laser range-finder we can assume that all rays emerge at one projection center, and visible points in a "full sphere of the visible 3D scene" are mapped onto the surface of a virtual sphere of some fixed radius (Note that the tripod of the range-finder will obstruct some area of the sphere.)

Panoramic Imaging: Sensor-Line Cameras and Laser Range-Finders   F. Huang, R. Klette, and K. Scheibe
© 2008 John Wiley & Sons, Ltd

DEFINITION **2.1.** *A panoramic camera or sensor performs a projection from 3D space onto a* capturing surface, *which is a geometric abstraction using the "most similar" basic surface shape.*

Basic surface shapes are rectangles, or surface patches of spheres, ellipsoids, paraboloids, cylinders or other geometric objects of comparable geometric simplicity.

In the case of a sensor-matrix or sensor-line camera, in an abstract sense we are projecting 3D data onto a planar rectangular window; the unbounded extension of this rectangular window identifies an *image plane*. In case of the (non-rotating) sensor line, the width of this window is defined by the width τ of one pixel only.

As another example, if the sensor is defined by an ideal circular rotation of a sensor-line camera (composing an image, column by column, during this rotation), the capturing surface is the surface of a straight cylinder of some finite height, and the unbounded extension of this finite cylinder defines an *image cylinder*.

In general, a capturing surface of a panoramic camera or sensor is a 2D manifold.[1] For the case of an (abstract) camera, for which we assume an optical system (and thus also an *optical axis*, which is centered with respect to the optical system), we also have the following:

DEFINITION **2.2.** *The* central point *of an image is the point where a camera's optical axis intersects the image plane.*

An (abstract) camera, which is basically characterized by central projection into a plane, is commonly called a "pinhole-type camera" (for more on this, see below).

If a panoramic sensor is defined by rotating a camera or an image plane (in an abstract geometric sense, characterizing a capturing surface this way) then the central point describes a curve on the capturing surface.

2.1.2 Pinhole Camera

The basic projection performed by a matrix-sensor camera is often modeled in photogrammetry by a *pinhole camera*; such a camera does not have a conventional glass lens, but an extremely small hole (e.g., 1/16 inch, which is as small as the tip of a ball-point pen) in a very thin material.

Pinhole cameras have a known history of about 2,500 years, with early reports on light projections on walls in darkened tents or rooms in China and Greece (Aristotle and Euclid), and they are still popular today for experimental photography or surveillance.[2] A simple pinhole camera (as used for the images shown in Figure 2.1) can be build out of a shoe box, using some tinfoil with a small hole (e.g., made by a needle) as a pinhole, and a piece of photographic paper in the image plane.

---

[1] For a definition of an nD manifold, $n \geq 2$, see, for example, Definition 7.10 in Klette and Rosenfeld (2004).

[2] See, for example, *http://photo.net/learn/pinhole/pinhole*, *http://www.pinholephotography.org/*, *http://www.pinhole. com/*, *http://users.rcn.com/stewoody/*, or *http://en.wikipedia.org/wiki/Pinhole_camera*.

Figure 2.1 Examples of pinhole photographs, illustrating typical results when following a basic pinhole camera approach only. Courtesy of Buybooks Marius, 2005.

Modern technology allows these cameras to record images at very low light levels, a full 90° image on a recording sensor matrix, and images are often sharper than with many camcorders. The ideal pinhole is circular, and the optimal pinhole diameter D_a (the *aperture*) was calculated as

$$D_a = 1.9 \cdot \sqrt{f\lambda}$$

by John Strutt, third Baron Rayleigh (1842–1919), where f is the distance between pinhole and image plane, and λ is the wavelength of light (e.g., 550 nm for yellow-green light).

Photogrammetry sometimes uses a pinhole camera model (with a circular pinhole) for discussing subjects in camera optics.

2.1.3 Ideal Pinhole Camera

Figure 2.2 shows ideal central projection, where the size of a pinhole was reduced to that of a dimensionless point, the *projection center*. (In other words, we have an infinitesimal aperture.) The capturing surface represents the photosensitive area, and it is a rectangle (approximating the sensor matrix) in the image or *projection plane*. This plane is actually behind the projection

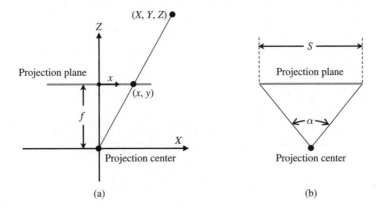

Figure 2.2 (a) A 2D sketch (top view) of projecting a 3D point (X, Y, Z) into (x, y). (b) Illustration for the definition of α.

center, and this is the common point of view in computer graphics. However, in computer vision or photogrammetry it proved to be convenient to illustrate and model the image plane in front of the projection center.

DEFINITION **2.3.** *An ideal pinhole camera is defined by central projection of 3D space into a rectangular window of size $A \times B$ in the projection plane.*

Assuming an xy coordinate system in the projection plane and an aligned XYZ coordinate system in 3D space, where the origin coincides with the projection center, such an ideal pinhole camera is only defined by its *focal length* $f > 0$, specifying the central projection (see Figure 2.2(a)) as a mapping

$$ x = \frac{fX}{Z} \quad \text{and} \quad y = \frac{fY}{Z}. \tag{2.1} $$

Both equations can be derived from similarities of triangles (also known as the ray theorem in elementary geometry).

In cases where the value of f is not (even in some abstract sense) defined by a focal length of a camera, we call it the *projection parameter*. Equations (2.1) remain valid.

An ideal pinhole camera has a *viewing angle* (see Figure 2.2(b)) of

$$ \alpha = 2 \arctan \frac{S}{2f}. $$

where S is equal to A or B (e.g., 24 mm in the vertical and 36 mm in the horizontal direction in the case of 35 mm photographic film). In computer graphics, the *viewing rectangle* is defined by the horizontal viewing angle, an aspect ratio A/B, and the focal length f.

The focal length f typically starts at 14 mm, and can go up to multiples of 100 mm. For example, for $S = 36$ mm and $f = 14$ mm, the horizontal viewing angle is about $\alpha = 104.25°$.[3]

Obviously, this ideal pinhole camera uses notions of optics in an abstract sense. The ideal pinhole camera also disregards the wave nature of light, assuming ideal geometric rays.

For a pinhole camera with an infinitesimal aperture it is common to assume that objects are in focus if they are "at infinity" (simply put: "very far away"). Projection rays, coming from a visible surface point "at infinity" and parallel to the optical axis, are bundled in a focal point which defines the position of the *focal plane* for objects at this distance.

The ideal pinhole camera assumes that objects are in focus, whatever their distance from the camera. If projecting a visible surface point at close range, in practice we would have to focus the camera at this range; the parameter f of the camera thus increases to $f + z$. In this book, we also call the resulting parameter focal length, and again use the symbol f.

We do not introduce the notion of a *camera constant*, which is also in popular use. The notion "focal length" is intuitively more appealing, but f is in general just a distance value characterizing a camera's central projection (calibrated or calculated with respect to a defined geometric model of the camera; it may be, in special circumstances, the "true" focal length f of the camera – see Section 2.2).

---

[3] For readers who prefer to define a *wide angle* accurately, let it be any angle greater than this particular $\alpha = 104.25°$, with 2π as an upper bound.

2.1.4 Sensor-Matrix Cameras

The quality of recorded photographic images has constantly improved since 1826, due to progress in optical lenses and the recording medium, with, for example, silver-based film being replaced since 1981 (Sony's Mavica) by digital technologies, defining *digital cameras*.

A *sensor-matrix camera* combines common camera optics with a grid (the "matrix") of phototransistors, using a CCD, CMOS, or further (novel) sensor principles. Basically, such a matrix (see Figure 2.3) is a field of rectangular light-sensitive cells, each of size $\tau_w \times \tau_h$ where τ_w and τ_h are in micrometers (e.g., about $2\,\mu$m). Ideally, the aspect ratio τ_w/τ_h between the width and height of a sensor cell should be equal to 1. In this book we assume square sensor cells of size $\tau \times \tau$.

2.1.5 Sensor-Line Cameras

As a special option, the sensor matrix might just be of size $1 \times M$, which defines a *sensor-line camera*. For example, $M \geq 10,000$ sensor cells is possible today at reasonable cost for professional photographic equipment.

Originally, sensor-line cameras were built for aerial mapping; see Figure 2.4 for an example. The WAAC used has three sensor-lines mounted on one focal plate (i.e., behind the optical system); each has about 5,000 CCD sensor elements for capturing gray-level image data. Figure 2.4 shows part of the Hauraki Gulf; note that the reflection captured on the water is different from that which a sensor-matrix camera with a single projection center would capture from about the same elevation.

Sensor-line cameras are today very popular in various applications of computer vision, such as industrial inspection. Figure 2.5 shows a research setup for inspecting pipe welding lines.

Sensor-matrix cameras, sensor-line cameras, pinhole cameras, and ideal pinhole cameras are characterized by having a planar capturing surface (in the first two cases, obviously, only

Figure 2.3   CCD sensor matrix of a digital camera.

Figure 2.4 A scan in February 2001 of Auckland's central business district (CBD), using one of the three WAAC sensor lines (see also Chapter 1). The aircraft was at about 3,000 m elevation, and its egomotion is eliminated from the image by geometric rectification of all the contributing line images. Courtesy of Ralf Reulke, DLR Berlin-Adlershof.

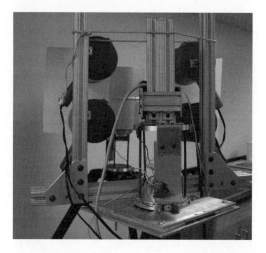

Figure 2.5 A research setup in 2006 to investigate the use of a sensor-line camera for inspecting pipe welding lines, with translational movement of the camera for longitudinal lines, or rotational movement for transverse lines. Courtesy of Andreas Koschan, University of Tennessee, Knoxville.

in some abstract sense). Distinguishing between planar and non-planar capturing surfaces, simplifies the expression of certain points of view and or results in this book. The authors decided for this reason to use the term "pinhole camera" to identify a basic camera architecture characterized (in an abstract sense) by central projection and a planar capturing surface.[4]

---

[4] Alternatively, *pinhole imaging* could also just identify central projection (through a small pinhole), for example, also on a cylindrical capturing surface. But we will not do so.

2.2 Optics

Real physical imaging with light involves some energy transfer. In optics, this is described by electromagnetic radiation in the *optical waveband*, which ranges from ultraviolet (UV) to infrared (IR). Therefore, imaging depends on the laws of propagation of electromagnetic waves, and the energy transport requires a finite (non-vanishing!) aperture of the optical system. The ideal pinhole camera needs to be extended to a camera model with non-vanishing aperture if more detailed understanding of the imaging process is required.

2.2.1 Waves and Rays of Light; Law of Refraction

For the spectral range of visible light, the wavelength of electromagnetic radiation is very small (approximately 0.5 μm). Therefore, we may in many cases neglect (in a first step) the wave nature of light. For the limit case of vanishing wavelength, the propagation of light can be described in terms of rays. These rays of light represent the direction of energy flux of the electromagnetic waves, and they are normal to the surfaces of wave fronts.

The propagation of light rays within refracting media (e.g., air or lenses) is described by the law of refraction (illustrated in Figure 2.6(a))

$$n \sin \alpha = n' \sin \alpha'$$

where n and n' denote refraction indices. Obviously, the law of refraction defines a non-linear mapping, and thus we cannot expect to have a linear imaging transform (i.e., a central projection) with rays of light. However, for small angles, between incoming rays and the normal of the surface of the lens (at the point where the ray hits the lens), we can approximate the sine function by the angle α itself. In this case we actually have approximately a linear imaging situation.[5]

2.2.2 Ideal Gaussian Optics

Under small-aperture conditions, a pencil of rays, which radiates from an object point **P** in a 3D scene (see Figure 2.6(b)) is transformed by the optical system into a set of rays, which (ideally)

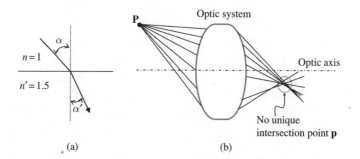

Figure 2.6 (a) Illustration of the law of refraction, and (b) its meaning for an optical system.

---

[5] For example, for a difference angle of 7°, the difference between $\sin \alpha$ and α is about 1%.

should all intersect again at exactly one point **p** thus forming a pencil of rays again. Such a unique intersection would be the image of **P**. In this ideal case we speak about *convergence of rays*. Basically, the optic system defines in this case a linear mapping of **P** into **p**, and we speak about *linear imaging*.

The linear imaging assumption is valid for optical rays within a small volume (known as *paraxial region*), centered around the optical axis of the optical system. Imaging laws in the paraxial region are described by linear *Gaussian optics*. Ideal Gaussian optics serves as a reference for real physical imaging. Deviations from Gaussian optics are considered to be image errors (or aberrations), although these are, of course, not errors in the sense of manufacturing tolerances but are due to the underlying physical laws.

The design of optical systems aims to enlarge the paraxial region, and, for this reason, more than just one single lens is normally used in such an optical system. The number of lenses increases with image quality demands.[6]

Now assume ideal Gaussian optics. Note that a surface point reflects a bundle of light rays into the 3D scene, and some of those may actually be in the paraxial region of a given optical system, even if the surface point is further away from the optical axis, but then may also vignetted.

We specify below three *characteristic Gaussian rays*[7] which describe typical refractions of rays within an optical system. In order to do so, we will utilize a few basic notions, which are also illustrated in Figure 2.7. We start with the following:

DEFINITION **2.4.** *The* optical axis *is the straight line which is incident with all the centers of all (rotationally symmetric) surfaces of lenses of a given optical system.*

Assume a bundle of rays, coming from infinity (i.e., parallel rays), which are transformed by the optical system into a (dimensionless) point **p** "behind the optical system", thus forming a pencil of rays behind the optical system. If the incoming bundle of parallel rays is also parallel to the optical axis, then point **p** equals point **F'** in Figure 2.7. Light may also pass the optical system in the opposite direction, thus defining point **F** by mapping a bundle of rays, all parallel to the optical axis, through the optical system.

DEFINITION **2.5.** *The* focal points **F** *and* **F'** *are the defining points of those bundles of rays, all parallel to the optical axis, with resulting pencils of rays, each incident with focal point* **F** *and* **F'***, respectively.*

The geometric refraction is modeled to occur at two *principal planes H* and *H'*, having *principal points* **H**$_+$ and **H'**$_+$ as intersection points with the optical axis. The principal planes *H* and *H'* are (abstract) parallel planes within the optical system: a recorded image in one plane would be always identical to the image in the other plane, due to an assumed parallel projection between both planes.

---

[6] A few years ago, glass lens production was basically by mechanical grinding, and produced (concave or convex) spherical lenses. New technologies allow relatively inexpensive production lines for aspherical lenses, which help to optimize optical system design. This has recently become noticeable in the quality of glass lenses in non-professional digital cameras, and already in cameras of mobile phones.

[7] Note that these are rays in optics, and there are also "Gaussian rays" in physics, defined by the propagation of electromagnetic waves through given media.

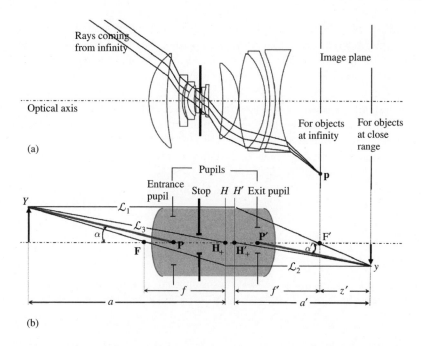

Figure 2.7 Geometric sketch of an optical system: (a) rays passing through a multi-lens optical system; (b) illustration of abstract definitions for such an optical system. A 3D point (X, Y, Z) is mapped into a 2D point (x, y, z_0), where z_0 defines the position of the image plane.

In the case of an optical system, formed by a single thin lens only, both principal planes can be assumed to be identical, forming *the* principal plane H, with a uniquely defined principal point H_+.

DEFINITION **2.6.** *The* focal length f or f' *is the distance between* **F** *and* **H**$_+$, *or between* **F'** *and* **H**$'_+$, *respectively.*

We now apply the above notions to identify three characteristic rays of Gaussian optics. All three rays project from 3D scene space, through the optical system, then towards the image, and are finally assumed to continue to infinity. The first ray $\mathcal{L}_1$ is assumed to be parallel to the optical axis; it passes through focal point **F'**. The second ray $\mathcal{L}_2$ is assumed to be incident with focal point **F**; it refracts such that it leaves the optical system parallel to the optical axis. The ray $\mathcal{L}_3$ passes through the principal point **H**$_+$ and continues from principal point **H**$'_+$; it leaves the optical system "parallel to itself", and can be assumed to follow the optical axis between **H**$_+$ and **H**$'_+$. See Figure 2.7(b) for those three characteristic rays.

2.2.3 Pupil, Aperture, and f-Number

Every technical implementation of an optical system defines some limitations for projected ray bundles (e.g., size and shape of lenses, camera stops). These limitations have effects on forming ideal ray pencils, and thus on the resulting image.

An *entrance pupil* (denoted *EP*) in the object space of the optical system defines a restriction on incoming rays. The entrance pupil is a virtual aperture, the object side image of the physical stop (i.e., iris), and the image of the stop in the image space is called the *exit pupil* (denoted *XP*).

The *central ray* for these pupils (drawn as a bold gray line in Figure 2.7) proceeds from an object point Y (here simply denoted by its Y coordinate only, as usual in a camera YZ coordinate system) to the center of the entrance pupil (point $\mathbf{P}$), and in image space from the center of the exit pupil (point $\mathbf{P'}$) to the image point specified by its y coordinate. This is the most important ray for tracing an optical system; together with the marginal rays (i.e., those limited by aperture) it describes a tube (or a bundle of rays) that are passing through the optical system.

Let D_{XP} and D_{EP} be the diameters of exit and entrance pupil, respectively. Both define the *pupil magnification ratio*

$$\beta_P = \frac{D_{XP}}{D_{EP}}.$$

The finite aperture of projected pencils of rays is characterized by

$$K = \frac{f'}{D_{EP}}$$

which is known as the *f-number*.

When rotating a sensor-matrix or sensor-line camera for stitching (in the simple case) or combining, column by column (in our case), panoramas, it is important that the center of the entrance pupil describes a circular rotation for avoiding parallax effects.

2.2.4 Relation to the Pinhole Camera

Gaussian optics describes imaging, and can be identified by describing the geometry of the three characteristic rays in relation to both principal planes H and H'. Obviously, these two planes, with principal points $\mathbf{H_+}$ and $\mathbf{H'_+}$, describe a refraction which differs from the ideal central projection as assumed for the ideal pinhole camera.

Also, the pupils define cones of light, and rays, emerging from any object point, are entering or exiting the optical system within the given limitations of those cones only.

Assume a pupil magnification ratio $\beta_p \neq 1$. Then the bold gray central ray (see Figure 2.7(b)) describes two angles with the optical axis, a pupil entrance and a pupil exit angle α and α', respectively; see Figure 2.8. The upper case is the reality, and the lower case is our goal. To achieve this goal, we have to shift both the exit pupil and the image plane horizontally along the optical axis such that the mathematical projection center $\mathbf{P^*}$ is equal to the optical projection center $\mathbf{P}$.

The *camera constant c* (defined by the distance between mathematical projection center $\mathbf{P^*}$ and image plane) can be estimated as

$$c = f' \left(1 - \frac{\beta}{\beta_P}\right)$$

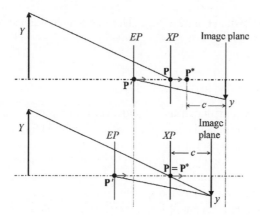

Figure 2.8 Illustration of mapping Y (of point (X, Y, Z)) into y (into image point (x, y, z_0)); the mathematical projection center $\mathbf{P}^*$ either differs from the optical projection center (top), or both are equal (bottom). Planes EP and XP as in Figure 2.7. Courtesy of K. Lenhardt.

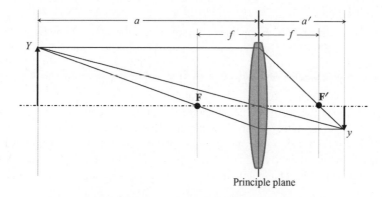

Figure 2.9 Geometric entities of a thin lens.

where β denotes the *image magnification*

$$\beta = \frac{-z'}{f'}.$$

Obviously, it follows that $c = f' + z'$ if $\beta_P = 1$. The distance z' between focal plane for objects at infinity, and focal plane for objects at close range (see Figure 2.7(b)) needs to be calculated by camera calibration.

A lens is *thin* if its thickness is very small compared to the parameters f or f'. Figure 2.9 illustrates a thin lens. In this case, we consider only a single principal plane, with $f = f'$. The magnitudes of the displayed distances a, a', and f satisfy the following relationship (known as the *Gaussian lens equation*):

$$\frac{1}{a} + \frac{1}{a'} = \frac{1}{f}$$

where $a' = f' + z'$, as shown in Figure 2.7(b). (The Gaussian lens equation often uses symbols b and g instead of a and a'.)

2.3 Sensor Models

A *sensor* is in general anything that measures and records or reports data. For example, a single CCD or CMOS cell is a sensor, and this is actually the motivation for the terms "sensor matrix" and "sensor line". However, besides using these two terms frequently in this book, we identify the notion *sensor* from now on by default not with such "micro-sensors" (also called *sensor cells*) but rather with "macro-sensors" (also called *devices*), which may be engineered using one or more cameras, and some other tools, such as one or more turntables, a mirror, a slider, and so forth. For example, a catadioptric camera system combines a camera with a mirror, and this is a (panoramic) sensor in our terminology. We also consider a range-finder to be a (panoramic) sensor.

2.3.1 Rotating Sensor-Line Cameras

A prototype of a rotating CCD sensor-line panorama camera was shown by DLR at Photokina 1998; it captured 8,000 RGB pixels in one line. Panoscan in California began manufacturing rotating sensor-line cameras (their MK-1) in January 1999; see Figure 2.10(a). With such a panoramic camera it is possible to have very-high resolution data compared to sensor-matrix cameras. The trilinear sensor of the MK-1 contains sensor elements for 7,072 color pixels, and during one 360° rotation the camera captures up to 22,400 columns (which is 453 MB). Figure 2.10(b) shows a rotating sensor-line camera (Eyescan) which was developed between 1999 and 2001 at DLR Berlin in Germany (in cooperation with KST GmbH, Dresden).

Such cameras are based on a rotating *trilinear sensor-line*: three (e.g., CCD or CMOS) lines (e.g., for red, green and blue channels) form one linear sensor-cell array, which is mounted on a

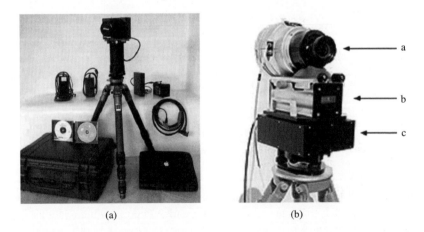

(a)   (b)

Figure 2.10   (a) Panoscan MK-1 (1999) with a PowerBook. (b) Eyescan M3 (2001) which uses a laptop for control and data storage.

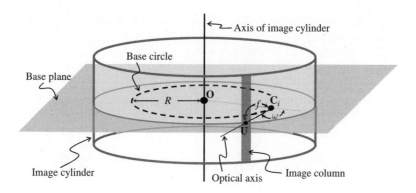

Figure 2.11   Basic entities of a rotating line camera.

planar surface behind the optical system of the camera.[8] As the sensor-cell array is rotated 360°
along with the camera (on an ideal circle), it describes (in some abstract sense) a cylindrical
surface (see Figure 2.11). The *image cylinder* describes the mathematical abstract location of
those rotating trilinear sensor lines.

DEFINITION **2.7.** *The* size *of a panoramic image* captured by a rotating sensor-line camera
is described by the parameters H and W, where the height *H is the number of pixel sensors
in the line, and the* width *W is the total number of lines captured in order to generate this
panoramic image.*

Ideally, the projection center of the sensor line rotates through points $\mathbf{C}_i$ (for $i \in \mathbb{Z}$) which
are all incident with a *base circle* (see Figure 2.11) in the *base plane*. A projection center $\mathbf{C}_i$
(for some i) identifies a position of the rotating sensor-line camera. The rotation axis is the axis
of the image cylinder, and this is assumed to be perpendicular to the base plane, intersecting
this plane at point $\mathbf{O}$.

The optical axis of the rotating sensor-line camera is assumed to remain incident with the
base plane, at all of its positions during the rotation. If the optical axis is incident with the
straight line passing through points $\mathbf{O}$ and $\mathbf{C}_i$, the angle between $\overline{\mathbf{OC}_i}$ and the optical axis
specifies the camera's viewing direction at position $\mathbf{C}_i$. Let $\mathbf{U}$ denote the intersection of the
optical axis and the cylindrical surface.

DEFINITION **2.8.** *The base circle has a radius R, which is called the* off-axis distance. *The
optical axis of a camera at position* $\mathbf{C}$ *forms the* principal angle ω *with the surface normal of
the image cylinder (passing through* $\mathbf{C}$*). Let* $\mathbf{U}$ *be the point where the optical axis intersects
the cylindrical surface. The Euclidean distance between* $\mathbf{C}$ *and* $\mathbf{U}$ *identifies the* projection
parameter f *of the camera at position* $\mathbf{C}$.

In the ideal case, the projection parameter f and the principal angle ω are assumed to remain
constant during a rotation of a sensor-line camera (i.e., during the recording of one panoramic
image).

---

[8] As discussed above, a given image plane can only be a "focal plane" for some of the 3D scene points. However, we
use the notion "focal plane" for this planar surface, where the sensor lines are mounted.

In the degenerate case of $R = 0$, we assume that the base plane is defined perpendicular to the rotation axis of the sensor, and ω becomes meaningless in this case.

Of course, when dealing with real rotating sensor-line cameras in real-world applications, we have to understand deviations from these assumptions.

Three Camera Components

A rotating sensor-line camera may consist of three major components: a camera head (see 'a' in Figure 2.10(b)), an optical bench (b), and a drive (c).

The camera head consists of optical and electronic components, fiber optics (elements for bidirectional data transfer), and the trilinear CCD sensor array.

The optical bench supports the camera head and is used for physical movement of the camera. It enables adjustment of the optical axis of the camera head parallel to the base plane, and aims in general to have the projection center of the camera exactly on the rotation axis (see Figure 2.11).

However, considering only this situation with $R = 0$ would actually restrict the potential use of such a camera. Shifting the camera head away from the rotation axis by a distance R, and tilting the camera's optical axis (with respect to the normal of the image cylinder; see the dashed line in Figure 2.11), by an angle ω, also allows stereo panoramas to be captured (as explained in Chapter 4).

Mechanical locking mechanisms on the optical bench are used to control the different capturing modes (namely, single-center or stereo modes). To use the camera in stereo mode it is possible to unlock the mechanism and allow a shift of the camera head by a distance R, which is restricted by the maximum length of the optical bench used.

The drive rotates the optical bench together with the camera head (for a continuous CCD line rotation). The drive contains a measuring system for angular increments, which allows the user to specify a variable angular interval and number of columns to be captured during the optical bench rotation. The drive also supports camera calibration (this is explained in Chapter 5).

Obviously, the shift of the camera head away from the rotation axis is limited by the optical bench used, and an experimental rotating sensor-line camera is illustrated in Figure 2.12, specially configured to allow larger distances R (up to 1 meter) for recording stereo panoramic images in various 3D scene geometries. The optimization of R and ω, depending on scene geometry, is one of the subjects of this book.

Angular Increment and Instantaneous Field of View

Let $\mathbf{C}$ denote the camera's projection center; this is often given a subscript i to indicate the projection center when capturing column i of the panorama, with $0 \le i \le W - 1$. If a camera is rotated through 360° (by a turntable) at uniform speed, then those projection center positions are uniformly distributed on the base circle. In such a case, an angle γ describes the constant angular distance between any pair of adjacent projection centers on the base circle with center at $\mathbf{O}$.

DEFINITION **2.9.** *The angle γ is called the* angular unit *or the* angular increment *of the rotating sensor-line camera.*

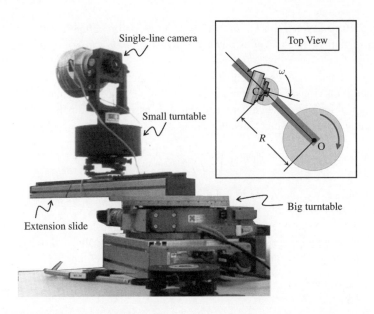

Figure 2.12 An experimental rotating sensor-line camera configuration from 2002, using an Eyescan M3 and an additional extension slider.

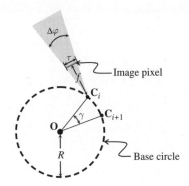

Figure 2.13 Angular increment γ and instantaneous field of view $\Delta\varphi$.

In Figure 2.13, this angular increment is illustrated by the angle formed by $\mathbf{C}_i$ and $\mathbf{C}_{i+1}$, with

$$\gamma = \angle \mathbf{C}_i \mathbf{O} \mathbf{C}_{i+1}.$$

Due to practical limitations, we always have a finite number W of projection center positions, and thus $\gamma > 0$. Obviously, we have the following:

$$\gamma = \frac{360°}{W}.$$

Figure 2.13 also illustrates another important parameter of a sensor-line camera:

DEFINITION **2.10.** *The horizontal field of view of a sensor-line camera defines an angle* $\Delta\varphi$ *called* the instantaneous field of view, *which depends upon the chosen focal length* f.

(The value of $\Delta\varphi$ also depends on sensor pixel size τ, but this remains, of course, constant for a given camera.) Obviously, the instantaneous field of view does not change as long as we do not alter the focal length of the camera, and thus it is assumed to remain constant during one rotation of the sensor-line camera. We have that

$$\Delta\varphi = 2 \cdot \arctan\left(\frac{\tau}{2f}\right) \approx \frac{\tau}{f}, \qquad \text{for } f \gg \tau. \tag{2.2}$$

When deciding how many image columns should be recorded during (say) a full $360°$ rotation, the goal is to ensure $\tau \times \tau$ (square) pixels.[9] In other words, the goal is to ensure uniform sampling of 3D scene data, ideally described by a "gap-less" and "non-overlapping" situation for subsequent image columns. In the case of $R = 0$, the value of the angular increment can be set equal to the instantaneous field of view. Then we have that

$$W = \frac{360°}{\Delta\varphi}.$$

For the general case of $R > 0$ and $\omega > 0$, this sampling issue (i.e., the determination of W) will be discussed in detail in Chapter 7.

Field of View and Focus

The horizontal field of view (H_{FoV}) of a rotating sensor-line camera can be obtained by the formula

$$H_{FoV} = \gamma(W - 1) + \Delta\varphi.$$

This is also true for the general case when both R and ω are greater than zero or when we have different scales of sampling between horizontal and the vertical directions. For the special case when $\gamma = \Delta\varphi$, we have

$$H_{FoV} = \Delta\varphi W.$$

The vertical field of view V_{FoV} is defined by the focal length and the physical length of the sensor line. (This length is the number H of pixels times τ.) We have that

$$V_{FoV} = 2 \cdot \arctan\left(\frac{H\tau}{2f}\right).$$

The focal plane is assumed (by rules of Gaussian optics) to record a "crisp" (i.e., focused) projection **p** of a 3D point **P**; **p** is at distance a' from principal plane H' (see Figure 2.7). A panoramic camera may allow the focal plane to be shifted by a distance z' into a few

---

[9] The definition of a square pixel is only valid for a planar image surface. Imagine that the cylindrical capturing surface is unrolled and flattened (see Chapter 3 for details); this also defines a planar surface, but this is geometrically different from the original non-Euclidean capturing surface.

Focal length (mm)	Focus position				
	0	1	2	3	4
35	1.0	1.5	2.2	3.6	>9.9
45	1.9	3.1	4.6	8.5	>70
60	3.3	4.9	8.0	14.7	>122.0
100	9.2	14.6	22.0	40.0	>336.0
180	30.0	45.0	70.0	130.0	>1080.0

Table 2.1   Ideal object distances a (in meters) for the panoramic camera Eyescan M3 at its predefined focus positions.

prespecified *focus positions* named $0, 1, \ldots, n_{as}$, within an accuracy of some micrometers. These focus positions define physical shifts of the CCD sensor line, with the aim of supporting object distances ranging from close-range to far-range (with "at infinity" for $z' = 0$).

The aim when selecting a focus position is that the focus should ensure a reproducible image accuracy of one pixel, for example, to be verified in laboratory experiments. (Zoom lenses are not able to fulfill such a photogrammetric requirement; this uncertainty is known as the *hysteresis effect*.)

Each optical system has a *depth of the field of view* (D_{FoV}) which depends on its aperture. As a rule of thumb, the D_{FoV} is specified by an interval (defined by lower and upper distance limits)

$$I_{D_{FoV}} = \left[\frac{2a}{3}, \frac{5a}{3} \right]. \tag{2.3}$$

For example, for $a = 8$ m we have that $I_{D_{FoV}} \approx [5.3\,\text{m}, 13.3\,\text{m}]$. Table 2.1 provides values of a for an example of a panoramic camera; these values depend upon the focal length of the optical system used.[10]

The fifth column in Table 2.1 (i.e., focus position 4) shows values for the focus position "at infinity" (called *hyperfocal length*), which supports $z' = 0$ as the minimum object distance for acceptable sharp imaging. The hyperfocal length a_h is geometrically defined by the lens equation, an acceptable circle of confusion σ, and the f-number K (the relation between optical diameter and focal length):

$$a_h = \frac{f^2}{K\sigma} + f.$$

Example: Eyescan M3

Examples of panoramic images on cylindrical capturing surfaces, shown in this book, were captured with an Eyescan M3, where a single image is several hundreds of megapixels, up to

---

[10] Equation (2.3) is approximate; using the lens equation it actually follows that D_{FoV} has to be symmetric. However, this equation takes into account the psychological effect that objects further away only require a low resolution, and can be still perceived as being high-contrast, compared to objects closer to the camera (viewer).

Focal length (mm)	Image size (GByte)	Resolution (degrees)	Resolution (mm at 10 m)
25	1.28	0.0160	2.80
35	1.79	0.0115	2.00
45	2.30	0.0089	1.66
60	3.07	0.0067	1.16
100	5.12	0.0040	0.70
180	9.21	0.0022	0.39

Table 2.2 Image resolution for Eyescan M3, depending on the optical system ("lens") used, for sensor pixel size $\tau = 7$ μm. The values in the last column show the geometric resolution at 10 m distance (edge length of a square projected into a single pixel).

multiples of gigapixels. This camera possesses a 70 mm Kodak CCD trilinear sensor (meaning that it has one sensor line for each of the three R, G, and B color channels). Three color values form a value of one pixel. Each image has a vertical resolution of $H = 10,200$ pixels (for each of the three color channels), and during one 360° rotation we have up to $W = 160,000$ image columns, depending on the lens used.

We give a few examples of technical parameters for the Eyescan M3. The trilinear sensor line of 70 mm length, containing 10,200 pixels, defines a vertical pixel size of

$$\tau = \frac{70\,\text{mm}}{10,200} \approx 7\,\mu\text{m}.$$

Table 2.2 shows dependencies between image resolution and the lens used. The image size in this table is based on having 48 bits (i.e., 3×16 for RGB images) for each pixel, and a full 360° scan. (Actually, the electronic unit used had a 14-bit analog–digital converter, but data were stored in the common 16-bit format.)

Rotating sensor-line cameras are examples of *omnidirectional imaging systems*. 'Omnidirectional' implies here equal visual sensitivity "in all directions". For a rotating sensor-line camera (along an ideal circle), the obvious choice for "all directions" are the normals of that circle. (Catadioptric camera systems are another example of omnidirectional imaging; here, the obvious choice of "all directions" are all normals to a half sphere.)

2.3.2 Rotating Multi-Line Cameras

So far we have assumed that there is exactly one sensor line in the rotating camera, centered at its optical axis. (This also covers the case of a trilinear sensor, with three individual sensor lines for red, green, and blue channels, but considering all three sensor cells for one individual pixel as being at the same abstract location).

In fact, from the start, when sensor-line cameras were first designed for space missions, they were designed in a way that multiple sensor lines were mounted on a focal plate (see the comment in Section 2.1.5).

For example, in the WAAC there are three sensor lines, one mounted as for the single-sensor-line case: centered at the optical axis, looking along the optical axis. This line has been called the *nadir line*. The nadir line would look "straight down" during a flight. Symmetrically to the nadir line, two more lines are mounted on the focal plate, designed for "looking forward" and "looking backward", respectively. The directions, from positions of pixels on those two symmetric lines through the mathematical projection center P^* (see Figure 2.8), form angles ω and (symmetrically) $-\omega$ with the optical axis.

When rotating this camera (say, on a tripod), the projection geometry of those two symmetric lines is not the same as rotating a single sensor line with principal angle ω or $-\omega$. Of course, this camera can be used for implementing the single multi-sensor-line approach by just using the nadir line, and having the whole camera rotated by principal angle ω with respect to the optical axis.

A model of such a multi-sensor-line camera is the *Wide Angle Optical Stereo Scanner* (WAOSS) or *High-Resolution Stereo Camera* (HRSC) of DLR in Germany, which combines nine sensor lines on the focal plate, and was particularly designed for Mars missions.

2.3.3 Rotating Sensor-Matrix Cameras

Obviously, a multi-sensor-line camera may be generalized to a matrix-sensor camera. A cylindrical panorama can now be scanned progressively by any chosen sensor-column i of a rotating sensor-matrix camera (not by stitching images, but by composing a panoramic image out of image columns, each image column taken by the same sensor column i but at different rotational angles); see Figure 2.14(a). In this subsection, we assume that the sensor-matrix columns are all parallel to the rotation axis.

We proceed as described for the single-line camera except that the captured image line is now taken from a fixed sensor column of the sensor matrix. Note that any column of the sensor matrix thus defines one specific panoramic image. The choice of the column corresponds (somehow; see below) to the choice of angle ω in case of the single-line camera. Thus, we continue to use the symbol ω for the angle identified by the selected column, assuming that the angle ψ between the surface normal of the sensor matrix and the normal of the base circle (at a given position of the camera) defines an additional parameter.

Similar to the (single) sensor-line camera case, if the projection center of the matrix camera coincides with the rotation center $\mathbf{O}$ (see Figure 2.15), the situation degenerates to a single projection center for all contributing lines. Ignoring optical distortions, in this case any sensor

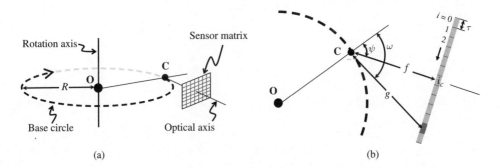

Figure 2.14 (a) Rotation of a sensor-matrix camera for panoramic imaging. (b) Illustration of the calculation of the projection parameter g and principal angle ω for a selected column of the sensor matrix.

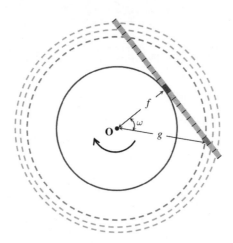

Figure 2.15   A sketch of a multi-sensor-line or matrix-sensor camera; a (single) sensor-line camera has
the sensor line at the position of the dark gray pixel. Rotation of this camera for $\psi = 0$. sensor lines define
a principal angle ω and a projective constant g (radius of dashed circle); different values of g define
different "projective scalings" of the panorama.

column can be chosen, and (up to different scalings!) it would always be basically the same
generated panoramic image (assuming a static scene and constant illumination).

In general, when a matrix camera is rotated with $R \geq 0$, the scanned panoramas depend upon
the selected generating column. Thus, unlike the case of a rotating single sensor-line camera,
a sensor-matrix camera of image resolution $M \times H$, where M is the width and H the height
of the matrix, is able to generate up to M different panoramas of size $W \times H$ (e.g., after a
complete 360° scan). Furthermore, all these panoramas have different projection parameters g
(note: we have $g = f$ for a sensor column chosen exactly in the middle of the sensor matrix)
and principal angles ω, depending on the column chosen.

The panorama composed by the ith column of the sensor matrix has the following parameter
values: R is equal to the distance between the optical center of the matrix camera and the rotation
axis; the values of the projection parameter and the principal angle are given by

$$g = \sqrt{f^2 + (i - i_c)^2 \tau^2}$$

and

$$\omega = \psi + \arctan\left(\frac{(i - i_c)\tau}{f}\right)$$

respectively, where f is the focal length of the sensor-matrix camera, i_c is the central sensor
column[11] of the matrix camera, τ is the size of a single sensor cell (assuming perfectly square
pixels), and ψ is the angle between the optical axis of a matrix camera and the normal of the
base circle passing through the optical center (see Figure 2.14(b)). If $\psi = 0$, we may generate
$M/2$ (± 1) different panoramas due to symmetries.

---

[11] The central sensor column of a matrix camera is the column that the optical axis of the camera passes through.

2.3.4 Laser Range-Finder

Until now, laser scanners have followed one of three design principles. Historically the first option was the *time of flight* principle, whereby the 3D coordinates of a surface point are derived by measuring the time delay between a pulse transmission and the detection of its reflected signal, and the direction of the transmitted pulse. Such a time-of-flight system allows the (unambiguous within certain accuracy limits) measurement of distances up to several hundreds of meters. Dense surface scans (measuring distances to thousands or millions of surface points) can take up to several hours.

As a second option, the *phase measurement principle* is often applied to medium-range scanners. The phase shift of a modulated wave is measured when transmitting and receiving this wave; the phase shift depends on the distance. In combination with the time-of-flight technique, uniqueness (note that phase repeats at 2π) is ensured. High acquisition rates and high densities of measured 3D surface points are supported by phase shift systems. See Figure 2.16 for an example of measured surface points, where gray levels correspond to distance.

A third laser-based approach is similar to the technique of structured lighting (commonly used in computer vision). The *triangulation laser scanner* technique uses a point laser to scan the 3D scene. A camera observes the projected light dot in the scene. The triangle, defined by point laser, camera, and projected light dot, allows the position of the illuminated surface point to be determined under the assumption that the relative position of camera has been calibrated with respect to the point laser. Instead of a single laser dot, a laser stripe is commonly used to accelerate the acquisition process. Very high accuracy (i.e., of some micrometers), but a limited scan range of some meters are the main characteristics of a triangulation laser scanner.

Table 2.3 briefly summarizes the differences among time-of-flight, phase-shift, and triangulation-based systems in terms of accuracy, range, and scan rate. Table 2.4 lists a few laser range-finders (as available in 2003) and their advantages and disadvantages. Figure 2.17 illustrates conventional LRFs with respect to their deviations in range measurements. In this

Figure 2.16 Part of a spherical panoramic range-scan (range is visualized by gray-levels) captured with a laser range-finder (a courtyard in Neuschwanstein castle in Germany). A panoramic camera is visible on the left defining a location for a second panoramic scan (a cylindrical color image).

Design principle	Range (m)	Accuracy (mm)	Scan rate (points per second)
Time of flight	<1500	<20	up to 12,000
Phase shift	<100	<10	up to 625,000
Triangulation	some meters	<0.1	up to 10,000

Table 2.3 Measurement characteristics of three design principles for a laser scanner.

System	Advantages	Disadvantages
Callidus	Very large FOV	Very coarse vertical resolution
Cyrax2500	Good accuracy	Small scanning window
S25	Very high accuracy for short ranges	Does not work in sunlight; not suited for long ranges
GS100	Large FOV	Large noise
Riegl Z210	High ranges possible Large FOV	Low accuracy
Riegl Z420i	Very high ranges possible; large FOV	Large noise
Z + F IMAGER 5003	Very high scanning speed; large FOV	Low edge quality; limited angular resolution

Table 2.4 A few LRFs with their advantages and disadvantages. FOV stands for "field of view". From Boehler et al. (2003).

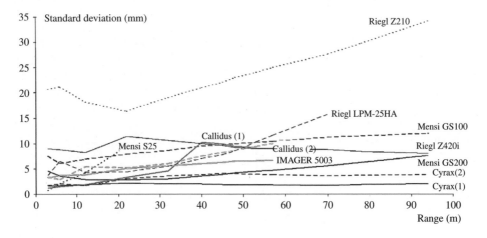

Figure 2.17 Standard deviation of range data, at distances between 2 and 100 m (from Boehler et al., 2003).

simple test, a planar surface was scanned at varying distances from the LRF. The resulting deviation of measured range data is a reliable source of information for the relative accuracy of range measurements. These measurements were repeated and averaged for three different surface paints of the used planar surface, with intensities of about 80%, 50%, or 8% (note that a surface paint defining 100% would have been "totally white", and a surface paint defining 0% would have been "totally black").

Example: Panoramic Laser Range-Finder IMAGER 5003

Laser scanners may also differ with respect to their scan geometry (i.e., how the scan rays progress during a single scan of a 3D scene). The range-scan examples shown in this book were recorded by a panoramic Z+F IMAGER 5003 scanner. In this case, the scan geometry

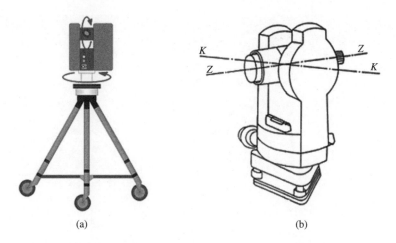

(a)   (b)

Figure 2.18   Sketch of scan geometry defined by two scan directions (vertical and horizontal): (a) laser range-scanner; (b) theodolite with two rotation axes, the pitch axis, traditionally called K (German: *Kippachse*), and the roll axis, called Z (German: *Zielachse*).

is defined as follows: angular increments between scan rays are uniformly defined in two dimensions: vertically by a rotating deflecting mirror, and horizontally by rotating the whole measuring system. The vertical scan range of the IMAGER 5003 is 310° (which leaves 50° uncovered), and the horizontal scan range is 360°. This scan geometry is similar to that known for theodolites (see Figure 2.18(b)), which are traditional instruments for measuring (manually) both horizontal and vertical angles.

Actually a horizontal scan range of 180° is sufficient to measure each visible 3D point once, because the LRF scans overhead. Scanning the full 360° horizontally can be used for calibration; in this case, each visible 3D point is scanned twice.

Figure 2.19   Raw data (without redundancy) of an uncalibrated LRF scan. Calibration is needed to specify the exact motion of the mirror or the LRF (one row in the shown image corresponds to one position of the mirror).

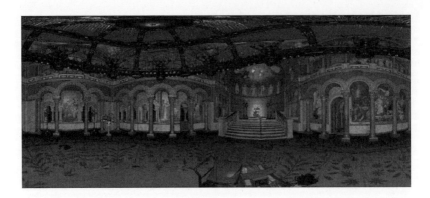

Figure 2.20   Data of a calibrated LRF image in spherical coordinates.

Figure 2.19 shows an uncalibrated raw data set without redundancy (i.e., 310° times 180°) captured by using the panoramic scan geometry described. Figure 2.20 shows the data set after calibration (to be discussed in Chapter 5); now both "half-images" are unified in one panoramic scan.

2.4 Examples and Challenges

Figure 2.21 shows an example of a panoramic image. The image was taken near the Harbour Bridge in Auckland in 2001; the distance to the CBD is approximately 2 km. A 60 mm lens was used for this image of size 56,580 × 10,200 pixels, which is about 15 m × 3 m at 100 dpi. The size of this image is 3 GB. Figure 2.22(b) illustrates the image's resolution; it is possible to recognize an advertisement in the CBD.

Correct geometric modeling of captured scenery is difficult to achieve for such a very high-resolution panoramic image. Basically, having a capturing surface in the form of an ideal straight cylinder would resolve the difficulties. But, in reality, several mechanical errors (e.g., non-uniform rotations, tumbling of axis, eccentricities) can occur.

Furthermore, a rotating line camera does not take images (obviously) in a single shot. The panoramic image is "composed", line by line, and the acquisition of, for example, 160,000 lines may require 10 minutes or more in bright daylight. Moving objects in a recorded scene will

Figure 2.21   This panoramic 380° image was taken in 2001 at Auckland's Northcote Point, and shows a view near the Harbour Bridge towards Auckland's CBD. The labeled rectangular window of this image is shown again in Figure 2.22(a) (see Plate 23).

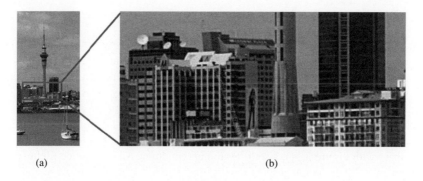

(a)   (b)

Figure 2.22   (a) The rectangular window in Figure 2.21 showing a part of Auckland's CBD. (b) Zooming in further reveals this detail of 500 × 200 pixels (see Plate 24).

Figure 2.23   Two windows within one panoramic image captured for surveillance of Braunschweig airport: airport tower with distortions caused by vibrations (left), and people ("ghosts") move while the rotating line sensor is scanning in their direction (right) (see Plate 25).

appear geometrically distorted. A rotating line camera is basically designed for recording static scenes (but effects of dynamic scenes sometimes attract special attention from photographers, for example how waves on the water appear when captured with a rotating sensor-line camera). Moving leaves or grass are examples of textures in landscapes which will appear with some kind of "motion pattern".

Also, a rotating sensor-line camera should not move during an image acquisition scan (note the apparent vibrations on the bridge while capturing the image shown in Figure 2.21). The use of an *inertial measuring unit* might be helpful in measuring movements of the camera for later compensation. Figure 2.23 illustrates distortions which occurred while the line scanner was operating on an airfield, impacted by various vibrations. (An image-based algorithm to correct such kind of distortions is described in Section 5.2.5.)

Some applications have recently suggested that sensor-line cameras may also be efficiently used for other types of camera movements, not only in aircraft, satellites or on a turntable. However, these applications are not yet "saturated", and we will not discuss them in this book.

2.5 Exercises

2.1. Calculate the hyperfocal length for an ideal Gaussian optics with $f = 35$ mm and a diameter of 6.25 mm for the entrance pupil with an acceptable circle of confusion of $\sigma = 0.025$ mm. What is the geometric resolution (in millimeters) of a pixel at this distance?

2.2. Prove that the formula for calculating the horizontal field of view is also true for the case when $R > 0$ and $\omega > 0$.

2.3. [Possible lab project] Repeat the experiment of Exercise 1.3, but now position the camera in such a way that the projection center is definitely not on the axis of the image cylinder (compare with Figure 2.11). Use for R the maximum possible distance (defined by your equipment). Stitch the images again with some (freely available) software and report on the errors that occurred during this process. Would it be possible to stitch those images more accurately by modeling the sensor parameters in your own stitching software?

2.6 Further Reading

The Slovak mathematician and physicist Josef Petzval (1807–1891) studied the pin-hole camera, and his results were later improved by the British Nobel laureate Lord Rayleigh (see Lord Rayleigh, 1891).

On the optics of sensor systems in general see Hecht and Zajac (1974) and Hecht (2002), and see Hornberg (2006) in particular for an explanation of optics in relation to a pinhole camera model (see, for example, the chapter about optics by Karl Lenhardt, or the online version there of (Lenhardt, 2005)).

Sensor-line cameras ("linear pushbroom cameras") have been discussed in Gupta and Hartley (1997) for possible applications in computer vision. For examples of rotating sensor-line cameras, see the Wikipedia entry on *panoscan*, and links provided on that page. For technical details of Eyescan M3, see Scheibe et al. (2001).

Rotating line cameras have been described at an abstract level in Huang et al. (2002); in particular, the parameters R and ω are characterized with respect to optimizing those two parameters for stereo viewing of a scene which is characterized by closest and furthest distance between camera and objects of interest (see Chapters 6 and 7 in this book).

For time-of-flight and phase-shift data, as listed in Table 2.3, see Sgrenzaroli (2005). On the panoramic Z+F IMAGER 5003 scanner, see Heinz et al. (2001). The accuracy of laser scanners is discussed in Zhang et al. (2005).

Schneider and Maas (2004) discuss issues of non-uniform rotations, tumbling of axes, or eccentricities which may occur for a rotating line camera. A CCD line scanner, freely movable in 3D space, is discussed in Griesbach et al. (2004) and Reulke et al. (2003) which leads to even more challenges for geometric rectification of the sequence of recorded image lines.

WAAC was discussed (with complete technical data sheet) in Sandau and Eckardt (1996) and Reulke and Scheele (1998). For a successor (MFC), see Börner et al. (2008). The technique of 3-line pushbroom scanning was invented in the late 1970s in the Mapsat (USGS) and Stereosat (JPL) projects; O. Hofmann (at Messerschmidt-Bolkow-Blohm, MBB) also proposed an airborne 3-line pushbroom scanner at that time. The concept was further developed at DLR, in collaboration with associated companies such as MBB, or DASA.

3

Spatial Alignments

This chapter considers the positioning of panoramic sensors or panoramas in 3D space, defining and using world or local (e.g., camera or sensor) coordinate systems. The chapter also specifies coordinates and locations of capturing surfaces or panoramic images. The chapter starts with a few fundamentals for metric spaces, coordinate system transforms, or projections from 3D space into panoramic images, briefly recalling mathematical subject areas such as linear algebra, projective geometry, and surface geometry.

3.1 Mathematical Fundamentals

It is assumed that the reader already knows basic mathematical concepts such as typically taught in first or second year university classes. This section recalls such concepts and prepares for the formal discussion of panoramic sensors, but will not start at basic mathematical levels. For example, it is assumed that the reader is already familiar with basic vector and matrix algebra, or homogeneous coordinates (introduced in the 19th century by A. F. Möbius, and today of crucial importance, for example, in computer graphics or vision).

This section also specifies the (default) notation used in this book.[1] For example, by default we use Greek letters for angular values, and bold letters for points, vectors, or matrices. $\mathbb{R}$ is the set of all reals and $\mathbb{Z}$ is the set of all integers.

3.1.1 Euclidean Spaces and Coordinate Systems

A *Euclidean space* $\mathbb{R}^n$, with $n \geq 1$, is defined by *Euclidean distances* $D_e(\mathbf{p}, \mathbf{q})$ between its points $\mathbf{p} = (p_1, \ldots, p_n)$ and $\mathbf{q} = (q_1, \ldots, q_n)$, and *Euclidean angles*. Let $||\mathbf{p}|| = D_e(\mathbf{p}, \mathbf{o}) \geq 0$, where $\mathbf{o} = (0, \ldots, 0)$ defines the *origin* of the space. For this *norm* $||\mathbf{p}||$ we have that

$$\cos \alpha = \frac{\mathbf{p} \cdot \mathbf{q}}{||\mathbf{p}|| \, ||\mathbf{q}||}$$

---

[1] See also the list of symbols at the front of the book.

Panoramic Imaging: Sensor-Line Cameras and Laser Range-Finders   F. Huang, R. Klette, and K. Scheibe
© 2008 John Wiley & Sons, Ltd

defines the Euclidean angle α, uniquely in the interval $[0, \pi)$, between rays starting at **o** and passing through **p** or **q**. (Thus, the definition of an angle is actually a conclusion from the definition of distance.) The *inner product*

$$\mathbf{p} \cdot \mathbf{q} = p_1 q_1 + \ldots + p_n q_n$$

is defined for points or vectors **p** and **q**.

By default, we use Cartesian xy coordinates for the 2D plane $\mathbb{R}^2$, Cartesian XYZ coordinates in 3D space $\mathbb{R}^3$, and Cartesian ij coordinates for discrete coordinates in the discrete 2D plane $\mathbb{Z}^2$. (A coordinate system is *Cartesian* if any pair of coordinate axes forms a right angle.) xy or XYZ coordinates can be expressed with arbitrary accuracy, but ij coordinates only within the selected discrete scale of both axes.

The World

A panoramic sensor records 3D data of a scene. Together with time, this defines a four-dimensional (4D) space. This space may also be assumed to be Euclidean by treating all four coordinates equally with respect to distance and angle definitions. However, in this book we aim to model of static scenes only. Therefore, it is sufficient to deal only with a 3D space, without an additional coordinate for time.

DEFINITION **3.1.** *The* world *is a Euclidean 3D space* $\mathbb{R}^3$, *equipped with a Cartesian XYZ* world coordinate system *with origin denoted as* **W**; *all three coordinate axes are scaled uniformly with respect to a selected physical distance unit.*

The uniformly applied distance unit might be, for example, millimeters or micrometers. (Sometimes, the focal length f of the camera used is selected as the base unit.) In general, we will not list physical units, as long as actual sizes in 3D scenes are not of relevance for the discussed parameters or formulas in a given context. (It simplifies implementations if all involved objects are measured for the same physical unit.) For example, a sensor matrix, a position of a laser-range finder, or an image cylinder (e.g., of a rotating sensor-line camera) are all objects in the world, and thus should be measured uniformly in the XYZ world coordinate system by the selected unit.

By $\mathbf{P}_w = (X_w, Y_w, Z_w)$ we indicate that this point is defined in the world coordinate system, with origin at **W**.

The XYZ (or $X_w Y_w Z_w$) world coordinate system is normally defined by an application context (e.g., the origin **W** at a selected "landmark" in the scene), or, during sensor calibration, by a calibration pattern.

We consider the XZ *plane* to be the *ground plane* (i.e., the plane on which the objects in the world are standing), with the Y-axis pointing downward. The XY *plane* is sometimes also called the *front plane*.

Inhomogeneous and Homogeneous Coordinates

Typically we use inhomogeneous Cartesian XYZ or xy coordinates, if sufficient for clarity. However, this is sometimes inconvenient. For example, 4D homogeneous coordinates (X, Y, Z, W), with $W \neq 0$, identify the point $(X/W, Y/W, Z/W)$ in 3D inhomogeneous Cartesian coordinates, and, as is well known, the use of such 4D homogeneous coordinates allows coordinate transforms to be represented in a uniform (and, thus, simple) way.

Points, Coordinates, and Distances

Points in 3D space are typically given in (inhomogeneous) Cartesian coordinates and denoted, for example, by $\mathbf{P} = (X, Y, Z) = (P_x, P_y, P_z)$. Points in 2D space are denoted, for example, by $\mathbf{p} = (x, y) = (p_x, p_y)$ or $\mathbf{p}_i = (x_i, y_i)$ when multiple points are considered at the same time, where $i \in \mathbb{Z}$. Besides those inhomogeneous coordinates, we also occasionally use homogeneous coordinates $\mathbf{P} = (X, Y, Z, 1)$ or $\mathbf{p} = (x, y, 1)$.

A distance between two points (say, $\mathbf{P}$ and $\mathbf{Q}$ in Euclidean 3D space) is defined to be the Euclidean distance

$$D_e(\mathbf{P}, \mathbf{Q}) = \sqrt{(P_x - Q_x)^2 + (P_y - Q_y)^2 + (P_z - Q_z)^2}.$$

Spherical coordinates are formally represented by two angular values φ (longitudinal angle) and ϑ (latitudinal angle) and a distance value D (measured from the origin to a given point). Assume a point $\mathbf{P}$ in 3D Cartesian coordinates. Its representation in spherical coordinates[2] is then given as follows:

$$P_x = D \sin \vartheta \cdot \sin \varphi,$$
$$P_y = D \sin \vartheta \cdot \cos \varphi, \qquad (3.1)$$
$$P_z = D \cos \vartheta.$$

For a 3D point on the surface of a cylinder of radius D, we have that

$$P_x = D \sin \varphi,$$
$$P_y = P_y, \qquad (3.2)$$
$$P_z = D \cos \varphi$$

where the cylinder is about the y-axis.

A point in 3D space $\mathbb{R}^3$ can thus be represented by Cartesian coordinates P_x, P_y, and P_z, by polar (or spherical) coordinates D, φ, and ϑ, or cylindrical coordinates D, φ, and P_y. Note that the coordinate system chosen does not define whether the space $\mathbb{R}^3$ of interest is Euclidean or not: this is decided by the metric and the angle definition used. For example, Cartesian 3D coordinates and the use of any Minkowski metric L_n (except $n = 2$) would define a non-Euclidean metric space $\mathbb{R}^3$.

Orientation of Coordinate Systems

A *right-handed 3D Cartesian coordinate system* is defined as follows: if the thumb, index finger and middle finger of the right hand are held so that they form three right angles, then

---

[2] The value of the longitudinal angle depends on whether the angle is counted starting at Y- or X-axis. The Y-axis is our default, and we proceed clockwise because measuring devices used in our applications all rotated clockwise.

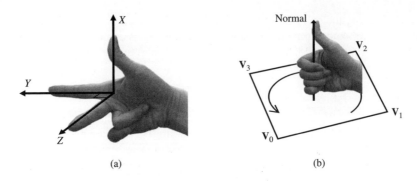

(a)   (b)

Figure 3.1   (a) Right-handed Cartesian coordinate system. (b) Illustration of the 'thumbs up' rule.

the thumb indicates the X-axis, the index finger the Y-axis, and the middle finger the Z-axis; see Figure 3.1(a).

For example, the determination of a *normal vector*[3] of polygons in computer graphics applications typically follows the right-hand rule: when an observer looks at a polygon from that side such that the vertex sequence of the polygon forms a counterclockwise loop (which identifies the *front side* of the polygon), then the normal vector points towards the observer. This can be illustrated by a closed right fist, with the thumb up: the four remaining fingers point in counterclockwise orientation, while the thumb shows the direction of the normal vector (see Figure 3.1(b)).

By default, we assume a right-handed world coordinate system.

Local Coordinate Systems

We also assume *local* Cartesian XYZ coordinate systems in the world while using a camera or sensor at a particular location. These systems again span the $\mathbb{R}^3$ of our world. A point in $\mathbb{R}^3$ can thus be represented in different coordinate systems. A standard 'allocation' procedure is as follows: first identify the world coordinate system, then measure points within a camera or sensor coordinate system, and finally map local representations of those points into the uniform world coordinate system.

DEFINITION **3.2.** *A camera defines a right-handed XYZ camera coordinate system, with its origin* **C** *at the projection center; the Z-axis coincides with the optical axis of the camera (pointing into the 3D scene).*

A sensor defines a right-handed XYZ sensor coordinate system with origin at a point **O***; axes are aligned with the sensor in some particular way to be specified in each case. If there is one defining axis of the sensor (which is not an optical axis), then this is assumed to be the Y-axis.*

By $\mathbf{P}_c = (X_c, Y_c, Z_c)$ we indicate that this point is defined in a camera coordinate system, with origin at **C**, and by $\mathbf{P}_o = (X_o, Y_o, Z_o)$ that it is defined in a sensor coordinate system, with

---

[3] For a definition of the surface normal, see next subsection.

origin at **O**. Camera and sensor coordinates are defined with respect to the same physical unit (such as millimetres or micrometers) as the world coordinate system.

A sensor (such as a rotating sensor-line camera) may be of a "more complex architecture", compared to a camera, and the definition of a local sensor coordinate system is then often a matter of deciding how to align axes to the sensor.

A rotating sensor-line camera does not have just one optical axis; however, the rotation axis is uniquely defined, and this is here taken as the Y-axis of the sensor. This also corresponds to our choice of taking the XZ plane as being the ground plane. For the rotating sensor-line camera, the XZ plane is assumed to be parallel to the base plane of this sensor.

If there is no rotation axis, then the Y-axis is often also chosen to be parallel to a linear component of the camera or sensor (e.g., to a side of the rectangular sensor matrix of a camera), for simplifying calculations of projection formulas.

The third axis (the X-axis) is finally defined such that a right-handed coordinate system is formed. For the rotating sensor-line camera, the optical axis of the initial camera position may be used for identifying the Z-axis of this sensor's XYZ coordinate system.

Sometimes we also assume for simplicity that local and world coordinate systems coincide (e.g., if the sensor is only used within one 3D scene at a single position). There is only an affine transform between both coordinate systems, and their parameters can often also be assumed to be known.

Euclidean and Non-Euclidean Subspaces

Subspaces of our (Euclidean) world are again Euclidean if and only if they are linear manifolds (i.e., straight lines, planes, or $\mathbb{R}^3$ itself). This allows distances and angles to be defined as in the entire space (i.e., as being Euclidean distances and angles). The 3D local spaces of camera or sensor are also Euclidean; these 3D manifolds coincide with the world.

In the previous chapter, we identified capturing surfaces as special types of 2D manifolds, which also includes non-linear subspaces. For example, the distance between two points on a cylindrical surface, defined by a shortest arc connecting these two points within (!) the cylindrical surface, is in general larger than the Euclidean distance between those two points in the 3D Euclidean space. This identifies a cylindrical surface as a non-Euclidean subspace of the world. Similarly, a spherical surface is non-Euclidean.

3.1.2 2D Manifolds and Surface Normals

Cylindrical or spherical surfaces are examples of 2D manifolds, which can be parameterized by 2D coordinates. By default we use uv coordinates for such non-Euclidean surfaces.

Representations of 2D Manifolds

A surface can be represented analytically either by an equation $E(X, Y, Z) = 0$ or in parametric form,

$$\Gamma(u, v) = (F(u, v), G(u, v), H(u, v)).$$

For the latter case we assume that the functions F, G, and H have partial derivatives

$$F_u(u, v) = \frac{\partial F(u, v)}{\partial u}, \quad F_v(u, v) = \frac{\partial F(u, v)}{\partial v}$$

(and similarly for G and H). As an example, consider the surface Γ of the sphere $X^2 + Y^2 + Z^2 - r^2 = 0$. A possible parameterization for this surface (equation (3.1), but with different symbols) is as follows:

$$\Gamma(\alpha, \beta) = (r \cos \alpha \cos \beta, r \cos \alpha \sin \beta, r \sin \alpha)$$

where $0 < \alpha < \pi$ and $0 < \beta < 2\pi$. Here α is the latitudinal angle, and β is the longitudinal angle. (Note that this parameterization does not represent the entire surface; one semicircle joining the poles is not included.)

As a second example, let Γ be defined by the equation $w = F(u, v)$ for $(u, v) \in B \subseteq \mathbb{R}^2$. Then Γ has the parameterization $\{(u, v, F(u, v)) : (u, v) \in B\}$. Such a Γ is called a *Monge patch*; it is *smooth* if and only if F is continuously differentiable for all $(u, v) \in B$.

Surface Area, Gradient, and Normal

Let Γ be a Monge surface patch defined by $w = F(u, v)$.[4] The *area* of $\Gamma = \{(u, v, F(u, v)) : (u, v) \in B\}$ is as follows:

$$\mathcal{A}(\Gamma) = \int_B \sqrt{1 + (F_u(u, v))^2 + (F_v(u, v))^2} \, d(u, v). \tag{3.3}$$

The vector $(F_u(u, v), F_v(u, v))^{\mathrm{T}}$ is called the *gradient* of Γ at $(u, v) \in B$. Let

$$\mathbf{n}_+(u, v) = (-F_u(u, v), -F_v(u, v), 1)^{\mathrm{T}}$$

and

$$\mathbf{n}_-(u, v) = (F_u(u, v), F_v(u, v), -1)^{\mathrm{T}}.$$

These vectors are the *normals* to Γ at (u, v), differing only by the chosen (positive or negative) direction. The choice of normal is often a subjective one; for surfaces of solid objects, we assume that the normal points outward from the solid object.

From equation (3.3) it follows that we also have the following:

$$\mathcal{A}(\Gamma) = \int_B \|\mathbf{n}_+(u, v)\|_2 \, d(u, v) = \int_B \|\mathbf{n}_-(u, v)\|_2 \, d(u, v).$$

This formula can be used, for example, to estimate the area of triangulated surfaces, by estimating normals at all vertices of the given triangulation.

---

[4] To be mathematically correct, such a Monge patch is defined for (u, v) in a closed bounded measurable set $B \subset \mathbb{R}^2$, where the first-order partial derivatives of F exist and are continuous on a set B_1 such that B is contained in the topological interior of B_1.

3.1.3 Vectors, Matrices, and Affine Transforms

Vectors and Matrices

Vectors are generally denoted by bold lower-case letters, and matrices by bold capitals. However, there are exceptions where we also use bold capitals for 3D vectors to be consistent with our point notation, because we often switch from a 3D point to the vector represented by this point, and vice versa.

The norm $||\mathbf{a}||$ specifies the length of a vector $\mathbf{a}$; for example, for a 3D vector $\mathbf{a} = (X, Y, Z)^{\mathsf{T}}$ we have that

$$||\mathbf{a}|| = \sqrt{X^2 + Y^2 + Z^2}.$$

The *unit vector* or the *direction* of a vector $\mathbf{a}$ is defined as follows:

$$\mathbf{a}^{\circ} = \mathbf{a}/||\mathbf{a}||.$$

It follows that $\mathbf{a} = ||\mathbf{a}|| \cdot \mathbf{a}^{\circ}$. Unit vectors along the X-, Y-, or Z-axis are denoted by $\mathbf{x}^{\circ} = (1, 0, 0)^{\mathsf{T}}$, $\mathbf{y}^{\circ} = (0, 1, 0)^{\mathsf{T}}$, or $\mathbf{z}^{\circ} = (0, 0, 1)^{\mathsf{T}}$, respectively.

Besides inhomogeneous representations of vectors, we also occasionally use homogeneous representations $\mathbf{a} = (X, Y, Z, 0)^{\mathsf{T}}$ of a 3D vector, or $\mathbf{b} = (x, y, 0)^{\mathsf{T}}$ of a 2D vector. Note that homogeneous coordinates allow a clear distinction to be made between points (use of 1 as the last coordinate) and vectors (use of 0).

Affine Transform between 3D Coordinate Systems

An *affine transform* is a linear mapping in n-dimensional Euclidean space, for $n \geq 1$.

We often need to map a 3D point from one coordinate system to another (e.g., from sensor into world coordinate system), where these two coordinate systems are related by a rotation and a translation (possibly also by a subsequent scaling; but note that we assume equal scaling in world and local coordinate systems). This mapping is a special case of an affine transform.

For example (without scaling), in 3D we use a 3×3 rotation matrix $\mathbf{R}$ and a 3×1 translation vector $\mathbf{t}$; a point $\mathbf{P}_w = (X_w, Y_w, Z_w)$ is mapped via

$$(X_c, Y_c, Z_c)^{\mathsf{T}} = \mathbf{R} \cdot (X_w, Y_w, Z_w)^{\mathsf{T}} + \mathbf{t}$$

into a 3D point $\mathbf{P}_c = (X_c, Y_c, Z_c)$, and that is from world coordinates into camera coordinates. Homogeneous coordinates allow such an affine transform to be represented by one 3×4 matrix $\mathbf{A}$, with

$$(X_c, Y_c, Z_c)^{\mathsf{T}} = \mathbf{A} \cdot (X_w, Y_w, Z_w, 1)^{\mathsf{T}}$$
$$= [\mathbf{R}\,\mathbf{t}] \cdot (X_w, Y_w, Z_w, 1)^{\mathsf{T}}.$$

Alternatively, the matrix $\mathbf{A}$ can be defined as a 4×4 matrix, such that the above transformation can be rewritten as

$$(X_c, Y_c, Z_c, 1)^{\mathsf{T}} = \mathbf{A} \cdot (X_w, Y_w, Z_w, 1)^{\mathsf{T}}$$
$$= \begin{bmatrix} \mathbf{R} & \mathbf{t} \\ \mathbf{0}^{\mathsf{T}} & 1 \end{bmatrix} \cdot (X_w, Y_w, Z_w, 1)^{\mathsf{T}}$$

where $\mathbf{0}^{\mathsf{T}}$ is a 3×1 zero vector.

Graphic systems (such as OpenGL) uniformly use 4×4 matrices in homogeneous coordinates for all the required coordinate transforms in 3D (inhomogeneous) space. This allows all projective situations (such as mapping a point to infinity) to be dealt with in one uniform and simple algebraic system.

3D Rotations in 3D Coordinates

Rotation angles about three different axes are denoted by ψ, ϕ, and κ, respectively. The rotation about the X-axis with angle ψ is defined by the rotation matrix

$$\mathbf{R}_x(\psi) = \begin{pmatrix} 1 & 0 & 0 \\ 0 & \cos(\psi) & -\sin(\psi) \\ 0 & \sin(\psi) & \cos(\psi) \end{pmatrix},$$

the rotation about the Y-axis with angle ϕ is defined by the rotation matrix

$$\mathbf{R}_y(\phi) = \begin{pmatrix} \cos(\phi) & 0 & \sin(\phi) \\ 0 & 1 & 0 \\ -\sin(\phi) & 0 & \cos(\phi) \end{pmatrix}$$

and the rotation about the Z-axis with angle κ by the rotation matrix

$$\mathbf{R}_z(\kappa) = \begin{pmatrix} \cos(\kappa) & -\sin(\kappa) & 0 \\ \sin(\kappa) & \cos(\kappa) & 0 \\ 0 & 0 & 1 \end{pmatrix}.$$

Sometimes we use different notations for specifying a rotation (e.g., $\mathbf{R}_\varphi$, a deliberate rotation of the sensor about the (local) Y-axis for scanning a panorama). In any case, all rotation matrices are denoted by $\mathbf{R}$. In general, if there is no special indication of a particular rotation axis, then a given rotation matrix is assumed to be $\mathbf{R} = \mathbf{R}_x(\psi) \cdot \mathbf{R}_y(\phi) \cdot \mathbf{R}_z(\kappa)$. The resulting matrix $\mathbf{R}$ is then given as

$$\begin{pmatrix} c\phi \cdot c\kappa & s\phi \cdot s\kappa & s\phi \\ c\psi \cdot s\kappa + s\psi \cdot s\phi \cdot s\kappa & c\psi \cdot c\kappa - s\psi \cdot s\phi \cdot s\kappa & -s\psi \cdot s\phi \\ s\psi \cdot s\kappa - c\psi \cdot s\phi \cdot s\kappa & s\psi \cdot c\kappa + c\psi \cdot s\phi \cdot s\kappa & s\psi \cdot c\phi \end{pmatrix}$$

where ψ, ϕ, and κ are the rotation angles about the X-, Y-, and Z-axis, respectively, and c and s stand for the cosine and sine function, respectively.

Use of Quaternions for Modeling Rotations

Rotations about a specified axis in 3D space are frequently required, and the use of quaternions is beneficial in implementing them efficiently. Quaternions are an extension of complex numbers. They are 4-tuples, similar to vectors in a four-dimensional space $\mathbb{R}^4$, but with special algebraic rules. We denote them by

$$\bar{q} = (Q, X, Y, Z) \qquad \text{with } Q \in \mathbb{R} \text{ and } (X, Y, Z) \in \mathbb{R}^3$$

(not to be confused with 4D homogeneous coordinates). This combines a scalar Q (this characterizes the rotation angle below) with a vector $\mathbf{q} = (X, Y, Z)^{\mathrm{T}}$ (this characterizes the rotation axis below).

The product $(Q_1, X_1, Y_1, Z_1) \circ (Q_2, X_2, Y_2, Z_2) = (Q_3, X_3, Y_3, Z_3)$ of two quaternions is defined as follows:

$$Q_3 = Q_1 Q_2 - X_1 X_2 - Y_1 Y_2 - Z_1 Z_2,$$
$$X_3 = Q_1 X_2 + X_1 Q_2 - Y_1 Z_2 + Z_1 Y_2,$$
$$Y_3 = Q_1 Y_2 + X_1 Z_2 + Y_1 Q_2 - Z_1 X_2,$$
$$Z_3 = Q_1 Z_2 - X_1 Y_2 + Y_1 X_2 + Z_1 Q_2.$$

This product is not commutative. For two quaternions $\bar{p} = (P, \mathbf{p}^{\mathrm{T}})$ and $\bar{q} = (Q, \mathbf{q}^{\mathrm{T}})$, defined by two 3D vectors $\mathbf{p}$ and $\mathbf{q}$, the product is also equal to

$$\bar{p} \circ \bar{q} = \begin{pmatrix} PQ - \mathbf{p} \cdot \mathbf{q} \\ P \cdot \mathbf{q} + Q \cdot \mathbf{p} + \mathbf{p} \times \mathbf{q} \end{pmatrix}$$

where $\mathbf{p} \cdot \mathbf{q}$ and $\mathbf{p} \times \mathbf{q}$ are the normal inner and outer vector products, respectively.

A rotation with angle ϕ and about an axis identified by unit vector $\mathbf{e}$ is represented by the quaternion

$$\bar{e} = \left(\cos(\tfrac{\phi}{2}), \mathbf{e}^{\mathrm{T}} \sin(\tfrac{\phi}{2}) \right) = (e_0, e_1, e_2, e_3).$$

This is a unit quaternion (i.e., with $|\bar{e}| = \sqrt{e_0^2 + e_1^2 + e_2^2 + e_3^2} = 1$). The rotation, identified by ϕ and $\mathbf{e}$, of a vector $\mathbf{p}$ into a vector $\mathbf{q}$, expressed as quaternions $\bar{p} = (0, \mathbf{p}^{\mathrm{T}})$ and $\bar{q} = (0, \mathbf{q}^{\mathrm{T}})$, is then represented (when applying the multiplication rule as specified above) by the product of quaternions

$$\bar{q} = \bar{e} \circ \bar{p} \circ \bar{e}^*, \tag{3.4}$$

with $\bar{e}^*$ as the *conjugated quaternion* of $\bar{e}$ defined as follows:

$$\bar{e}^* = \left(\cos(\tfrac{\phi}{2}), -\mathbf{e}^{\mathrm{T}} \sin(\tfrac{\phi}{2}) \right) = (e_0, -e_1, -e_2, -e_3).$$

This finally allows the products in equation (3.4) to be expressed in matrix form as follows, also specifying a 3×3 matrix $\mathbf{R_e}(\phi)$ this way:

$$\bar{q}^{\,\mathrm{T}} = \begin{pmatrix} 0 \\ \mathbf{q} \end{pmatrix} = \begin{pmatrix} 1 & 0 & 0 & 0 \\ 0 & 1 - 2(e_2^2 + e_3^2) & 2(e_1 e_2 - e_0 e_3) & 2(e_1 e_3 + e_0 e_2) \\ 0 & 2(e_1 e_2 + e_0 e_3) & 1 - 2(e_1^2 + e_3^2) & 2(e_2 e_3 - e_0 e_1) \\ 0 & 2(e_1 e_3 - e_0 e_2) & 2(e_2 e_3 + e_0 e_1) & 1 - 2(e_1^2 + e_2^2) \end{pmatrix} \bar{p}^{\,\mathrm{T}}$$

$$= \begin{pmatrix} 1 & \mathbf{0}^{\mathrm{T}} \\ \mathbf{0} & \mathbf{R}_{\phi, \mathbf{e}} \end{pmatrix} \begin{pmatrix} 0 \\ \mathbf{p} \end{pmatrix}$$

with $\mathbf{0} = (0, 0, 0)^{\mathrm{T}}$. Thus, for any rotation angle ϕ and rotation axis $\mathbf{e}$, just calculate the matrix $\mathbf{R_e}(\phi)$, and the algebra of quaternions is (only) used for the derivation of the final equation.

3.2 Central Projection: World into Image Plane

With equation (2.1) we had a first brief result on mapping the XYZ world into the xy image plane of an ideal pinhole camera. The xy plane is assumed to be parallel to the XY plane, and axes x and y are parallel to axes X and Y, respectively.

3.2.1 Symmetric Perspective Projections

A *perspective* or *central projection* in 3D space may be defined by one 4×4 homogeneous matrix $\mathbf{M}$ as follows:

$$\mathbf{M} = \begin{pmatrix} f & 0 & 0 & 0 \\ 0 & f & 0 & 0 \\ 0 & 0 & f & 0 \\ 0 & 0 & 1 & 0 \end{pmatrix}.$$

The transform

$$\begin{pmatrix} x \\ y \\ z \\ w \end{pmatrix} = \mathbf{M} \cdot \begin{pmatrix} X \\ Y \\ Z \\ 1 \end{pmatrix} \tag{3.5}$$

in homogeneous coordinates results in the projection equation

$$(x, y, z, w)^{\mathrm{T}} = (fX, fY, fZ, Z)^{\mathrm{T}}$$

and, in inhomogeneous coordinates,

$$(x', y', z')^{\mathrm{T}} = \left(\frac{x}{w}, \frac{y}{w}, \frac{z}{w} \right)^{\mathrm{T}} = \left(\frac{fX}{Z}, \frac{fY}{Z}, f \right)^{\mathrm{T}}.$$

Central projection was discussed in Chapter 2, and is recalled here simply for the sake of comparison with asymmetric perspective projections in this more applied context.

3.2.2 Asymmetric Perspective Projections

Multiplication of different affine transform matrices with the central projection matrix $\mathbf{M}$ results in an *asymmetric perspective projection*. An asymmetric perspective transform models,

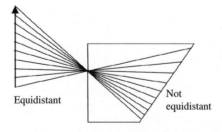

Figure 3.2 Asymmetric projection which causes some scaling in the image.

for example, a shearing or shifting of the image of a sensor-matrix camera (e.g., tilting the sensor matrix, or shifting the sensor's nodal point, which may define the central point in an image); see Figure 3.2.

3.3 Classification of Panoramas

This section classifies recorded panoramas into various types due to basic scanning properties, but also, more importantly, due to mutual spatial alignments, and also due to spatial relationships to the world.

3.3.1 Views and Projection Centers

Each panorama is identified with one *view* of the world, which is mathematically characterized by a *location* (a viewpoint in a 3D world coordinate system), a *view vector* (defined for an initial state of the sensor), an *up-vector* (e.g., for leveling purposes), and the sensor geometry (i.e., sensor parameters).[5] This book also adopts the term *attitude* from flight navigation (also common in engineering in general), which is equivalent to the term *pose* which is of common use in computer vision (and in mathematical models in general).

DEFINITION **3.3.** *The* attitude *or* pose *(of a sensor or camera) is defined by location, view vector, and up-vector.*

Thus, attitude (or pose) and sensor geometry together define a view. By adding time (typically, an interval of time rather than an exact point in time) we would obtain a unique identification of a recorded panorama up to given accuracy limits (in space and time).

DEFINITION **3.4.** *A panorama may be generated as an individual event; this is a* single-view panorama. *A set of interrelated panoramas, taken for different views, defines a* multi-view panorama *of some finite cardinality (= numbers of contributing panoramas).*

For example, stereo viewing demands at least two panoramas (and this is *one* multi-view panorama, and obviously a set of cardinality 2).

View calibration, in particular *attitude* or *pose registration*, of a panoramic sensor is used for calculating defining parameters of participating panoramas, typically within a single world coordinate system.

A panorama may be *composed* of multiple partial scans (i.e., images or range-scans), each defined by an individual projection center. For example, for rotating (single- or multi-)sensor-line cameras, these centers describe the base circle of radius R, as described in Chapter 2.

---

[5] In this book, a view is generally defined by more than just one projection center; in discussing the rotating sensor-line camera in the previous chapter we illustrated that a single panoramic image may be generated by using multiple projection centers. Also, a change of a parameter, which defines the viewing geometry (e.g., the focal length of the camera), specifies a "new way of viewing", thus a new view.

DEFINITION **3.5.** A single-center panorama *is defined by contributing images which all have the same (i.e., constant) projection center. Otherwise, it is a* multi-center panorama. *In both cases we assume a constant sensor geometry; view vectors and up-vectors may change.*

The constant projection center of a single-center panorama is also called the *nodal point*. In this case we have off-axis distance $R = 0$.

For example, a pinhole camera with a wide viewing angle defines a single-center panorama, and a translational move of a pinhole camera (e.g., images taken from an airplane) allows an image mosaic to be generated as an example of a multi-center panorama. A set of several image mosaics defines a multi-view panorama.

In the case of a single-center panorama, the view is defined by the nodal point (the logical choice for identifying the position), an initial viewing direction, an up-vector, and parameters defining the sensor geometry such as focal length or numbers of contributing images.

By default, cylindrical panoramas are assumed to be multi-center (i.e., $R > 0$); a single-center panorama is the exception due to the difficulty of ensuring $R = 0$ in practice.

In the case of a multi-center panorama, the view has a more complex description; for example (see Figure 1.5(b)), the position (one of the three components of the attitude) is specified in some way, and this might be with respect to all the contributing projection centers, or by selecting one of these (e.g., the initial projection center).

DEFINITION **3.6.** *A cylindrical panoramic image*[6] *is a function* $E_\mathcal{P}(R, f, \omega, \gamma, S)$ *where R, f, ω, and γ are the defining parameters (introduced in Chapter 2) of a rotating sensor-line camera and S is the scene captured.*

In general we will suppress S, and write $E_\mathcal{P}(R, f, \omega, \gamma)$.

3.3.2 Refined Classification

We defined cylindrical panoramas in Chapter 1. For the following characterizations we assume that each (cylindrical) panorama is uniquely associated with an *axis* (i.e., the axis of the image cylinder), and with a *center* (i.e., the center of the base circle of the sensor).

Multi-view Panoramas (General Case)

The most general case is a multi-view panorama, without limiting pose or sensor geometries. The only restrictions are that the set of panoramas is finite, and all are cylindrical. See Figure 3.3(a) for a sketch of this general case.

Parallel-Axis or Leveled Panoramas

A set of cylindrical panoramas whose associated axes are all parallel defines a *parallel-axis panorama*; see Figure 3.3(b). In particular, if the axes are all orthogonal to the sea level (e.g., by using a "bull's-eye" for leveling), then they are called *leveled panoramas*.

---

[6] We follow Klette et al. (1998) where images have been identified with irradiances; thus we use the symbol E (common in physics for denoting irradiance) also as the symbol for an image.

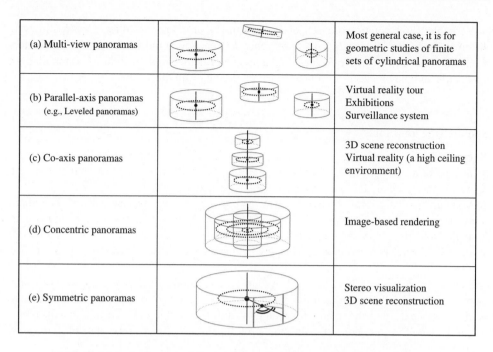

(a) Multi-view panoramas		Most general case, it is for geometric studies of finite sets of cylindrical panoramas
(b) Parallel-axis panoramas (e.g., Leveled panoramas)		Virtual reality tour Exhibitions Surveillance system
(c) Co-axis panoramas		3D scene reconstruction Virtual reality (a high ceiling environment)
(d) Concentric panoramas		Image-based rendering
(e) Symmetric panoramas		Stereo visualization 3D scene reconstruction

Figure 3.3   Different types of multi-view panoramas and their applications.

Leveled panoramas are often used for visualization or reconstruction of large scenes. A perfect leveling reduces the affine transform between two sensor coordinate systems to just a translation, with no need for a rotation. Captured scenes correspond to human visual experience. The size of overlapping fields of view is maximized in leveled panoramas. Leveled panoramas also simplify epipolar geometry; this will become clear when we discuss stereo analysis later in this book.

Coaxial Panoramas

A set of cylindrical panoramas, whose associated axes all coincide, defines a *coaxial panorama*; see Figure 3.3(c). If the sensor geometry for capturing two coaxial panoramas is identical (the only differences being the heights of centers), then the epipolar geometry is quite simple. Epipolar lines are corresponding image columns in this case. This simplifies stereo matching. In addition, the implementation of such a configuration is reasonably straightforward (i.e., after positioning a rotating camera on a tripod).

Concentric Panoramas

A set of panoramas where not only their axes but also their associated centers coincide is called a *concentric panorama*; see Figure 3.3(d). Of course, concentric panoramas can also be leveled (to the sea level), and then all the advantages of leveled panoramas apply as well.

A complete 360° rotation of a sensor-matrix camera, of image resolution $H \times M$, allows M different panoramas of size $H \times W$ to be generated, each with individual camera parameters (i.e., a defined projective parameter g and a unique principal angle ω). All M panoramas are concentric and may be used for image-based rendering techniques.

Symmetric Panoramas

Two concentric panoramas, $E_{\mathcal{P}_R}(R, f, \omega, \gamma)$ and $E_{\mathcal{P}_L}(R, f, (2\pi - \omega), \gamma)$, are called a *symmetric panorama* or a *symmetric pair*; see Figure 3.3(e). Their principal angles are symmetric with respect to the associated normal vector of the base circle.

A symmetric pair is also a (rectified) *stereo panorama* – by analogy with (rectified) binocular stereo images for pinhole cameras. Plate 7 illustrates part of an anaglyphic panorama. (Note that, due to the cylindrical sensor surface, the projection of a horizontal straight line appears curved when displaying the panorama ("incorrectly") on a planar surface.)

Stereo panoramas can be generated by only one rotational scan using a two-sensor-line camera. Figure 3.4 shows a ("historic") prototype of a stereo panoramic camera.[7] The camera is fixed on an extended slider, and it is facing towards the turntable. Some experiments reported in this book were conducted using this prototype to acquire various types of multi-view panorama.

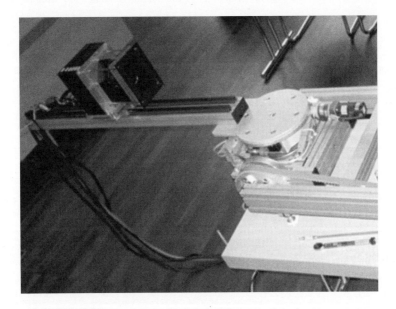

Figure 3.4 Stereo panoramic camera at the Institute of Space Sensor Technology and Planetary Exploration, DLR Berlin, 2000.

---

[7] This camera was based on the WAAC.

3.4 Coordinate Systems for Panoramas

This section discusses coordinate systems for cylindrical panoramas, provides some basics for coordinate system transforms, and concludes with a brief subsection for the case of a laser range-finder.

Sometimes we have to address points on capturing surfaces (which are 2D manifolds) uniquely by (easy calculable) 2D coordinates. For pixel identification it is also desirable to have a simple (integer-only) addressing scheme. Before considering cylindrical panoramas in detail, we start with the case of a Euclidean capturing surface, and compare it with a spherical capturing surface.

3.4.1 Planar Capturing Surface

In case of a (for example, wide-angle) central projection onto an image plane (as assumed for an ideal pinhole camera) the projection is defined by equation (2.1), from camera XYZ space into an xy plane, with origin C of the XYZ space at the projection center, and origin of the xy plane at the mathematical center of the image.

The xy image plane is a Euclidean subspace of the world (i.e., metric relations are as in world coordinates). Figure 3.5 shows this camera coordinate system (three black arrows) and two copies of the xy coordinate system, one in front of the projection center, and one (mirrored) behind, both at distance f from the principal plane.

Figure 3.5 illustrates an ideal pinhole camera; the distance f between image plane and projection center is simply called the *focal length*. From a central perspective point of view,

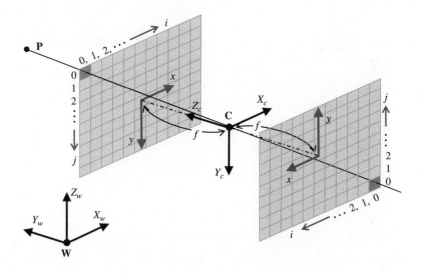

Figure 3.5 Cartesian xy coordinates in the image plane, and a discrete ij image coordinate system. The figure also shows a camera coordinate system, with origin at projection center C. The bottom left-hand tripod indicates the world coordinate system. The point P maps into the upper left-hand corner of the copy of the image in front of the projection center, and thus into the lower right-hand corner of the copy of the image behind the projection center.

it does not matter whether we consider the image plane in front or behind the focal point, and we choose the plane in front. The origin of the xy plane (i.e., the center of the focal plane) is assumed to coincide with the central point, the intersection of the optical axis with the xy plane.

Integers i and j are used for identifying columns and rows in the image, respectively. We assume that $0 \le i \le M - 1$ and $0 \le j \le H - 1$. These discrete ij coordinates refer to pixels in this image, and, in the example shown, the origin is at the upper left corner, with the i-axis to the right and the j-axis pointing downward.

Assuming square sensor elements of width or height τ (e.g., $\tau = 0.007$ mm is a common pixel size for current CCD technology), we can easily transform ij pixel coordinates into xy coordinates in the image plane. All calculations are in Euclidean space.

The origin of the xy coordinate system is defined by the intersection of the camera's optical axis (i.e., the z-axis of the camera coordinate system) and the image plane. This intersection point is referred to as *image center* or *principle point*.

The image center of a planar capturing surface can either be determined (calibrated) by measurements, or identified by an (ideal) pixel position. We specify an image center by real coordinates x_0 and y_0 if determined by measurements, or by integer coordinates i_c and j_c otherwise.

For example, Figure 3.6 shows a particular situation and image pixel $(i, j) = (9, 3)$. We assume that $\tau = 0.5$ mm, $x_0 = 2.5$ mm, $y_0 = 2.5$ mm, $i_c = 5$ (i.e., the sixth column), and $j_c = 5$ (i.e., the sixth row). The conversion from the ij to the xy coordinates is as follows:

$$(x, y) = (i\tau - x_0, \, j\tau - y_0) = (2, -1)$$

or

$$(x, y) = ((i - i_c)\tau, \, (j - j_c)\tau) = (2, -1).$$

The image center (x_0, y_0) can be defined by any real coefficients (the measurements). For instance, if we assume $(x_0, y_0) = (2.43, 2.55)$ and $(i_c, j_c) = (5, 5)$, then the above two formulas for calculating (x, y) will not lead to the same answer.

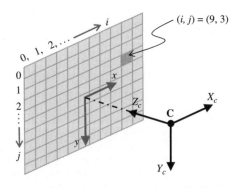

Figure 3.6   Illustration for examples of image centers.

3.4.2 Spherical Capturing Surfaces

In general, the coordinates of a 3D point $\mathbf{P} = (P_x, P_y, P_z)$ (either in world or local coordinates) are given by spherical coordinates

$$\varphi = \arctan\left(\frac{P_x}{P_y}\right)$$

with $\varphi \in [0, 2\Pi)$, and either

$$\vartheta = \arctan\left(\frac{P_x}{P_z}\right)\sin(\varphi)$$

or

$$\vartheta = \arctan\left(\frac{P_y}{P_z}\right)\cos(\varphi)$$

or

$$\vartheta = \arccos\left(\frac{P_z}{\sqrt{P_x^2 + P_y^2}}\right)$$

with $\vartheta \in [0, \Pi)$. D is simply the distance from $\mathbf{P}$ to the origin of the coordinate system. For a spherical capturing surface, we have that D is constant, and (ϑ, φ) defines a 2D coordinate system for this non-Euclidean surface.

We could consider discrete coordinates ij, defined by

$$i = \mathrm{round}(\varphi/\Delta\varphi)$$

and

$$j = \mathrm{round}(\vartheta/\Delta\vartheta)$$

for some scaling parameters $\Delta\varphi$ and $\Delta\vartheta$. However, a "pixel" ij would describe a spherical rectangle of varying size (as is well known from spherical coordinates on the globe).

3.4.3 Cylindrical Capturing Surfaces

For a cylindrical capturing surface, see equations (3.2): here we use (D, φ) to identify one surface point at a given elevation. Because the radius D is constant, the elevation and φ together define a 2D coordinate system for all points on this non-Euclidean surface.

In general, the coordinates of a 3D point $\mathbf{P} = (P_x, P_y, P_z)$ in world coordinates are given by cylindrical coordinates

$$\varphi = \arctan\left(\frac{P_x}{P_z}\right).$$

(called the *abscissa* of the unrolled cylinder) with $\varphi \in [0, 2\Pi)$, and the elevation L defined with respect to the base plane. We have

$$L = D \cdot \frac{P_y}{P_x} \cdot \sin(\varphi)$$

or

$$L = D \cdot \frac{P_y}{P_z} \cdot \cos(\varphi)$$

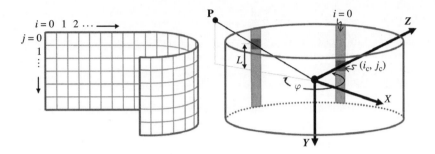

Figure 3.7 A partially unrolled cylinder (left) and a, single-center panorama projection model (right).

(L is called the *ordinate* of the unrolled cylinder). Note that the sign of L is the same as the sign of P_y.

Figure 3.7 illustrates the unrolling of a cylinder. We use integer coordinates i and j to identify columns and rows, respectively, with $0 \le i \le W - 1$ and $0 \le j \le H - 1$. This constitutes a simple scheme for addressing all pixels in a cylindrical panorama. The definition of the cylindrical coordinates are shown on the right in Figure 3.7. The initial column (i.e., $i = 0$) is set to the column where the Z-axis intersects with the cylinder. i_c may be any integer between 0 and $W - 1$, and $i_c = 0$ may act as default. For simplification we may also assume that $x_0 = 0$. Moreover, we assume that y_0 or j_c remains the same for all image columns. If H is an even number, then $j_c = H/2$, and if H is an odd number, then we have that $j_c = (H+1)/2$.

Consider the case of a single projection center where the image center is known to be (x_0, y_0). Let $x_0 = 0$. It follows that pixel coordinates i and j can be determined by

$$i = \text{round}(\varphi/\Delta\varphi)$$

and

$$j = \text{round}\left(\frac{L + y_0}{\tau}\right)$$

for some real $y_0 \ge 0$ (as identified by calibration, and assumed to be constant for all image columns).

For multi-center panoramas, the final image coordinates also depend on the position of the projection centers (e.g., defined by angle ω and off-axis distance R), and on further deviations from an "ideal sensor geometry" which we describe later.

3.5 Geometric Projection Formula for Cylindrical Panorama

In this section, we derive a projection formula for a cylindrical panorama in the ideal off axis case, which is categorized as multi-center panorama (i.e., both R and ω are greater than zero). Consider a known 3D point **P** with coordinates (X_w, Y_w, Z_w) in a world coordinate system. We wish to calculate the projection of **P** on the recorded cylindrical panorama $E_P(R, f, \omega, \gamma)$ in terms of ij image coordinates.

For the derivation, we make use of three different 3D coordinate systems for the world, the sensor, and the line camera. Points defined with respect to these coordinate systems are denoted by (X_w, Y_w, Z_w), (X_o, Y_o, Z_o), and (X_c, Y_c, Z_c), respectively. In particular, the camera

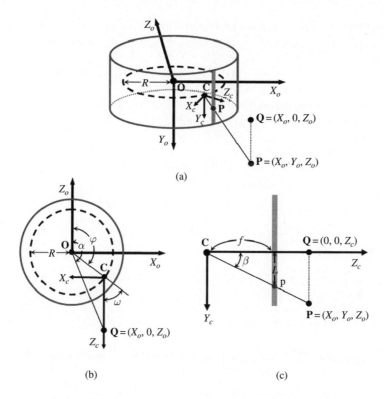

Figure 3.8   Projection geometry of a cylindrical panorama.

coordinate system is defined such that the Z_c-axis coincides with camera's optical axis and the Y_c-axis is parallel to the sensor's Y_o-axis. Figure 3.8(a) illustrates both the sensor and the camera coordinate systems.

The point **P** is first transformed from the world coordinate system into the sensor coordinate system before calculating its projection on the image. We denote the coordinates of **P** with respect to the sensor coordinate system as (X_o, Y_o, Z_o). We have

$$\begin{bmatrix} X_o \\ Y_o \\ Z_o \end{bmatrix} = [\mathbf{R}_{wo} \quad -\mathbf{R}_{wo}\mathbf{t}_{wo}] \begin{bmatrix} X_w \\ Y_w \\ Z_w \\ 1 \end{bmatrix}$$

where $\mathbf{R}_{wo}$ is a 3×3 rotation matrix and $\mathbf{t}_{wo}$ is a 3×1 translation vector. Both together describe the affine transform of the world coordinate system into the sensor coordinate system. The 3×4 matrix $[\mathbf{R}_{wo} - \mathbf{R}_{wo}\mathbf{t}_{wo}]$ is the *transformation matrix*, and its 12 parameters (which we will need in Chapter 5) are denoted as follows:

$$[\mathbf{R}_{wo} \quad -\mathbf{R}_{wo}\mathbf{t}_{wo}] = \begin{bmatrix} t_{11} & t_{12} & t_{13} & t_{14} \\ t_{21} & t_{22} & t_{23} & t_{24} \\ t_{31} & t_{32} & t_{33} & t_{34} \end{bmatrix}.$$

The projection of (X_o, Y_o, Z_o) can be expressed in image coordinates (i, j), and we determine i and j separately. To determine i, we first calculate the cylindrical coordinate φ of the point (X_o, Y_o, Z_o). Lets recall the definition of φ, that is, the angle between the Z-axis of the sensor coordinate system and the line segment $\overline{\text{OC}}$, where $\mathbf{C}$ is the camera projection center associated to image column i; see Figure 3.8(b).

Consider a 3D point $\mathbf{Q}$ be the projection of point $\mathbf{P}$ onto the XZ plane of the sensor coordinate system; see Figure 3.8(a). $\mathbf{Q}$ has sensor coordinates $(X_o, 0, Z_o)$. From the top view of the projection geometry (see Figure 3.8(b)) we conclude that

$$\varphi = \alpha - \angle\mathbf{COQ} = \alpha - \omega + \angle\mathbf{CQO}.$$

Thus we have that

$$\varphi = \begin{cases} \arctan\left(\dfrac{X_o}{Z_o}\right) - \omega + \arcsin\left(\dfrac{R\sin\omega}{\sqrt{X_o^2 + Z_o^2}}\right), & \text{if } Z_o \geq 0, \\[3mm] \pi + \arctan\left(\dfrac{X_o}{Z_o}\right) - \omega + \arcsin\left(\dfrac{R\sin\omega}{\sqrt{X_o^2 + Z_o^2}}\right), & \text{otherwise.} \end{cases}$$

If φ exceeds 2π then it is transformed modulo 2π into $[0, 2\pi)$. For the case of interest (i.e., for $Z_o \geq 0$) we can simplify the equation as follows:

$$\varphi = \arctan\left(\frac{X_o}{Z_o}\right) - \omega + \arcsin\left(\frac{R\sin\omega}{\sqrt{X_o^2 + Z_o^2}}\right)$$

$$= \arcsin\left(\frac{X_o}{\sqrt{X_o^2 + Z_o^2}}\right) - \omega + \arcsin\left(\frac{R\sin\omega}{\sqrt{X_o^2 + Z_o^2}}\right)$$

$$= \arcsin\left(\frac{X_o}{\sqrt{X_o^2 + Z_o^2}}\sqrt{1 - \frac{R^2\sin^2\omega}{X_o^2 + Z_o^2}} + \frac{R\sin\omega}{\sqrt{X_o^2 + Z_o^2}}\sqrt{1 - \frac{X_o^2}{X_o^2 + Z_o^2}}\right) - \omega$$

$$= \arcsin\left(\frac{X_o\left(\sqrt{X_o^2 + Z_o^2 - R^2\sin^2\omega}\right) + Z_o R\sin\omega}{X_o^2 + Z_o^2}\right) - \omega. \tag{3.6}$$

To obtain i, we have

$$i = \text{round}\left(\frac{\varphi}{\Delta\varphi}\right) \tag{3.7}$$

where $\Delta\varphi$ is the angular increment (or angular unit) of the sensor.

Next, for the calculation of j we can make use of the angle β as shown in Figure 3.8(c). This is the angle between the Z-axis of the camera coordinate system, with origin at $\mathbf{C}$, and the line segment $\overline{\text{CP}}$.

Points $\mathbf{P}$ and $\mathbf{Q}$ have respective coordinates $(0, Y_c, Z_c)$ and $(0, 0, Z_c)$ with respect to the camera coordinate system, where $Y_c = Y_o$. From the side view of the camera coordinate system, with origin at $\mathbf{C}$ (see Figure 3.8(c)), the angle β can be calculated by

$$\beta = \arctan\left(\frac{Y_c}{Z_c}\right)$$

where

$$Z_c = \frac{\overline{OQ}\sin(\angle COQ)}{\sin\omega}.$$

Thus, we have that

$$\beta = \arctan\left(\frac{Y_o\sin\omega}{\sqrt{X_o^2 + Z_o^2}\,\sin\left(\omega - \arcsin\left(\frac{R\sin\omega}{\sqrt{X_o^2 + Z_o^2}}\right)\right)}\right)$$

$$= \arctan\left(\frac{Y_o\sin\omega}{\sqrt{X_o^2 + Z_o^2}\left(\sin\omega\cos\arcsin\left(\frac{R\sin\omega}{\sqrt{X_o^2 + Z_o^2}}\right) - \cos\omega\left(\frac{R\sin\omega}{\sqrt{X_o^2 + Z_o^2}}\right)\right)}\right)$$

$$= \arctan\left(\frac{Y_o}{\sqrt{X_o^2 + Z_o^2}\left(\sqrt{1 - \frac{R^2\sin^2\omega}{X_o^2 + Z_o^2}} - \cos\omega\frac{R}{\sqrt{X_o^2 + Z_o^2}}\right)}\right)$$

$$= \arctan\left(\frac{Y_o}{\sqrt{X_o^2 + Z_o^2 - R^2\sin^2\omega} - R\cos\omega}\right). \tag{3.8}$$

Finally, the image point's elevation L defined with respect to the base plane (having positive value downwards) can be determined as follows:

$$L = f\tan\beta$$

$$= \frac{fY_o}{\sqrt{X_o^2 + Z_o^2 - R^2\sin^2\omega} - R\cos\omega}.$$

The image coordinate j can be calculated by

$$j = \text{round}\left(\frac{L}{\tau}\right) + j_c \tag{3.9}$$

where j_c is the center of the ith image column and τ denotes the pixel size.

3.6　Rotating Cameras

In this section we briefly discuss cases of rotating sensor-line or sensor-matrix cameras. First we define two useful notions, continuing to ignore camera motion for this purpose.

3.6.1　Image Vectors and Projection Rays

We consider the local 3D coordinate system of a camera or sensor, without looking at relations within a world coordinate system. Just as the geographical coordinates of the Earth are independent of the Earth's motion in the universe, we also ignore the motion of the camera or sensor, confining ourselves to looking at the "small world" of one camera or sensor.

For a camera with a planar capturing surface, any xy coordinates in this surface and f describe the location of a pixel, which is also characterized by discrete coordinates (sometimes also called *indices*) i and j and pixel constant τ.

In the case of a camera with only one sensor line, the pixel index i is obsolete and may be used to describe the motion (e.g., rotation or translation) of this single-line camera.

DEFINITION **3.7.** *An* image vector v *points from a projection center* $\mathbf{C}$ *to an image point.*

Let $\mathbf{v}_{ij}$ be the image vector which describes a pixel position in the camera coordinate system defined by indices i and j. This ijth *image vector* equals

$$\mathbf{v}_{ij} = \begin{pmatrix} i\tau - x_0 \\ j\tau - y_0 \\ f \end{pmatrix}$$

for a matrix sensor, and

$$\mathbf{v}_j = \begin{pmatrix} 0 \\ j\tau - y_0 \\ f \end{pmatrix}$$

for a single-line camera, say at $i = 0$. Such an image vector may also be seen as a "vectorial image point" in the camera coordinate system.

In the following example we assume a matrix camera. The central point (x_0, y_0) of the image is not necessarily equal to the geometric center $(H/2, W/2)$ of the image. For example, the uppermost, leftmost pixel $(0, 0)$ of a matrix might be defined in a specific local camera coordinate system by

$$v_{00} = (-5\,\text{mm}, -5\,\text{mm}, 35\,\text{mm})^{\text{T}}$$

with a sensor matrix of a size of $10\,\text{mm} \times 10\,\text{mm}$, a focal length of $35\,\text{mm}$, and a pixel size of $\tau = 0.01\,\text{mm}$; thus with $1{,}000 \times 1{,}000$ pixels. An ideally positioned sensor matrix has its geometric center at the central point (i.e., at $x_0 = 5\,\text{mm}$ and $y_0 = 5\,\text{mm}$), and the optical axis would be (exactly) perpendicular to the sensor matrix. (A misalignment of the sensor matrix by "only" $0.1\,\text{mm}$ would result in a shift by 10 pixels.) Applying the transform in equation (3.10) below, we may rotate this ideal central point into world coordinates (i.e., as defined by this camera) as follows:

$$\begin{pmatrix} -5 \\ 35 \\ 5 \end{pmatrix} = \begin{pmatrix} 1 & 0 & 0 \\ 0 & 0 & 1 \\ 0 & -1 & 0 \end{pmatrix} \begin{pmatrix} -5 \\ -5 \\ 35 \end{pmatrix}.$$

An image vector $\mathbf{v}_{ij}$ and a scaling factor $\lambda > 0$ define a *projection ray* $\mathcal{L} = \lambda \cdot \mathbf{v}_{ij}$ (either in local camera or in world coordinates). $\lambda = \infty$ describes a projection ray of infinite length; each 3D point in the scene intersected by such an infinite $\mathcal{L}$ is theoretically visible for the camera; assuming totally transparent air and opaque objects, the closest visible 3D point is recorded at pixel ij. Continuing with our example, assume that a 3D point at world coordinates $(-2\,\text{m}, 14\,\text{m}, 2\,\text{m})$ (i.e., the distance equals $14\,\text{m}$ and the altitude equals $2\,\text{m}$) is projected into the central point; this means we have that

$$\begin{pmatrix} -2{,}000 \\ 14{,}000 \\ 2{,}000 \end{pmatrix} = \lambda \begin{pmatrix} 1 & 0 & 0 \\ 0 & 0 & 1 \\ 0 & -1 & 0 \end{pmatrix} \begin{pmatrix} -5 \\ -5 \\ 35 \end{pmatrix}. \tag{3.10}$$

It follows that $\lambda = 400$. The Euclidean distance $\lambda|\mathbf{v}|$ is the length $|\mathcal{L}|$ of the *actual projection ray* $\lambda \cdot \mathbf{v}_{ij}$, which starts at the projection center and ends at the projected 3D object point. In our example, this length is about $14.28\,\text{m}$; the projected 3D point has a Z coordinate in camera coordinates of $\lambda v_z = 14\,\text{m}$.

Now we apply equation (3.10) in its given form, to transform world into camera coordinates:

$$\begin{pmatrix} -5 \\ -5 \\ 35 \end{pmatrix} = \frac{1}{\lambda} \begin{pmatrix} 1 & 0 & 0 \\ 0 & 0 & -1 \\ 0 & 1 & 0 \end{pmatrix} \begin{pmatrix} -2{,}000 \\ 14{,}000 \\ 2{,}000 \end{pmatrix}.$$

World coordinates of the 3D object point are first transformed into camera coordinates $P_c = (-2{,}000, -2{,}000, 14{,}000)^\mathsf{T}$; the divisions P_{cx}/P_{cz} and P_{cy}/P_{cz} eliminate the scaling factor λ and we obtain that

$$x = \mathbf{v}_x = (P_{cx}/P_{cz})\mathbf{v}_z$$

and

$$y = \mathbf{v}_y = (P_{cx}/P_{cz})\mathbf{v}_z.$$

These are nothing more than the general projection equations $x = f P_{cx}/P_{cz}$ and $y = f P_{cy}/P_{cz}$. Regarding pixel indices, we have that

$$i = (x + x_0)/\tau$$

and

$$j = (y + y_0)/\tau.$$

So far we have not considered camera motion. An object point with $X \neq 0$ would not be visible for a single-line camera (note that $\mathbf{v}_x$ is always zero in this case). Therefore we have to move such a camera (i.e., we are scanning a scene, line by line). The next subsections describe relations between image points and object points for a rotation with $R = 0$ (i.e., a single projection center), $R > 0$ (i.e., multiple projection centers), and also for the general case of $R > 0$ and $\omega > 0$.

3.6.2 Single-Center Panorama (Ideal Case)

We consider a rotating sensor-line or sensor-matrix camera with a single projection center (i.e., $R = 0$) and focal length f. We assume a local 3D sensor coordinate system where the y-axis coincides with the rotation axis (pointing downward). We define rotation angles $\varphi(k) = k\gamma$ for a given constant angular increment γ, and any $k \geq 0$ defines a new image.

Each rotation angle (or each image) defines a rotation matrix $\mathbf{R}_{\varphi(k)}$ in the local sensor coordinate system, and a new attitude of the rotating camera.[8]

Let $\mathcal{L}$ be a projection ray for image vector $\mathbf{v}_{ij}$, for some $\lambda > 0$, which characterizes a pixel position in the kth image, defined by indices i and j. We have that

$$\mathcal{L} = \lambda \mathbf{R}_{\varphi(k)} \begin{pmatrix} i\tau + x_0 \\ j\tau + y_0 \\ f \end{pmatrix} = \lambda \begin{pmatrix} f \sin\varphi(k) + \cos\varphi(k)(i\tau + x_0) \\ j\tau + y_0 \\ f \cos\varphi(k) - \sin\varphi(k)(i\tau + x_0) \end{pmatrix}.$$

For example, when rotating a (monochromatic) single-line camera, generating pixels with indices i and j, we have that

$$\mathcal{L} = \lambda \mathbf{R}_{\varphi(i)} \begin{pmatrix} 0 \\ j\tau + y_0 \\ f \end{pmatrix} = \lambda \begin{pmatrix} f \sin\varphi(i) \\ j\tau + y_0 \\ f \cos\varphi(i) \end{pmatrix}.$$

A point in 3D world coordinates $\mathbf{P}_w$ is then given by

$$\mathbf{P}_w = \mathbf{t}_0 + \mathbf{R}\mathbf{R}_\varphi \lambda \mathbf{v}. \tag{3.11}$$

We recall that $\mathbf{t}_0 + \mathbf{R}$ is a transformation between coordinate systems, $\mathbf{t}_0$ is the location of our sensor coordinate system in the world, and in the case of a single projection center, it is equal to the projection center. $\mathbf{R}$ combines the rotations $\mathbf{R}_\psi$, $\mathbf{R}_\phi$, and $\mathbf{R}_\kappa$, with $\mathbf{R}_\psi = \mathbf{R}_\psi - 90°$.

It follows that that both cases (i.e., a rotating sensor-matrix and a rotating single-line camera) lead basically to the same final coordinate equation defining the spatial relation between object and image point. The only difference is defined by values of $\mathbf{v}_x$. The question arises as to why $\mathbf{v}_x$ is assumed to be zero for a single-line camera. The simple answer is that calculations are easier this way! Actually, a panchromatic "single-line" contains three sensor lines (for R, G, and B) with some spacing in the $\mathbf{v}_x$ direction, typically of some micrometers. Also we have to consider some central point displacement as illustrated in Figure 3.9.

An uncalibrated image has a significant color shift of several pixels (due to shifted locations of the three sensor lines, also in the ideal case of no central point displacement). Later in the book we specify a pre-calibration step for correcting color shift, also for the case of $R > 0$, but with some assumptions for this general off-axis case. Therefore we can continue with the terminology of a single-line camera also for a color sensor, identifying one RGB triplet with exactly one position, with $\mathbf{v}_x = 0$.

---

[8] In the case of a rotating sensor-line camera, the rotation matrix $\mathbf{R}_{\varphi(k)}$ identifies the kth row of the recorded panoramic image.

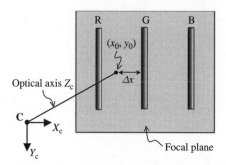

Figure 3.9 Illustration of an RGB sensor line, which is actually composed of three sensor lines. All three lines are mounted on the focal plate. The figure shows a case where the center of the combined red, green and blue line has a distance Δx from the central point.

3.6.3 Multi-Center Panorama with $\omega = 0$

We consider the case of a rotating single-sensor-line camera, but now also allowing $R > 0$ (possibly just due to incorrect positioning of the optics of the camera with respect to the rotation axis, with an actual intention of having $R = 0$). In this case it follows that

$$\mathbf{P}_w = \mathbf{t}_0 + \mathbf{R}\mathbf{R}_{\varphi(i)} \left(\lambda \mathbf{v}_j + R\mathbf{z}^\circ \right). \tag{3.12}$$

Recall that $\mathbf{z}^\circ$ is the unit vector along the z-axis of the sensor. We have that

$$\mathbf{P}_w = \mathbf{t}_0 + \mathbf{R}\mathbf{R}_{\varphi(i)} \left[\lambda \begin{pmatrix} 0 \\ j\tau + y_0 \\ f \end{pmatrix} + \begin{pmatrix} 0 \\ 0 \\ R \end{pmatrix} \right].$$

3.6.4 Multi-Center Panorama with $\omega \neq 0$

We also discuss the standard rotational motion model allowing that $\omega \neq 0$; this requires an off-axis approach with $R > 0$. Symmetric values $+\omega$ and $-\omega$ are convenient choices for creating a symmetric pair of panoramas (e.g., to acquire stereo data, or for stereo visualization).

In Equation (3.12), the viewing direction needs to be changed due to angle $\omega \neq 0$; the coordinate transform is now given in the following equation:

$$\mathbf{P}_w = \mathbf{t}_0 + \mathbf{R}\mathbf{R}_{\varphi(i)} \left(\lambda \mathbf{R}_\omega \mathbf{v}_j + R\mathbf{z}^\circ \right). \tag{3.13}$$

Filling in more details, we also have that

$$\mathbf{P}_w = \mathbf{t}_0 + \mathbf{R}\mathbf{R}_{\varphi(i)} \left[\lambda \mathbf{R}_\omega \begin{pmatrix} 0 \\ j\tau + y_0 \\ f \end{pmatrix} + \begin{pmatrix} 0 \\ 0 \\ R \end{pmatrix} \right].$$

The matrix $\mathbf{R}_\omega$ specifies the additional rotation of the sensor line due to $\omega \neq 0$.

3.6.5 General Case of a Rotating Sensor-Line Camera

Figure 3.10 illustrates the general case of a rotating RGB line camera. In applications we also have to model the following deviations[9] which are apparent in real life:

- At any discrete moment i in time, the sensor line is tilted (within the local coordinate system) by three angles which define over time a time-dependent rotation matrix $\mathbf{R}_i(\alpha, \beta, \delta)$; this defines the *inner attitude* of a sensor line about all three axes with respect to the central point x_0, y_0.
- The red and blue 'sub-lines' have an offset Δ with respect to the central point on the green line.
- The optical axis is rotated by ξ about the X_o-axis.
- The optical axis is rotated by the fixed principal angle ω about the Y_o-axis.
- The sensor line is rotating with (an eccentricity, or a desired) off-axis distance $R > 0$.

The inner attitude $\mathbf{R}_i(\alpha, \beta, \delta)$, the central point (x_0, y_0), and the offset Δ now allow us to decide about different ways for modeling the rotation and shift of the sensor line. For example, we may rotate it first, and then shift it to its final position, or vice versa. Whatever model is selected, six degrees of freedom are sufficient to describe the positioning of the sensor line in any case.

Figure 3.11 illustrates differences between two types of rotations, either about the projection center, or about the focal point. In the first case, it would not matter whether the sensor line

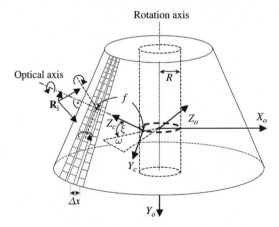

Figure 3.10   The $X_o Y_o Z_o$ coordinate system of the rotating line camera: the optical axis identifies the central point x_0, y_0 and is tilted by $\mathbf{R}_\xi$ and $\mathbf{R}_\omega$; each red-green-blue line has a constant distance Δ between central point (on green line) and red or blue line. The tilt of the sensor line with respect to the optical axis is specified by $\mathbf{R}_i(\alpha, \beta, \delta)$.

---

[9] Optical distortions are dealt with for a given camera in a pre-calibration process which is not part of the geometric correction process.

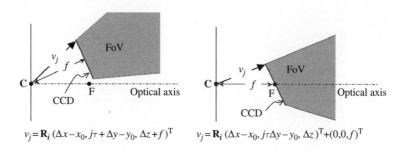

$$v_j = \mathbf{R}_i\,(\Delta x - x_0, j\tau + \Delta y - y_0, \Delta z + f)^{\mathrm{T}} \qquad v_j = \mathbf{R}_i\,(\Delta x - x_0, j\tau\Delta y - y_0, \Delta z)^{\mathrm{T}} + (0,0,f)^{\mathrm{T}}$$

Figure 3.11 Illustration of the difference between a rotation of the sensor line around the projection center (left), and about the focal point F (right).

or the whole sensor (i.e., the camera system) is rotated. These changes of inner attitude and sensor attitude would just be dual operations.

We model the rotation with respect to the focal point. Attempting to be closer to the actual physical processes, we also decide first to rotate the sensor line, and then to shift it into the focal point F. (This is the work flow when a focal plane is physically mounted in a camera.)

In conclusion, the image vector $\mathbf{v}_j$ is split into two terms as follows:

$$\mathbf{v}_j = \mathbf{v}_{j,\Delta} + \mathbf{v}_f = (\Delta_x - x_0, j\tau + \Delta_y - y_0, 0)^{\mathrm{T}} + (0, 0, f)^{\mathrm{T}}. \qquad (3.14)$$

Altogether, the coordinate transform is now

$$\mathbf{P}_w = \mathbf{t}_0 + \mathbf{R}\mathbf{R}_{\varphi(i)}\left[\lambda\mathbf{R}_\xi\mathbf{R}_\omega[\mathbf{R}_i\mathbf{v}_j + f\mathbf{z}^\circ] + R\mathbf{z}^\circ\right]. \qquad (3.15)$$

In more detailed form, this equation also reads as follows:

$$\mathbf{P}_w = \mathbf{t}_0 + \mathbf{R}\mathbf{R}_{\varphi(i)}\left[\lambda\mathbf{R}_\xi\mathbf{R}_\omega\left[\mathbf{R}_i\begin{pmatrix}\Delta_x - x_0 \\ j\tau + \Delta_y - y_0 \\ \Delta_z\end{pmatrix} + \begin{pmatrix}0 \\ 0 \\ f\end{pmatrix}\right] + \begin{pmatrix}0 \\ 0 \\ R\end{pmatrix}\right]. \qquad (3.16)$$

This more detailed form of equation (3.15) is applied repeatedly later on.

3.7 Mappings between Different Image Surfaces

Mappings of spherical or cylindrical panoramas into a plane are of interest, for example, for visualization purposes.

3.7.1 Reprojection onto a Straight Cylinder

Sensor calibration involves the discussion of optical distortions, such that projection can be assumed to be onto an ideal capturing surface. For example, when capturing a single-center panorama with a rotating sensor-line camera, we assume that a focal plane (defined by a calibrated focal length) rotates around a rotation axis, thus forming an image cylinder. However, due to distortions, this can only (approximately) be described by forming a cone rather than a straight cylinder.

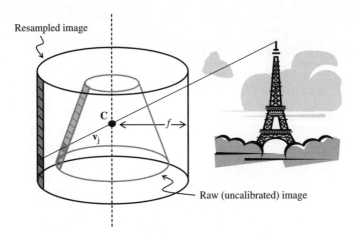

Figure 3.12 Mapping of a single-center panorama, captured on a conic surface, onto a cylindrical surface, where **C** denotes the projection center.

For purposes of correction, we project the cone onto, say, the smallest cylinder such that it contains the cone, and all projected points (on the cone) can be mapped onto this cylinder (details follow in Chapter 5); see Figure 3.12. The radius of the specified (ideal) cylinder is then the focal length f, and this cylinder is used in all subsequent calculations: the ijth *image vector* is defined by the reference point of pixel location (i, j) on the ideal cylinder (i.e., in the resampled image).

For calculating cylindrical image coordinates, the image vector $\mathbf{v}_j$ (see equation (3.11)) has to be transformed by the tilt parameters $\mathbf{R}_x(\xi)$ and $\mathbf{R}_\omega$ for the principal axis, the interior sensor attitude $\mathbf{R}_i(\alpha, \beta, \delta)$ and the offset parameter Δ, and the rotation $\mathbf{R}_\varphi$ to $\mathbf{v}'_j = \mathbf{v}_j(\xi, \omega, \alpha, \beta, \delta, \Delta, \varphi)$:

$$\mathbf{R}^{-1} \cdot (\mathbf{t} - \mathbf{t}_0) = \mathbf{R}_\varphi(\lambda \cdot \mathbf{R}_\xi \cdot \mathbf{R}_\omega[\mathbf{R}_i \cdot \mathbf{v}_{j,\Delta} + \mathbf{v}_f] + R\mathbf{z}^\circ) \tag{3.17}$$

$$= \lambda \cdot \mathbf{v}'_j + \mathbf{R}_\varphi \cdot R\mathbf{z}^\circ. \tag{3.18}$$

Now we project each pixel onto a straight (hollow) cylinder following the relation

$$\mathbf{R}_{\varphi'}\left[\lambda'\begin{pmatrix} 0 \\ j' \cdot \tau' \\ f' \end{pmatrix} + \begin{pmatrix} 0 \\ 0 \\ R \end{pmatrix}\right] = \lambda \cdot \mathbf{v}'_j + \mathbf{R}_\varphi \cdot R\mathbf{z}^\circ. \tag{3.19}$$

The left-hand side of equation (3.19) corresponds to a straight (hollow) cylinder, whereby λ' is a scaling factor of the object coordinates. The parameter f' is the new virtual focal length; R is the new radius of this cylinder; f' is freely selectable. The parameter τ' is the selectable pixel size, and $\Delta\varphi'$ the new angular resolution of this cylinder. To retain the image ratio, the pixel size is given by

$$\tau' = \frac{\tau \cdot f'}{f} \cdot \frac{\Delta\varphi'}{\Delta\varphi}. \tag{3.20}$$

In the case of $R = 0$, the cylindrical image coordinates are

$$i' = \arctan\left(\frac{v'_{jx}}{v'_{jz}}\right) \cdot \frac{1}{\Delta\varphi'}, \tag{3.21}$$

$$j' = \frac{f \cdot v'_{jy}}{\sqrt{v'^2_{jx} + v'^2_{yz}}} \cdot \frac{1}{\tau'}, \tag{3.22}$$

where i' and j' represent the position of a pixel on the rectified cylinder.

If $R \neq 0$ then the projection onto a straight hollow cylinder could be calculated by solving equation (3.19) for $R \neq 0$. However, this would result in the loss of the important relation between object coordinates and image coordinates (in contrast to the case $R = 0$). Thus, the projection is possible in principle for $R > 0$, but not really meaningful.

3.7.2 Cylindrical Panorama onto Sphere

Plate 8 shows an example of the projection of three cylindrical panoramic images onto a sphere. Because of the relatively small vertical field of view of the rotating sensor-line camera used it was necessary to combine three scans. The upper (green) and the lower (red) scan were acquired with the panoramic sensor having a tilted rotation axis $\mathbf{R}_\xi$. The angle of the rotation axis can vary between $-45°$ and $+45°$. For a tilted rotation axis, the unrectified image coordinates are basically *conic coordinates*.[10]

Similar to the reprojection onto an ideal cylindrical surface, the transformed image vector $\mathbf{v}'$ has to be projected onto a sphere. The left-hand side of equation (3.23) corresponds to a sphere with radius D and the right-hand side involves the transformed image vector $\mathbf{v}' = \mathbf{R}_\xi \mathbf{v}$:

$$D\begin{pmatrix} \sin\theta \cdot \sin\varphi \\ \sin\theta \cdot \cos\varphi \\ \cos\theta \end{pmatrix} = \lambda \cdot \mathbf{v}'_j, \tag{3.23}$$

$$i' = \arctan\left(\frac{v'_{jx}}{v'_{jz}}\right) \cdot \frac{1}{\Delta\varphi'},$$

$$j' = \arccos\left(\frac{v'_{jy}}{\sqrt{v'^2_{jx} + v'^2_{jz}}}\right)\frac{1}{\vartheta'}.$$

3.7.3 Cylindrical Panorama onto Tangential Plane

Figure 3.13 illustrates a projection of a section of an image cylinder into a tangential plane. Pixel (i, j) in the ith column of the panoramic image is mapped into pixel (i', j') in the tangential plane, using the equations

---

[10] These coordinates are given by unrolling the cone into a plane, with subsequent deformation into a rectangle.

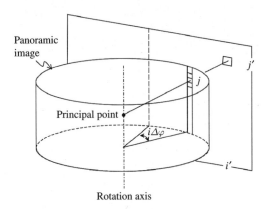

Figure 3.13   Projection of an image cylinder into a tangential plane.

$$i' = f \tan(i\Delta\varphi)\frac{1}{\tau}, \tag{3.24}$$

$$j' = \frac{j}{\cos(i\Delta\varphi)}. \tag{3.25}$$

Plate 9 illustrates a result of such a central-perspective rectification, by comparing an unrolled cylinder with the projection onto a tangential plane. Obviously, less than 180° of the panoramic image can be projected. Besides a "straightening" of incorrectly curved lines in the unrolled (isogonal) cylinder, the projected image also shows the widening effect of central projection for objects further away from the central point.

3.8 Laser Range-Finder

We transform all LRF data at a given attitude into a polar coordinate system with a horizontal range of 360° and a vertical range of 180°. At this step, all LRF calibration data are available and required. Each 3D point, measured with the LRF in the scene, is described either in local polar coordinates by the triple (D, ϑ, φ), or in local Cartesian coordinates as a vector $\mathbf{p} = (p_x, p_y, p_z)$. The two representations are related to each other as follows:

$$p_x = D \sin\vartheta \cos\varphi,$$

$$p_y = D \sin\vartheta \sin\varphi,$$

$$p_z = D \cos\vartheta.$$

The laser scanner rotates clockwise. The first scan line starts at the positive y-axis in the local LRF coordinate system, at the horizontal angle of 0 degrees. Figure 3.14 illustrates the local coordinate system of the LRF.

The attitude of the LRF with respect to a reference vector $\mathbf{t}$ in the world coordinate system is defined by one rotation matrix $\mathbf{R} = \mathbf{R}_\Omega \cdot \mathbf{R}_\phi \cdot \mathbf{R}_\kappa$ and a translation vector $\mathbf{t}_0$:

$$\mathbf{t} = \mathbf{t}_0 + \mathbf{R} \cdot \mathbf{p}. \tag{3.26}$$

In conclusion of this short section, for the LRF we have listed only a few formulas here which were also given earlier for the rotating sensor-line camera.

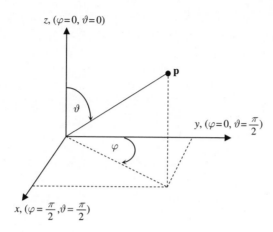

Figure 3.14   Range-finder coordinate system: the z-axis points to the zenith; point **p** is defined by slant ϑ, φ, and distance D to the origin.

3.9 Exercises

3.1. Assume that a local left-handed *XYZ* Cartesian coordinate system needs to be mapped into a right-handed *XYZ* Cartesian coordinate system. What kind of affine transform allows this in general?

3.2. Draw a right-handed *XYZ* coordinate system onto a box (e.g., of matches). Perform the following three motions:

  (1) Place the box onto a table. Now rotate the object first by 90° about its x-axis, then by 90° about its y-axis, and finally by 90° about its z-axis. Take a photo of the box in its final attitude.
  (2) Place the box onto the table, exactly as for (1). Now rotate first about the z-axis, then about the y-axis, and finally about the x-axis.
  (3) Place the box onto the table, exactly as for (1). The table defines our world coordinate system: the edges of the table specify the *XY* coordinate axes, and the normal of the table defines the direction of the *Z*-axis. Now rotate the object first by 90° about the *X*-axis, then by 90° about the *Y*-axis, and finally by 90° about the *Z*-axis. Take a photo of the final attitude of the box.

  Is motion (1) or motion (2) equivalent to motion (3)? Express your result in a general statement in mathematical terms.

3.3. [Possible lab project] Implement your own program which projects all your images of Exercise 1.3 onto a cylindrical surface. Compare and discuss differences between results using the available program and your own implementation.

3.4. [Possible lab project] Implement a program which projects any < 180° fraction of a cylindrical image (i.e., which is given as being on the surface of a straight cylinder; you may use a resulting image from Exercise *3.3.*, or a, normally 360°, panorama from the net) onto a plane which is tangential to the cylinder.

In particular, allow the tangential plane to be rotated by the same angle as used during your image capturing experiments in Exercise 1.4. Compare original (i.e., recorded) images with your projected images.

3.5. [Possible lab project] This lab project requires a sensor-line camera, or you may also use input images as available on the website for this book. The images are recorded in the described setup with an optical axis tilted to the rotation axis by angle ξ ($-45°$ bottom [floor], $0°$ [normal panorama], and $+45°$ top [ceiling]). Implement a program which projects all images into a sphere, for a full spherical angle of 4π. Then project only the ceiling into a tangential plane which is defined as the top of the cylinder. Why is it now possible to project the full $360°$ into the tangential plane, contrary to Exercise 3.4.?

3.10 Further Reading

For homogeneous coordinates and their use in computer graphics, see Hill and Kelley (2007).

There are various online textbooks on the subject of linear algebra (such as listed on *http://en.wikipedia.org/wiki/Linear_algebra*).

For projective geometry, see Coxeter (2003) in general, and Hartley and Zisserman (2000) in particular in the context of computer vision.

For surface geometry in the context of 3D image analysis, see Klette and Rosenfeld (2004); see also books cited therein on differential geometry for general references.

Ishiguro et al. (1992) details the essential features of a multi-center panoramic image acquisition model, and the approach has since been further analyzed, for example, by Peleg and Ben-Ezra (1999) and Huang et al. (2001c).

The majority of applications of panoramic 3D scene visualizations seem to be based on or assume single-center panoramas; see Zheng and Tsuji (1992), Chen (1995) and Huang et al. (2001a) for some early examples.

A projection equation for "linear pushbroom cameras" (i.e., sensor-line cameras in the terminology of this book) was given in Gupta and Hartley (1997).

The concentric case was discussed in Shum and He (1999). Coaxial panoramas are widely used for 3D scene reconstruction, and also for catadioptric approaches (Southwell et al., 1996; Nene and Nayar, 1998; Petty et al., 1998).

For stereo matching in symmetric panoramas, see, for example, Barnard and Fischler (1982), Ohta and Kanade (1985), Cox (1994) and Šára (1999).

4

Epipolar Geometry

Epipolar geometry characterizes the geometric relationships between projection centers of two cameras, a point in 3D space, and its potential position in both images. A benefit of knowing the epipolar geometry is that, given any point in either image (showing the projection of a point in 3D space), epipolar geometry defines a parameterized (i.e., Jordan) curve ℓ (the epipolar curve) of possible locations of the corresponding point (if visible) in the other image. The parameter t of the curve ℓ(t) then allows in computational stereo to move along the curve when testing image points for actual correspondence. Interest in epipolar geometry is also motivated by stereo viewing.[1] There is a "preferred" epipolar geometry which supports depth perception, and this is defined by parallel epipolar lines; these lines are a special case of epipolar curves.

4.1 General Epipolar Curve Equation

The epipolar geometry of two sensor-matrix cameras defines epipolar lines, but these are in general not all parallel in each of the two image planes. The epipolar geometry of two rotating sensor-line cameras defines Jordan curves on cylinders, and we will also identify the case where two panoramic images support depth perception.

Computational stereo with large panoramic images (of the order of $10,000 \times 100,000$ pixels) definitely requires the knowledge of epipolar curves to allow reasonably sized search spaces for corresponding points. This ensures not only efficiency in stereo matching, but also a decrease in the probability of false matching.

This section states an epipolar curve equation theorem and its proof for the general case of multi-view panoramas. Specific cases (such as leveled or symmetric panoramas) are discussed in Section 4.2.

---

[1] Sir Charles Wheatstone demonstrated his "stereopticon" in 1838 for line drawings, and a year later for photographs. Different techniques have been used since then for depth perception by showing one image to the left, and the other image to the right eye, either in parallel or in alternation.

Panoramic Imaging: Sensor-Line Cameras and Laser Range-Finders   F. Huang, R. Klette, and K. Scheibe
© 2008 John Wiley & Sons, Ltd

We consider an arbitrary pair of multi-view panoramas $E_{\mathcal{P}_1}(R_1, f_1, \omega_1, \gamma_1)$ and $E_{\mathcal{P}_2}(R_2, f_2, \omega_2, \gamma_2)$, where R, f, ω, and γ are the sensor parameters specifying off-axis distance, effective focal length, principal angle, and angular increment. Subscripts 1 and 2 are used to indicate that parameters may differ between panoramas. We also call sometimes panorama $E_{\mathcal{P}_1}$ the *source image*, and panorama $E_{\mathcal{P}_2}$ the *destination image*, and use "source" or "destination" for other components in the same sense. In this chapter, we assume that the same sensor-line camera is used to acquire any arbitrary panoramic pair. This implies that the pixel sizes of both the source and the destination images are identical with dimensions of τ (width) $\times \tau$ (height).

A multi-view panoramic pair defines a *general epipolar curve equation* because the epipolar geometry of other types of cylindrical panoramic pairs can be derived from this equation. An image point $\mathbf{p}_1$ on $E_{\mathcal{P}_1}$ is given. This image point is a projection of an (unknown) point in 3D space, which (assuming it is visible) will project into the *corresponding* image point $\mathbf{p}_2$ in $E_{\mathcal{P}_2}$. Without knowing the projected 3D point, the knowledge about sensor parameters and $\mathbf{p}_1$ allows the possible locations of $\mathbf{p}_2$ to be specified.

The origin of the source sensor coordinate system is denoted by $\mathbf{O}_1$, and of the destination sensor coordinate system by $\mathbf{O}_2$. The 3×3 rotation matrix $\mathbf{R}_{o_1 o_2}$ and the 3×1 translation vector $\mathbf{t}_{o_1 o_2}$ specify orientation and position of the destination sensor coordinate system with respect to the source sensor coordinate system. The rotation matrix $\mathbf{R}_{o_1 o_2}$ is decomposed into its three row vectors $[\mathbf{r}_1^T \mathbf{r}_2^T \mathbf{r}_3^T]^T$, and the translation vector $\mathbf{t}_{o_1 o_2}$ is represented by its three elements $(t_x, t_y, t_z)^T$.

The curves on the image cylinder will be described by cylindrical coordinates as described in Chapter 3. An image point on a panorama $E_{\mathcal{P}}(R, f, \omega, \gamma)$ will be represented by coordinates (φ, L, D_0). We recall that φ indicates the orientation, which lies in the range $[0, 2\pi)$; L indicates level or elevation, where we have $L = 0$ on the base plane and $L > 0$ downwards; and D_0 denotes the radius of the image cylinder. Figure 4.1 recalls the geometry of the imaging model. The value of D_0 is constant for all image points on the same panorama. We have

$$D_0 = \sqrt{R^2 + f^2 + 2Rf \cos \omega}.$$

An image point on the 2D cylindrical surface can also be represented by only the first two variables, φ and L. We call (φ, L) the image coordinates of an image point. The general epipolar curve theorem is stated as follows.

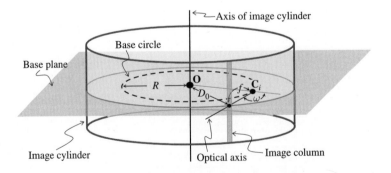

Figure 4.1   Geometry of the imaging model.

4.1. THEOREM. *Let (φ_1, L_1) and (φ_2, L_2) denote the image coordinates of the projection of a 3D point in the source image $E_{\mathcal{P}_1}(R_1, f_1, \omega_1, \gamma_1)$ and the destination image $E_{\mathcal{P}_2}(R_2, f_2, \omega_2, \gamma_2)$, respectively. Consider φ_1 and L_1 as being given. The corresponding epipolar curve on the destination image $E_{\mathcal{P}_2}$ can be represented by the parameterized function*

$$L_2(\varphi_2) = \frac{f_2 r_2^T \cdot l}{\sin(\varphi_2 + \omega_2) r_1^T \cdot l + \cos(\varphi_2 + \omega_2) r_3^T \cdot l - R_2 \cos \omega_2}$$

for $\varphi_2 \in [0, 2\pi)$, and the curve L_2 is only valid if the value of the denominator is greater than zero. The vector l is defined as follows:

$$l = u + \frac{R_2 \sin \omega_2 + \cos(\varphi_2 + \omega_2) r_1^T \cdot u - \sin(\varphi_2 + \omega_2) r_3^T \cdot u}{\sin(\varphi_2 + \omega_2) r_3^T \cdot v - \cos(\varphi_2 + \omega_2) r_1^T \cdot v} v$$

where u and v are two constant vectors defined as

$$u = \begin{bmatrix} R_1 \sin \varphi_1 - t_x \\ -t_y \\ R_1 \cos \varphi_1 - t_z \end{bmatrix} \quad and \quad v = \begin{bmatrix} \sin(\varphi_1 + \omega_1) f_1 \\ L_1 \\ \cos(\varphi_1 + \omega_1) f_1 \end{bmatrix}.$$

Furthermore, the matrix $[r_1^T r_2^T r_3^T]^T$ and the vector $[t_x, t_y, t_z]^T$ specify the orientation and the position of the rotating sensor-line camera of $E_{\mathcal{P}_2}$ with respect to the sensor coordinate system of $E_{\mathcal{P}_1}$.

Proof An image point $\mathbf{p}_1$ with image coordinates (φ_1, L_1) on $E_{\mathcal{P}_1}$ defines a 3D projection ray, denoted by $\mathcal{L}$. Let $\mathbf{C}_1$ denote the projection center associated with image point $\mathbf{p}_1$. The projection ray $\mathcal{L}$ emits from $\mathbf{C}_1$ and is incident with the (projection) point $\mathbf{p}_1$; see Figure 4.2. The projection ray can first be expressed in a local camera coordinate system which has its origin at $\mathbf{C}_1$; in this case it is written as $\mathcal{L}_{c_1}$ to distinguish it from other representations. (Ray $\mathcal{L}$ may also be expressed in another coordinate system in a given context.) The projection ray $\mathcal{L}_{c_1}$ can be described by $\mathbf{0} + \lambda \mathbf{k}$, where $\mathbf{0}$ is a 3D zero vector, $\lambda \in \mathbb{R}$ is any scalar, and $\mathbf{k}$ is an image vector defined as

$$\mathbf{k} = \begin{bmatrix} 0 \\ L_1 \\ f_1 \end{bmatrix} \tag{4.1}$$

where f_1 is the effective focal length associated with the source panoramic image $E_{\mathcal{P}_1}$.

The projection ray $\mathcal{L}_{c_1}$ is then transformed from the camera coordinate system to the sensor coordinate system (with origin at $\mathbf{O}_1$) of the source panoramic image $E_{\mathcal{P}_1}$, denoted by $\mathcal{L}_{o_1}$. We have

$$\mathcal{L}_{o_1} = \begin{bmatrix} \mathbf{R}_{o_1 c_1}^{-1} & \mathbf{t}_{o_1 c_1} \end{bmatrix} \begin{bmatrix} \mathcal{L}_{c_1} \\ 1 \end{bmatrix}$$

$$= \begin{bmatrix} \mathbf{R}_{o_1 c_1}^{-1} & \mathbf{t}_{o_1 c_1} \end{bmatrix} \begin{bmatrix} \mathbf{0} + \lambda \mathbf{k} \\ 1 \end{bmatrix}$$

$$= \mathbf{t}_{o_1 c_1} + \lambda \mathbf{R}_{o_1 c_1}^{-1} \mathbf{k} \tag{4.2}$$

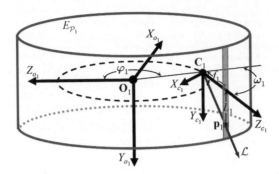

Figure 4.2   A 3D projection ray $\mathcal{L}$ defined by the image point $\mathbf{p}_1$.

where the rotation matrix $\mathbf{R}_{o_1c_1}$ and translation vector $\mathbf{t}_{o_1c_1}$ specify the orientation and position of the camera coordinate system (with origin at $\mathbf{C}_1$) with respect to the sensor coordinate system (with origin at $\mathbf{O}_1$), respectively. All the elements of $\mathbf{R}_{o_1c_1}$ and $\mathbf{t}_{o_1c_1}$ can be expressed by φ_1 and the sensor parameters R_1, ω_1, and γ_1 of the panoramic image $E_{\mathcal{P}_1}$.

The angle between the Z_{o_1}-axis of the sensor coordinate system and the line segment $\overline{\mathbf{O}_1\mathbf{C}_1}$ is denoted by φ_1 (see Figure 4.2). The rotation matrix is

$$\mathbf{R}_{o_1c_1} = \left[\begin{array}{ccc} \cos(\varphi_1+\omega_1) & 0 & -\sin(\varphi_1+\omega_1) \\ 0 & 1 & 0 \\ \sin(\varphi_1+\omega_1) & 0 & \cos(\varphi_1+\omega_1) \end{array} \right]. \tag{4.3}$$

The translation vector is

$$\mathbf{t}_{o_1c_1} = \left[\begin{array}{c} R_1 \sin\varphi_1 \\ 0 \\ R_1 \cos\varphi_1 \end{array} \right]. \tag{4.4}$$

To calculate the potential positions of the corresponding point of $\mathbf{p}_1$ in the destination panorama, a further transform of the projection ray $\mathcal{L}_{o_1}$ is necessary. We transform it into the panorama coordinate system (with origin at $\mathbf{O}_2$) of the destination panoramic image $E_{\mathcal{P}_2}$; see Figure 4.3. The transformed projection ray is denoted by $\mathcal{L}_{o_2}$. We have the following:

$$\begin{aligned} \mathcal{L}_{o_2} &= \begin{bmatrix} \mathbf{R}_{o_1o_2} & -\mathbf{R}_{o_1o_2}\mathbf{t}_{o_1o_2} \end{bmatrix} \begin{bmatrix} \mathcal{L}_{o_1} \\ 1 \end{bmatrix} \\ &= \begin{bmatrix} \mathbf{R}_{o_1o_2} & -\mathbf{R}_{o_1o_2}\mathbf{t}_{o_1o_2} \end{bmatrix} \begin{bmatrix} \mathbf{t}_{o_1c_1} + \lambda\mathbf{R}_{o_1c_1}^{-1}\mathbf{k} \\ 1 \end{bmatrix} \\ &= \mathbf{R}_{o_1o_2}\mathbf{t}_{o_1c_1} + \lambda\mathbf{R}_{o_1o_2}\mathbf{R}_{o_1c_1}^{-1}\mathbf{k} - \mathbf{R}_{o_1o_2}\mathbf{t}_{o_1o_2} \\ &= \mathbf{R}_{o_1o_2}\left(\mathbf{t}_{o_1c_1} - \mathbf{t}_{o_1o_2} + \lambda\mathbf{R}_{o_1c_1}^{-1}\mathbf{k}\right) \end{aligned} \tag{4.5}$$

where the rotation matrix $\mathbf{R}_{o_1o_2}$ and translation vector $\mathbf{t}_{o_1o_2}$ specify the orientation and position of the sensor coordinate system (with origin at $\mathbf{O}_2$) with respect to the sensor coordinate system (with origin at $\mathbf{O}_1$), respectively.

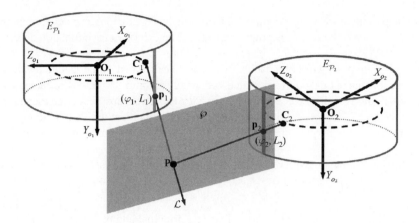

Figure 4.3   Pair of cylindrical panoramas in general position and orientation.

We expand the rotation matrix $\mathbf{R}_{o_1 c_1}$ and the translation vector $\mathbf{t}_{o_1 c_1}$ in equation (4.5), and substitute the image vector $\mathbf{k}$ as given in equation (4.1). The equation of the projection ray $\mathcal{L}_{o_2}$ is thus as follows:

$$
\mathcal{L}_{o_2} = \mathbf{R}_{o_1 o_2} \left(\begin{bmatrix} R_1 s\varphi_1 \\ 0 \\ R_1 c\varphi_1 \end{bmatrix} - \mathbf{t}_{o_1 o_2} + \lambda \begin{bmatrix} c(\varphi_1+\omega_1) & 0 & s(\varphi_1+\omega_1) \\ 0 & 1 & 0 \\ -s(\varphi_1+\omega_1) & 0 & c(\varphi_1+\omega_1) \end{bmatrix} \begin{bmatrix} 0 \\ L_1 \\ f_1 \end{bmatrix} \right)
$$

$$
= \mathbf{R}_{o_1 o_2} \left(\begin{bmatrix} R_1 s\varphi_1 \\ 0 \\ R_1 c\varphi_1 \end{bmatrix} - \mathbf{t}_{o_1 o_2} + \lambda \begin{bmatrix} s(\varphi_1+\omega_1) f_1 \\ L_1 \\ c(\varphi_1+\omega_1) f_1 \end{bmatrix} \right) \tag{4.6}
$$

where R_1, ω_1, and γ_1 are the camera off-axis distance, principal angle, and angular increment of the source panorama, respectively, and φ_1 and L_1 are the image coordinates of the given image point. Symbols s and c denote sin and cos functions, respectively, and are occasionally used for the sake of brevity.

Furthermore, in the theorem the rotation matrix $\mathbf{R}_{o_1 o_2}$ is decomposed into its three row vectors as $[\mathbf{r}_1^T \mathbf{r}_2^T \mathbf{r}_3^T]^T$, and the translation vector $\mathbf{t}_{o_1 o_2}$ is represented by three elements as $(t_x, t_y, t_z)^T$. The expression for the projection ray $\mathcal{L}_{o_2}$ in equation (4.6) is rearranged as follows:

$$
\mathcal{L}_{o_2} = \begin{bmatrix} \mathbf{r}_1^T \\ \mathbf{r}_2^T \\ \mathbf{r}_3^T \end{bmatrix} \left(\begin{bmatrix} R_1 s\varphi_1 \\ 0 \\ R_1 c\varphi_1 \end{bmatrix} - \begin{bmatrix} t_x \\ t_y \\ t_z \end{bmatrix} + \lambda \begin{bmatrix} s(\varphi_1+\omega_1) f_1 \\ L_1 \\ c(\varphi_1+\omega_1) f_1 \end{bmatrix} \right)
$$

$$
= \begin{bmatrix} \mathbf{r}_1^T \\ \mathbf{r}_2^T \\ \mathbf{r}_3^T \end{bmatrix} (\mathbf{u} + \lambda \mathbf{v}) \tag{4.7}
$$

$$
= \begin{bmatrix} \mathbf{r}_1^T \cdot \mathbf{u} + \lambda \mathbf{r}_1^T \cdot \mathbf{v} \\ \mathbf{r}_2^T \cdot \mathbf{u} + \lambda \mathbf{r}_2^T \cdot \mathbf{v} \\ \mathbf{r}_3^T \cdot \mathbf{u} + \lambda \mathbf{r}_3^T \cdot \mathbf{v} \end{bmatrix}
$$

where $\lambda \in \mathbb{R}$ is any scalar, and vectors $\mathbf{u}$ and $\mathbf{v}$ are specified in the theorem.

Let point $\mathbf{p}_2$ denote the corresponding point of $\mathbf{p}_1$ in the destination image $E_{\mathcal{P}_2}$. Point $\mathbf{p}_2$ is the projection of some (unknown) 3D point on ray $\mathcal{L}_{o_2}$. The epipolar curve function describes the projection of $\mathcal{L}_{o_2}$ onto $E_{\mathcal{P}_2}$, and can be expressed in the form $L_2(\varphi_2)$ where L_2 is a function of one variable φ_2. In other words, the values of L_2 can be calculated by the specified values of φ_2.

Let $\mathbf{C}_2$ denote the optical point associated with image column φ_2 of $E_{\mathcal{P}_2}$. For each column φ_2, the corresponding L_2 value can be found by the following two steps. First, find the intersection point, denoted by $\mathbf{P}$, of ray $\mathcal{L}_{o_2}$ with the plane $\wp$ which is incident with $\mathbf{C}_2$ and image column φ_2. Second, obtain L_2 by projecting point $\mathbf{P}$ into the image column φ_2' of panorama $E_{\mathcal{P}_2}$ (see Figure 4.3).

The equation of plane $\wp$ with respect to the sensor coordinate system (with origin at $\mathbf{O}_2$) is

$$- \cos(\varphi_2 + \omega_2)X + \sin(\varphi_2 + \omega_2)Z = R_2 \sin \omega_2 \tag{4.8}$$

where R_2, ω_2, and γ_2 are camera off-axis distance, principal angle, and angular increment of the destination panorama $E_{\mathcal{P}_2}$, respectively.

We substitute the X and Z coordinates of projection ray $\mathcal{L}_{o_2}$ in equation (4.7) into equation (4.8) and solve for λ. We obtain

$$- \cos(\varphi_2 + \omega_2)(\mathbf{r}_1^T \cdot \mathbf{u} + \lambda \mathbf{r}_1^T \cdot \mathbf{v}) + \sin(\varphi_2 + \omega_2)(\mathbf{r}_3^T \cdot \mathbf{u} + \lambda \mathbf{r}_3^T \cdot \mathbf{v}) = R_2 \sin \omega_2$$

and thus

$$\lambda = \frac{R_2 \sin \omega_2 + \cos(\varphi_2 + \omega_2)\mathbf{r}_1^T \cdot \mathbf{u} - \sin(\varphi_2 + \omega_2)\mathbf{r}_3^T \cdot \mathbf{u}}{\sin(\varphi_2 + \omega_2)\mathbf{r}_3^T \cdot \mathbf{v} - \cos(\varphi_2 + \omega_2)\mathbf{r}_1^T \cdot \mathbf{v}}$$

We define $\mathbf{l} = \mathbf{u} + \lambda \mathbf{v}$ (note: as in the theorem).

The intersection point $\mathbf{P}$ can now be calculated by substituting λ into equation (4.7). We denote the obtained coordinates of $\mathbf{P}$ as $(X_{o_2}, Y_{o_2}, Z_{o_2})^T$ with respect to the sensor coordinate system (with origin at $\mathbf{O}_2$). We have

$$\mathbf{P} = \begin{bmatrix} X_{o_2} \\ Y_{o_2} \\ Z_{o_2} \end{bmatrix} = \begin{bmatrix} \mathbf{r}_1^T \\ \mathbf{r}_2^T \\ \mathbf{r}_3^T \end{bmatrix} (\mathbf{u} + \lambda \mathbf{v}), \quad \text{for some value } \lambda$$

$$= \begin{bmatrix} \mathbf{r}_1^T \cdot \mathbf{l} \\ \mathbf{r}_2^T \cdot \mathbf{l} \\ \mathbf{r}_3^T \cdot \mathbf{l} \end{bmatrix}. \tag{4.9}$$

The point $\mathbf{P}$ is transformed into the camera coordinate system (with origin at $\mathbf{C}_2$) to calculate the projection of the image. We denote the coordinates of $\mathbf{P}$ after transformation as $(X_{c_2}, Y_{c_2}, Z_{c_2})^T$. We have the following:

$$\begin{bmatrix} X_{c_2} \\ Y_{c_2} \\ Z_{c_2} \end{bmatrix} = \begin{bmatrix} \mathbf{R}_{o_2 c_2} & -\mathbf{R}_{o_2 c_2} \mathbf{t}_{o_2 c_2} \end{bmatrix} \begin{bmatrix} X_{o_2} \\ Y_{o_2} \\ Z_{o_2} \\ 1 \end{bmatrix}$$

where the rotation matrix $\mathbf{R}_{o_2 c_2}$ and translation vector $\mathbf{t}_{o_2 c_2}$ specify the orientation and position of the camera coordinate system (with origin at $\mathbf{C}_2$) with respect to the sensor coordinate system (with origin at $\mathbf{O}_2$), respectively. After expansion we obtain the following:

$$
\begin{bmatrix} X_{c_2} \\ Y_{c_2} \\ Z_{c_2} \end{bmatrix} = \begin{bmatrix} c(\varphi_2+\omega_2) & 0 & -s(\varphi_2+\omega_2) \\ 0 & 1 & 0 \\ s(\varphi_2+\omega_2) & 0 & c(\varphi_2+\omega_2) \end{bmatrix} \begin{bmatrix} X_{o_2} - R_2 s\varphi_2 \\ Y_{o_2} \\ Z_{o_2} - R_2 c\varphi_2 \end{bmatrix}
$$

$$
= \begin{bmatrix} (X_{o_2} - R_2 s\varphi_2) c(\varphi_2+\omega_2) - (Z_{o_2} - R_2 c\varphi_2) s(\varphi_2+\omega_2) \\ Y_{o_2} \\ (X_{o_2} - R_2 s\varphi_2) s(\varphi_2+\omega_2) + (Z_{o_2} - R_2 c\varphi_2) c(\varphi_2+\omega_2) \end{bmatrix}
$$

$$
= \begin{bmatrix} X_{o_2} c(\varphi_2+\omega_2) - Z_{o_2} s(\varphi_2+\omega_2) + R_2 s\omega_2 \\ Y_{o_2} \\ X_{o_2} s(\varphi_2+\omega_2) + Z_{o_2} c(\varphi_2+\omega_2) - R_2 c\omega_2 \end{bmatrix}. \tag{4.10}
$$

Note that the value of X_{c_2} is equal to zero (as expected). We omit the details of further derivations here because they are all straightforward formula transformations.

To obtain the value of L_2, the point $(X_{c_2}, Y_{c_2}, Z_{c_2})^{\mathsf{T}}$ is projected into image column φ_2 by the following projection formula:

$$
L_2 = \frac{f_2 Y_{c_2}}{Z_{c_2}} = \frac{f_2 Y_{o_2}}{X_{o_2} \sin(\varphi_2+\omega_2) + Z_{o_2} \cos(\varphi_2+\omega_2) - R_2 \cos\omega_2} \tag{4.11}
$$

where X_{o_2}, Y_{o_2}, and Z_{o_2} can be substituted following equation (4.9). Thus we obtain the following epipolar curve equation (as stated in the theorem):

$$
L_2 = \frac{f_2 \mathbf{r}_2^{\mathsf{T}} \cdot \mathbf{l}}{\sin(\varphi_2+\omega_2)\mathbf{r}_1^{\mathsf{T}} \cdot \mathbf{l} + \cos(\varphi_2+\omega_2)\mathbf{r}_3^{\mathsf{T}} \cdot \mathbf{l} - R_2 \cos\omega_2}. \tag{4.12}
$$

Finally, according to the projection formula in equation (4.11), in cases where the value of Z_{c_2} is equal to zero, the value of L_2 is undefined. Moreover, according to the defined image projection geometry, when the value of Z_{c_2} is less than zero, the projection does not exist. Therefore, equation (4.12) is only valid if the Z_c coordinate of the 3D point is greater than zero (i.e., the denominator of L_2 needs to be greater than zero). $\qquad\square$

The general epipolar curve function stated in Theorem 4.1 for an arbitrary pair of cylindrical panoramas is rather complicated and lengthy. It is difficult to imagine the geometry of epipolar curves by reading the function. However, this function is useful for further mathematical analysis (as shown below), and characterizes epipolar geometry for any pair of cylindrical panoramas.

Example 4.1. Figure 4.4 illustrates an example of epipolar curves for the general case of a pair of panoramic images, demonstrating the geometric complexity of those curves.

The internal sensor parameters of the source panorama $E_{\mathcal{P}_1}(R_1, f, \omega_1, \gamma)$ and destination panorama $E_{\mathcal{P}_2}(R_2, f, \omega_2, \gamma)$ are as follows: $R_1 = 500\,\mathrm{mm}$, $\omega_1 = 45°$, $R_2 = 250\,\mathrm{mm}$, $\omega_2 = 65°$, $f = 35\,\mathrm{mm}$, and $\gamma = 0.36°$. The affine transform between both camera coordinate systems (associated with these two panoramas) can be described by a translation vector $\mathbf{t}_{o_1 o_2}$ and a rotation matrix $\mathbf{R}_{o_1 o_2}$. We set $\mathbf{t}_{o_1 o_2} = (2000, 300, 1500)^{\mathsf{T}}$ in millimeters and

$$
\mathbf{R}_{o_1 o_2} = \mathbf{R}_x(-1°)\mathbf{R}_y(-1°)\mathbf{R}_z(2°)
$$

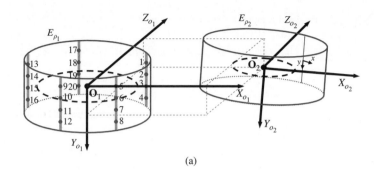

(a)

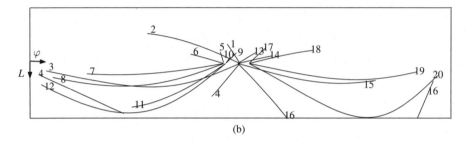

(b)

Figure 4.4 A pair of multi-view panoramas and epipolar curves in the destination image originating from 20 selected points in the source image $E_{\mathcal{P}_1}$.

where $\mathbf{R}_x$, $\mathbf{R}_y$, and $\mathbf{R}_z$ are rotation matrixes with respect to each of the three axes. Figure 4.4(a) shows the specified geometric relation between these two panoramas in 3D space.

Twenty numbered points have been selected in image $E_{\mathcal{P}_1}$. The corresponding epipolar curves in image $E_{\mathcal{P}_2}$, also numbered, are shown in Figure 4.4(b). An interesting observation is that epipolar curves do cross each other (in this case), and these intersections are not necessarily epipoles (e.g., curves 3 and 8). This is a new situation compared to the epipolar geometry of the pinhole camera model where epipolar lines only intersect at epipoles. However, if we only consider epipolar curves associated with image points on the same column of the source image $E_{\mathcal{P}_1}$ (e.g., curves labeled 17, 18, 19, and 20), then they only intersect (if at all) at epipoles.

Traditionally panoramas are assumed to be single-centered (i.e., single projection center), which means that the off-axis distance R is equal to zero. This is a special case of our general camera model, and we illustrate it in the next example.

Example 4.2. All the internal sensor parameters and the affine transform between both panoramas remain the same as in Example 4.1, except that we take $R_1 = R_2 = 0$. The resulting epipolar curves are shown in Figure 4.5. In this case, all epipolar curves intersect exactly at one point (i.e., the epipole).

The epipolar geometry of cylindrical single-center panoramas has been discussed in McMillan and Bishop (1995), Kang and Szeliski (1997) and Shum et al. (1998). In these approaches, each of the single-center panoramas is composed by stitching of planar

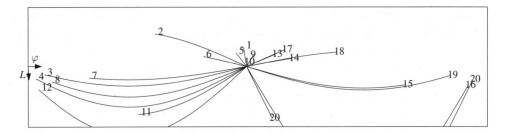

Figure 4.5   Epipolar curves for the single-center case (i.e., $R = 0$).

(i.e., pinhole) images. Negative results in stereo matching are linked in these cases to inaccurate epipolar geometry estimation. However, practical experiments show that a cylindrical panorama produced by rotating a camera (e.g., on a tripod) rarely satisfies the single-center assumption $R = 0$, not to mention the case of stitched images. Thus, negative results in stereo matching are totally foreseeable if using an ideal (oversimplified) single-center camera model to estimate the epipolar geometry in such cases.

The multi-view case should be used instead. Differences between cases shown in Figure 4.4(b) and Figure 4.5 illustrate the importance of the general model. As a consequence, it is necessary to calibrate and model the off-axis distance R, and (in general) also the principal angle ω for any stereo application that uses cylindrical panoramas.

4.2 Constrained Poses of Cameras

The general epipolar curve function of Theorem 4.1 is not a polynomial, and is lengthy and complicated due to the geometric complexity of the general cylindrical panoramic camera model. As a result, the estimation of the epipolar geometry from corresponding points of a multi-view panoramic pair may suffer severe numerical instability. This problem motivates us to constrain relative poses of panoramic cameras to some particular cases for practical purposes (e.g., to ensure computational simplicity or numeric stability).

This section discusses a few cases of constrained poses of panoramic pairs, where each case has its own practical merits and also existing applications. The cases selected include leveled, co-axial, concentric, and symmetric panoramic pairs as introduced in Chapter 3. We illustrate how Theorem 4.1 can be regarded as general guidance for computing epipolar curves for the whole family of cylindrical panoramas.

This section proceeds from general to more specific cases. We first derive the epipolar curve function for a leveled panoramic pair, and the remaining cases will then benefit from this derived epipolar curve function.

4.2.1 Leveled Panoramic Pair

In many applications, multiple panoramic images acquired at different viewpoints are necessary, for example, for 3D reconstructions of large historical sites or big shopping mall interiors, or for creating virtual tours of museums or famous scenic places. Epipolar geometry serves as a fundamental tool in various processes of those applications, such as image correspondence analysis, relative pose estimation, or novel view synthesis, where epipolar curves may be used to determine traversal paths or orders of image points.

An intuitive and practical way to reduce the dimensionality of the general epipolar curve function is to make all associated base planes of panoramas leveled with the sea level (e.g., by cameras on leveled tripods using a "bull's-eye" or more advanced leveling devices). In the model chapter, a multi-view panoramic pair whose associated axes are orthogonal to sea level was called a leveled panorama. We also call it a *leveled panoramic pair*.

Note that the elevations of cameras contributing to a leveled panoramic pair can be different. Under the assumption that a panoramic pair is perfectly leveled, the general epipolar curve function in Theorem 4.1 can be simplified. Recall that symbols s and c are used to denote sin and cos functions, respectively.

4.1. COROLLARY. *Let (φ_1, L_1) and (φ_2, L_2) be a pair of corresponding image points in a pair of leveled multi-view panoramas $E_{\mathcal{P}_1}(R_1, f_1, \omega_1, \gamma_1)$ and $E_{\mathcal{P}_2}(R_2, f_2, \omega_2, \gamma_2)$, respectively. Given φ_1 and L_1, the epipolar curve function in this case is as follows:*

$$L_2 = \frac{F(A - R_1 s(\varphi_2 + \sigma') - t_x c(\varphi_2 + \sigma) + t_z s(\varphi_2 + \sigma)) - f_2 t_y s(\varphi_2 + \sigma'')}{-B + E(R_2 c(\varphi_2 + \phi) + t_z) - G(R_2 s(\varphi_2 + \phi) + t_x)}$$

This function contains only one variable φ_2 that is in the range between 0 and 2π, and the epipolar curve is only valid if the value of this denominator is greater than zero. The angle ϕ determines the rotation with respect to the y-axis. All other symbols $A, B, E, G, F, \sigma, \sigma'$, and σ'' used are constants, and they are defined as follows:

$$A = R_2 \sin \omega_2,$$

$$B = R_1 \sin \omega_1,$$

$$E = \sin(\varphi_1 + \omega_1),$$

$$G = \cos(\varphi_1 + \omega_1),$$

$$F = \frac{L_1 f_2}{f_1},$$

$$\sigma = \omega_2 + \phi,$$

$$\sigma' = \sigma - \varphi_1,$$

$$\sigma'' = \sigma' - \omega_1.$$

Proof We derive this function from the general epipolar curve function in Theorem 4.1. The notation used in the proof has the same meaning as in the proof of Theorem 4.1.

In the leveled case, the 3×3 rotation matrix (that specifies the orientation of the sensor coordinate system of the destination image with respect to the sensor coordinate system of the source image) only consists of one variable that describes the rotation with respect to the Y-axis. We describe this rotation by angle ϕ. The three row vectors of the rotation matrix become $\mathbf{r}_1^T = (c\phi, 0, -s\phi)^T$, $\mathbf{r}_2^T = (0, 1, 0)^T$, and $\mathbf{r}_3^T = (s\phi, 0, c\phi)^T$.

Step by step, we substitute these row vectors of the rotation matrix into the general epipolar curve function of Theorem 4.1. First, we calculate $\mathbf{r}_1^T \cdot \mathbf{u}$, $\mathbf{r}_3^T \cdot \mathbf{u}$, $\mathbf{r}_1^T \cdot \mathbf{v}$, and $\mathbf{r}_3^T \cdot \mathbf{v}$. It follows that

$$\mathbf{r}_1^T \cdot \mathbf{u} = [c\phi, 0, -s\phi]^T \begin{bmatrix} R_1 s\varphi_1 - t_x \\ -t_y \\ R_1 c\varphi_1 - t_z \end{bmatrix}$$

$$= R_1 s\varphi_1 c\phi - t_x c\phi - R_1 c\varphi_1 s\phi + t_z s\phi$$

$$= R_1 s(\varphi_1 - \phi) - t_x c\phi + t_z s\phi;$$

$$\mathbf{r}_3^T \cdot \mathbf{u} = [s\phi, 0, c\phi]^T \begin{bmatrix} R_1 s\varphi_1 - t_x \\ -t_y \\ R_1 c\varphi_1 - t_z \end{bmatrix}$$

$$= R_1 s\varphi_1 s\phi - t_x s\phi + R_1 c\varphi_1 c\phi - t_z c\phi$$

$$= R_1 c(\varphi_1 - \phi) - t_x s\phi - t_z c\phi;$$

$$\mathbf{r}_1^T \cdot \mathbf{v} = [c\phi, 0, -s\phi]^T \begin{bmatrix} s(\varphi_1 + \omega_1) f_1 \\ L_1 \\ c(\varphi_1 + \omega_1) f_1 \end{bmatrix}$$

$$= f_1 s(\varphi_1 + \omega_1) c\phi - f_1 c(\varphi_1 + \omega_1) s\phi$$

$$= f_1 s(\varphi_1 + \omega_1 - \phi);$$

$$\mathbf{r}_3^T \cdot \mathbf{v} = [s\phi, 0, c\phi]^T \begin{bmatrix} s(\varphi_1 + \omega_1) f_1 \\ L_1 \\ c(\varphi_1 + \omega_1) f_1 \end{bmatrix}$$

$$= f_1 s(\varphi_1 + \omega_1) s\phi + f_1 c(\varphi_1 + \omega_1) c\phi$$

$$= f_1 c(\varphi_1 + \omega_1 - \phi).$$

In order to calculate vector $\mathbf{l}$ in Theorem 4.1, we first calculate the following two partial terms of vector $\mathbf{l}$:

$$c(\varphi_2 + \omega_2) \mathbf{r}_1^T \cdot \mathbf{u} - s(\varphi_2 + \omega_2) \mathbf{r}_3^T \cdot \mathbf{u}$$

$$= R_1 s(\varphi_1 - \phi) c(\varphi_2 + \omega_2) - t_x c\phi c(\varphi_2 + \omega_2) + t_z s\phi c(\varphi_2 + \omega_2)$$

$$\quad - R_1 c(\varphi_1 - \phi) s(\varphi_2 + \omega_2) + t_x s\phi s(\varphi_2 + \omega_2) + t_z c\phi s(\varphi_2 + \omega_2)$$

$$= R_1 s(\varphi_1 - \phi - \varphi_2 - \omega_2) - t_x c(\phi + \varphi_2 + \omega_2) + t_z s(\phi + \varphi_2 + \omega_2);$$

$$s(\varphi_2 + \omega_2) \mathbf{r}_3^T \cdot \mathbf{v} - c(\varphi_2 + \omega_2) \mathbf{r}_1^T \cdot \mathbf{v}$$

$$= f_1 s(\varphi_2 + \omega_2) c(\varphi_1 + \omega_1 - \phi) - f_1 c(\varphi_2 + \omega_2) s(\varphi_1 + \omega_1 - \phi)$$

$$= f_1 s(\varphi_2 + \omega_2 - \varphi_1 - \omega_1 + \phi).$$

To simplify the equations, let $\sigma = \omega_2 + \phi$, $\sigma' = \sigma - \varphi_1$, and $\sigma'' = \sigma' - \omega_1$; moreover, let $A = R_2 s \omega_2$. Vector $\mathbf{l}$ is as defined in Theorem 4.1, and it is equal to

$$\mathbf{u} + \frac{R_2 s \omega_2 - R_1 s(\varphi_2 + \sigma') - t_x c(\varphi_2 + \sigma) + t_z s(\varphi_2 + \sigma)}{f_1 s(\varphi_2 + \sigma'')} \mathbf{v}$$

$$= \begin{bmatrix} R_1 s\varphi_1 - t_x \\ -t_y \\ R_1 c\varphi_1 - t_z \end{bmatrix} + \frac{A - R_1 s(\varphi_2 + \sigma') - t_x c(\varphi_2 + \sigma) + t_z s(\varphi_2 + \sigma)}{s(\varphi_2 + \sigma'')} \begin{bmatrix} s(\varphi_1 + \omega_1) \\ \frac{L_1}{f_1} \\ c(\varphi_1 + \omega_1) \end{bmatrix}.$$

Let

$$\mathbf{l} = \frac{1}{s(\varphi_2 + \sigma'')} \begin{bmatrix} l_x \\ l_y \\ l_z \end{bmatrix}.$$

We have

$$l_x = R_1 s\varphi_1 s(\varphi_2 + \sigma'') - t_x s(\varphi_2 + \sigma'') + A s(\varphi_1 + \omega_1)$$
$$- R_1 s(\varphi_2 + \sigma') s(\varphi_1 + \omega_1) - t_x c(\varphi_2 + \sigma) s(\varphi_1 + \omega_1)$$
$$+ t_z s(\varphi_2 + \sigma) s(\varphi_1 + \omega_1),$$

$$l_y = -t_y s(\varphi_2 + \sigma'') + \frac{L_1}{f_1} \left(A - R_1 s(\varphi_2 + \sigma') \right.$$
$$\left. - t_x c(\varphi_2 + \sigma) + t_z s(\varphi_2 + \sigma) \right),$$

$$l_z = R_1 c\varphi_1 s(\varphi_2 + \sigma'') - t_z s(\varphi_2 + \sigma'') + A c(\varphi_1 + \omega_1)$$
$$- R_1 s(\varphi_2 + \sigma') c(\varphi_1 + \omega_1) - t_x c(\varphi_2 + \sigma) c(\varphi_1 + \omega_1)$$
$$+ t_z s(\varphi_2 + \sigma) c(\varphi_1 + \omega_1).$$

The numerator of L_2 in Theorem 4.1 becomes

$$f_2 \mathbf{r}_2^{\mathsf{T}} \cdot \mathbf{l} = \frac{f_2}{s(\varphi_2 + \sigma'')} (0, 1, 0) \begin{bmatrix} l_x \\ l_y \\ l_z \end{bmatrix}$$

$$= \frac{f_2 l_y}{s(\varphi_2 + \sigma'')}$$

$$= \frac{F \left(A - R_1 s(\varphi_2 + \sigma') - t_x c(\varphi_2 + \sigma) + t_z s(\varphi_2 + \sigma) \right) - t_y f_2 s(\varphi_2 + \sigma'')}{s(\varphi_2 + \sigma'')}.$$

Note that we have $F = L_1 f_2 / f_1$. Furthermore, the denominator of L_2 in Theorem 4.1 becomes

$$s(\varphi_2 + \omega_2)\mathbf{r}_1^{\mathrm{T}} \cdot \mathbf{l} + c(\varphi_2 + \omega_2)\mathbf{r}_3^{\mathrm{T}} \cdot \mathbf{l} - R_2 c\omega_2$$

$$= \frac{s(\varphi_2 + \omega_2)}{s(\varphi_2 + \sigma'')}[c\phi, 0, -s\phi]^{\mathrm{T}}\begin{bmatrix} l_x \\ l_y \\ l_z \end{bmatrix}$$

$$+ \frac{c(\varphi_2 + \omega_2)}{s(\varphi_2 + \sigma'')}[s\phi, 0, c\phi]^{\mathrm{T}}\begin{bmatrix} l_x \\ l_y \\ l_z \end{bmatrix} - R_2 c\omega_2$$

$$= \frac{l_x s(\varphi_2 + \omega_2)c\phi - l_z s(\varphi_2 + \omega_2)s\phi}{s(\varphi_2 + \sigma'')}$$

$$+ \frac{l_x c(\varphi_2 + \omega_2)s\phi + l_z c(\varphi_2 + \omega_2)c\phi}{s(\varphi_2 + \sigma'')} - R_2 c\omega_2$$

$$= \frac{1}{s(\varphi_2 + \sigma'')}(l_x s(\varphi_2 + \sigma) + l_z c(\varphi_2 + \sigma)) - R_2 c\omega_2$$

$$= \frac{1}{s(\varphi_2 + \sigma'')}\big(R_1 s(\varphi_1)s(\varphi_2 + \sigma'')s(\varphi_2 + \sigma)$$

$$- t_x s(\varphi_2 + \sigma'')s(\varphi_2 + \sigma) + A s(\varphi_1 + \omega_1)s(\varphi_2 + \sigma)$$

$$- R_1 s(\varphi_2 + \sigma')s(\varphi_1 + \omega_1)s(\varphi_2 + \sigma)$$

$$- t_x c(\varphi_2 + \sigma)s(\varphi_1 + \omega_1)s(\varphi_2 + \sigma) + t_z s^2(\varphi_2 + \sigma)s(\varphi_1 + \omega_1)$$

$$+ R_1 c\varphi_1 s(\varphi_2 + \sigma'')c(\varphi_2 + \sigma) - t_z s(\varphi_2 + \sigma'')c(\varphi_2 + \sigma)$$

$$+ A c(\varphi_1 + \omega_1)c(\varphi_2 + \sigma) - R_1 s(\varphi_2 + \sigma')c(\varphi_1 + \omega_1)c(\varphi_2 + \sigma)$$

$$- t_x c^2(\varphi_2 + \sigma)c(\varphi_1 + \omega_1) + t_z s(\varphi_2 + \sigma)c(\varphi_1 + \omega_1)c(\varphi_2 + \sigma)$$

$$- R_2 c\omega_2 s(\varphi_2 + \sigma'')\big)$$

$$= \frac{1}{s(\varphi_2 + \sigma'')}\big(R_1 c(\varphi_2 + \sigma')s(\varphi_2 + \sigma'')$$

$$- t_x s(\varphi_2 + \sigma)s(\varphi_2 + \sigma'') - t_z c(\varphi_2 + \sigma)s(\varphi_2 + \sigma'')$$

$$+ R_2 s\omega_2 c(\varphi_2 + \sigma'') - R_1 s(\varphi_2 + \sigma')c(\varphi_2 + \sigma'')$$

$$- t_x c(\varphi_2 + \sigma)c(\varphi_2 + \sigma'') + t_z s(\varphi_2 + \sigma)c(\varphi_2 + \sigma'')$$

$$- R_2 c\omega_2 s(\varphi_2 + \sigma'')\big)$$

$$= \frac{1}{s(\varphi_2 + \sigma'')}\big(R_1 s(\sigma'' - \sigma') - t_x c(\sigma - \sigma'') + t_z s(\sigma - \sigma'') + R_2 s(\omega_2 - \varphi_2 - \sigma_2)\big)$$

$$= \frac{1}{s(\varphi_2 + \sigma'')}(-B - t_x G + t_z E + E R_2 c(\varphi_2 + \phi) - G R_2 s(\varphi_2 + \phi)).$$

Thus we have that

$$L_2 = \frac{F(A - R_1 s(\varphi_2 + \sigma') - t_x c(\varphi_2 + \sigma) + t_z s(\varphi_2 + \sigma)) - f_2 t_y s(\varphi_2 + \sigma'')}{-B + E(R_2 c(\varphi_2 + \phi) + t_z) - G(R_2 s(\varphi_2 + \phi) + t_x)}$$

as required.   □

If panoramic images are taken at identical elevation then they are *horizontally aligned panoramas*, and the epipolar curve equation for such a pair (which is also leveled) can be further simplified into

$$L_2 = L_1 \cdot \frac{f_2}{f_1} \cdot \frac{R_2 s \omega_2 - R_1 s(\varphi_2 + \sigma') - t_x c(\varphi_2 + \sigma) + t_z s(\varphi_2 + \sigma)}{-R_1 s \omega_1 + E(R_2 c(\varphi_2 + \phi) + t_z) - G(R_2 s(\varphi_2 + \phi) + t_x)}.$$

This equation can be derived from Corollary 4.1 by setting $t_y = 0$.

An example of a pair of panoramic images, being "almost"[2] horizontally aligned, is shown in Figure 4.6. These panoramas were captured at DLR by the high-resolution, three-line WAAC in a complete 360° progressive scan. Each sensor line has 5,184 elements. The associated principal angles for both images are equal to 25°. The effective focal length of the line camera is 21.7 mm for the center sensor line, and 23.94 mm for its two symmetric sensor lines. Both the effective focal length and the principal angle had been exactly calibrated using on-site DLR facilities.

The off-axis distances for the panoramic pair are both equal to 100 mm, and the associated rotation axes of this panoramic pair are 1 m apart; despite careful measurement, there might be

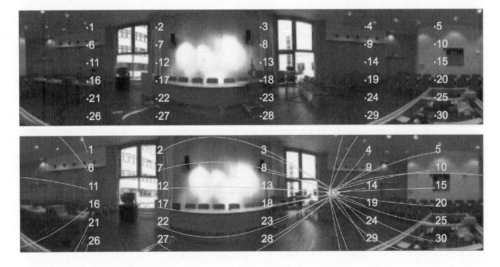

Figure 4.6 A pair of horizontally aligned panoramas. Source panorama with 30 test points labeled by "·" and indexed by numbers (top). Destination panorama superimposed with the corresponding epipolar curves (bottom).

---

[2] With a possible error in elevations due to a manual movement of the image acquisition system.

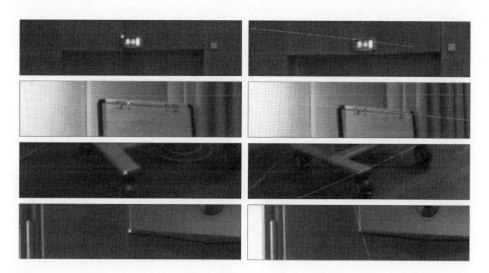

Figure 4.7 Close-ups of image points in the source panorama (left), and corresponding epipolar curves in the destination panorama (right) as calculated by the equation provided.

some error involved. The line camera took 22,000 line images each for two complete 360° scans at both positions. Most of the calculated epipolar curves pass through expected (corresponding) image points in the destination panorama. Close-up views of selected points (at "corners") and their corresponding epipolar curves are shown in Figure 4.7. This is an experimental illustration of the correctness of the provided epipolar curve equation for horizontally aligned leveled panoramas.

4.2.2 Co-axial Panoramic Pair

Two panoramas whose axes coincide form a *co-axial panoramic pair*. (The implementation of such a camera configuration is reasonably straightforward using a tripod.) The co-axis constraint eliminates two rotational and two translational parameters. The epipolar geometry of a co-axial panoramic pair is characterized by the following result:

4.2. COROLLARY. *Let* (φ_1, L_1) *and* (φ_2, L_2) *be a pair of corresponding image points in a co-axial pair of panoramas. Given* φ_1 *and* L_1, *we have*

$$L_2 = \frac{L_1\left(\frac{f_2}{f_1}\right)(R_2 s\omega_2 - R_1 s(\varphi_2 + \omega_2 + \phi - \varphi_1)) - t_y f_2 s(\varphi_2 + \phi - \varphi_1)}{-R_1 s\omega_1 - R_2 s(\varphi_2 - \omega_1 + \phi - \varphi_1)}$$

where the angle ϕ *determines the rotation with respect to the y-axis.*   .

Proof The equation can be derived from Corollary 4.1 by setting $t_x = 0$ and $t_z = 0$.   □

From this example we can conclude that epipolar curves of a co-axial pair can coincide with image columns for specific camera settings.

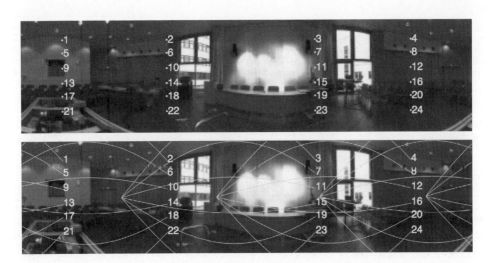

Figure 4.8   A pair of concentric panoramas. Source panorama with 24 test points labeled by "·" and indexed by numbers (top). Destination panorama superimposed with corresponding epipolar curves (bottom).

Now consider a pair of co-axial panoramas where associated centers coincide; this is a *concentric panoramic pair*. The epipolar curve equation of a concentric panoramic pair further simplifies to the following:

$$L_2 = \frac{L_1\left(\frac{f_2}{f_1}\right)(R_2 \sin \omega_2 - R_1 \sin(\varphi_2 + \omega_2 + \phi - \varphi_1))}{-R_1 \sin \omega_1 - R_2 \sin(\varphi_2 - \omega_1 + \phi - \varphi_1)}. \tag{4.13}$$

This is obtained by substituting $t_y = 0$ in the equation given in Corollary 4.2.

In practice, concentric panoramic pairs can be produced in a very simple way; for example, by rotating a matrix camera. In this case a set of panoramic images may be generated from different columns of the matrix images, where every fixed column generates exactly one of these panoramic images.

An example of a concentric panoramic pair taken by the WAAC is shown in Figure 4.8. The technical data are the same as for the leveled case, apart from the following modifications: the off-axis distance for the source panorama of the pair is equal to 100 mm, and 0 mm for the destination panorama of the pair. The calculated epipolar curves, superimposed on the destination image, again show incidence with corresponding points.

4.2.3 Symmetric Panoramic Pair

Two concentric panoramas whose associated off-axis distances, effective focal lengths, and angular units are pairwise identical, and whose principal angles sum to 360° (i.e., both angles can be denoted by ω and $-\omega$) are called a *symmetric panoramic pair*. These panoramic pairs have applications in stereo panoramic imaging (computational stereo and stereo viewing); they possess a simple epipolar geometry as characterized by the following result:

Figure 4.9 Anaglyph of a symmetric panoramic pair. It requires a presentation in color and anaglyphic eyeglasses for 3D viewing (see Plate 26).

4.3. COROLLARY. *Epipolar curves are straight lines and coincide with image rows, for any symmetric pair of panoramic images.*

Proof This fact follows from the epipolar curve equation in Corollary 4.2 by setting $t_y = 0$, $\phi = 0$, $f_1 = f_2$, $R_1 = R_2$, $\gamma_1 = \gamma_2$, and $\omega_2 = (2\pi - \omega_1)$. In this case, we have

$$
\begin{aligned}
L_2 &= L_1 \left(\frac{R_1 \sin(2\pi - \omega_1) - R_1 \sin(\varphi_2 - \varphi_1 + 2\pi - \omega_1)}{-R_1 \sin \omega_1 - R_1 \sin(\varphi_2 - \varphi_1 - \omega_1)} \right) \\
&= L_1 \left(\frac{-\sin \omega_1 - \sin(\varphi_2 - \varphi_1 - \omega_1)}{-\sin \omega_1 - \sin(\varphi_2 - \varphi_1 - \omega_1)} \right) \\
&= L_1
\end{aligned}
$$

regardless of the values of all other (as yet unspecified) parameters. This implies that the epipolar curve, corresponding to image point (φ_1, L_1), coincides with the image row L_2 on the destination image with $L_1 = L_2$. Thus, the epipolar curve is a straight line.   ☐

It follows from Corollary 4.3 that stereo-matching algorithms previously developed for pinhole binocular stereo are reusable. The corollary also characterizes symmetric pairs as being stereoscopically viewable (e.g., using anaglyphic techniques or polarized light) without further processing. An anaglyphic example of a symmetric pair is shown in Figure 4.9.

4.3 Exercises

4.1. Prove that $X_{c_2} = 0$ in equation (4.10).

4.2. [Possible lab project] The website for this book provides a symmetric panoramic stereo pair. As shown in this chapter, epipolar lines are just the image rows. However, this is an ideal assumption, and one line below or above should also be taken into account. Apply one of the stereo correspondence techniques, rated in the top 10 on the Middlebury stereo website *http://vision.middlebury.edu/stereo/*, to the symmetric stereo panorama provided, and estimate a sufficient length of the search interval for corresponding points (obviously, the full 360° panorama does not need to be searched for corresponding points), to ensure detection of corresponding points in the context of your recorded 3D scenes and the panoramic sensor used.

4.4 Further Reading

Epipolar geometry is widely studied and applied for pinhole cameras; see computer vision textbooks, such as Xu and Zhang (1996), Klette et al. (1998), Hartley and Zisserman (2000) and Faugeras and Luong (2001). There is increasing interest in stereo analysis for multiple planar images. Shape from motion, novel view synthesis (image-based rendering), ego-motion analysis, and object recognition and localization are examples of related applications; see, for example, Xu and Zhang (1996), Kanatani (1993), Laveau and Faugeras (1994), Zhang and Xu (1997), Evers-Senne et al. (2004) and Liu et al. (2005).

Compared with the pinhole case, much less work has so far been reported for panoramic cameras. Approaches computing epipolar curves in panoramic images were often focused on single-center panoramas; see, for example, McMillan and Bishop (1995) for image-based rendering applications. This chapter has presented a unified approach for computing epipolar curves for the general case of multi-view panoramas (as published in short form in Huang et al., 2001b). Epipolar geometry of cubic panoramas is studied in Kangni and Laganiere (2006). Jiang et al. (2006) discuss the epipolar geometry of a rotational stereo camera.

Seitz (2001) specifies all possible epipolar surfaces as a subset of doubly ruled surfaces[3] that allow a generation of images suitable for stereo vision. In our case of a cylindrical stereo panoramic pair, the epipolar surface is a half-hyperboloid.

There are many proposals for stereo algorithms; for some more historic examples, see Marr and Poggio (1979), Barnard and Fischler (1982), Ohta and Kanade (1985), Cox (1994) and Šára (1999), and for more recent work see references listed on the website *http://vision. middlebury.edu/stereo/*. Scheibe et al. (2006) also addresses stereo algorithms for panoramas. Zhu et al. (2005) use two gamma-ray sensor lines for stereo reconstruction (cargo inspection).

The co-axial constraint is also applied for catadioptric cameras (Southwell et al., 1996; Nene and Nayar, 1998; Petty et al., 1998; Gluckman and Nayar, 1999). For concentric panoramic pairs, see Ishiguro et al. (1992) and Shum and He (1999).

On the WAAC used in some of the experiments, see Sandau and Eckardt (1996) and Reulke and Scheele (1998).

---

[3] The only doubly ruled surfaces in 3D space are the plane, the hyperboloid, and the hyperbolic paraboloid (Hilbert and Cohn-Vossen, 1932).

5

Sensor Calibration

In this chapter we discuss the calibration of a rotating panoramic sensor (rotating sensor-line camera or laser range-finder). Having specified some preprocessing steps, we describe a least-squares error method which implements the point-based calibration technique (common in photogrammetry or computer vision, using projections of control points, also called calibration marks). This method has been used frequently in applications, and proved to be robust and accurate.

The chapter also discusses three calibration methods at a more theoretical level, for comparing calibration techniques for the general multi-center panorama case.[1] The aim of this discussion is to characterize the linear and non-linear components in the whole calibration process.

As pointed out at the end of Chapter 3, the geometry of the LRF can be understood (for calibration purposes) as a special case of the geometry of the rotating sensor-line camera. Therefore we prefer to use camera notation in this chapter. However, a section at the end of the chapter also discusses the specific errors to be considered for a laser range-finder.

5.1 Basics

Calibration of panoramic sensors builds on traditional camera calibration techniques (e.g., in computer vision or photogrammetry). This section briefly introduces the subject, and provides a few general definitions.

5.1.1 Camera Calibration

Photogrammetry or computer vision uses calibration marks, which are typically intersection points of lines, or of simple geometric shapes such as a disk or a square, possibly with some pattern; see Figures 5.1 and 5.2. In case of the checkerboard, vertices of white or black squares

---

[1] The first of these methods is a simplification of a point-based calibration technique.

Panoramic Imaging: Sensor-Line Cameras and Laser Range-Finders F. Huang, R. Klette, and K. Scheibe
© 2008 John Wiley & Sons, Ltd

Figure 5.1   Calibration area at the Institute for Photogrammetry at TFH Berlin (2005). Calibration marks are distributed within the scene.

Figure 5.2   A calibration checker board as used in 2006 in the research department at Daimler AG in Böblingen, Germany. The position of the board defines a (temporary) world coordinate system.

define the calibration marks. Marks have to remain in their positions for some time if camera attitude needs to be calibrated (with respect to a world coordinate system, defined by the positions of the given calibration marks). Moving the calibration rig during image capture allows camera parameters (such as focal length or central point), or relative positions between multiple cameras, to be calibrated.

Basically, a mathematical model, defining how points in the 3D scene (such as calibration marks) project into the image plane, is used to establish a system of *objective functions*, which need to be solved based on knowledge about measured positions of calibration marks, and the location of their projections in the image plane.

Optimization (e.g., by least-squares using an overdetermined equational system) allows quality improvement of camera calibration. The accuracy and the error sensitivity of estimated parameters depends on the degree of objective functions and permitted computation costs for their evaluation. It is of interest to study interrelations between simplification of objective functions (e.g., the reduction of degree towards linear) and accuracy of calibration.

A general calibration scenario is defined by the used calibration rigs or marks, localization of calibration marks in images, and the evaluation of objective functions used for the calculation of unknown parameters. The procedures involved (e.g., for the detection of traceable features, such as calibration marks, lines, or patterns, at subpixel accuracy) are often of interest in their own right. A checkerboard (see Figure 5.2) is today the default calibration tool for "normal" cameras, which are characterized by some linearity, defined by projection rays and an image plane.

Calibration of a "normal" sensor-matrix camera also has to deal with some degree of non-linearity, caused by geometric lens distortions, or deviations from linear Gaussian optics. Camera operations (e.g., shifting and tilting the sensor) may cause further non-linearity. Panoramic sensors are (typically) characterized by nonplanarity of capturing surfaces, and this introduces "new challenges" into the calibration task.

Criteria used for evaluating the performance of camera or sensor calibration include the number of images required, the difficulty of preprocessing steps, computational complexity of the procedures involved, degree of objective functions, amount of data required for calibration, (non-)dependence on initial values in the parameter estimation process, sensitivity of parameter estimation with respect to some type of noise or errors, or the reproducible verification of camera parameters.

5.1.2 Extrinsic and Intrinsic Parameters

It is common practice in photogrammetry or computer vision that the attitude (or pose) of a sensor is described by mapping its local Cartesian XYZ coordinate system into a (fixed) world $X_w Y_w Z_w$ coordinate system. Such an affine transform has six *degrees of freedom* because it may be specified by three translational and three rotational parameters.

DEFINITION **5.1.** Extrinsic parameters *define the pose of the camera or sensor in world coordinates.*

Extrinsic parameters of a camera or sensor correspond to the six degrees of freedom mentioned. In general, they are formally specified by identifying the rotation matrix and the translation of the affine transform concerned.

DEFINITION **5.2.** Intrinsic parameters *of a camera or sensor specify the mapping of 3D data onto the capturing surface of this device.*

The capturing surface is a virtual surface (see, for example, the discussion of a mapping onto a straight hollow cylinder above). There may be different parameters for different sensor geometries, and we do not specify them here.

Note that this mapping goes beyond geometric properties alone; it is also defined by photometric properties such as light sensitivity of individual sensor elements. Examples of intrinsic parameters are the dimensions of sensor arrays, the focal length, the central point (note that there may be several central points for various positions of a rotating sensor-line camera), the physical size τ_h or τ_v of individual sensor cells, the distribution of light sensitivity within a sensor array, radial distortion coefficients $\kappa_1, \kappa_2, \ldots$ of the optics used, or deviations of the geometry of the capturing surface from its ideal case (such as spherical, cylindrical, or planar).

One possibility is to carry along all the deviation parameters of a capturing surface. Another option is to calibrate a transform (using those deviation parameters) such that recorded panoramas are mapped by this geometric transform (and required resampling) onto an ideal capturing surface. The advantage is that a uniform model of the capturing surface simplifies further calculations. However, a disadvantage may be that, by resampling the original image twice (or even more often), the image's quality may deteriorate.

5.1.3 Registration and Calibration

A recorded panorama is taken for some pose of the sensor. Multi-view panoramas use multiple viewpoints[2] (e.g., required due to occlusions in the 3D scene).

DEFINITION **5.3.** Registration *is the process of determining the pose of a camera or sensor with respect to a (local or world) coordinate system.*

Obviously, registration in world coordinates is more general, and allows geometric relations to be specified between a panorama and world coordinates of registered objects or *landmarks* (i.e., particular positions in the world).

DEFINITION **5.4.** Calibration *is a process of determining (intrinsic and/or extrinsic) parameters of a camera or sensor.*

Often, calibration is based on some simplifying assumptions, such as modeling lens distortion with only one (traditionally called κ_1) or two parameters (called κ_1, κ_2). These simplified models assume ideal physical dependencies, for example that distortion increases with the radial distance from the principal point of the optics, but it is also known that such assumptions do not exactly match real imaging properties in general, and thus we have from time to

---

[2] Recall that a viewpoint is defined by the pose (i.e., the extrinsic parameters) and the specification of the camera or sensor (i.e., the intrinsic parameters).

time more sophisticated calibration techniques, taking even higher degrees of radial distortion into account. In fact, the best-known models today use polynomial terms of higher-order derivatives (up to some degree, say, 7); the coefficients vary between 7 (e.g., in the Helmert transformation) and 10 (e.g., in the Brown model).

Calibration involves radiometric or geometric specifications, and is done at the sensor's production site as well as at the time prior to recording a panorama. Radiometric calibration may, for example, identify the photometric accuracy and stability of every single CCD or CMOS sensor cell of the array or line sensor used.

In the case of a sensor-line camera with 10,000 or more pixel sensors, photometric or geometric calibration of intrinsic parameters is performed with extremely high accuracy requirements.

5.2 Preprocesses for a Rotating Sensor-Line Camera

Basic preprocessing operations are (i) the conversion of recorded data into a default image format, (ii) the correction of a color shift between red, green and blue channel, caused by the spatial alignment of the three RGB sensor lines, (iii) compensation of shading caused by optical components, and (iv) an optional data reduction from 16 to 8 bits with suitable scaling functions preserving contrast and information content of the image. However, it all starts with some *precalibration* at a specially designed calibration site, before such a high-accuracy sensor becomes available to the user.

5.2.1 Precalibration

This subsection is for information purposes only; obviously, the user of such a camera will not perform such a precalibration, but might be interested in some background (processes at the production site of the camera).

Collimator

A *collimator* (see Figure 5.3) allows precise illumination of individual (say, of all 10,200) sensor cells. Its use is very time-consuming. The use of a collimator is limited to measuring a camera system which is focused at infinity; this corresponds to the use of such a camera in an aircraft or satellite.[3] Of course, in close-range photogrammetry it is also important to use a near-range focus. The most important advantage of using a collimator is the opportunity of exact measurements of pixel viewing directions, independent of any model assumption. These measurements can be saved in a lookup table, or can be used to calculate new virtual pixel positions for an ideal pinhole-type model, or even transform the captured image into an image as taken by an ideal pinhole-type model.

---

[3] In principle, a collimator can also be defocused, from infinity to near-range focused optical sensors. However, this is almost limited to a very small range (i.e., defocusing the collimator corresponds to an effective shift of the sensor by some micrometers, and a refocus from infinity to, say, less than 1 meter, corresponds to an effective shift of some millimeters). Exact data depend both on the collimator and camera used.

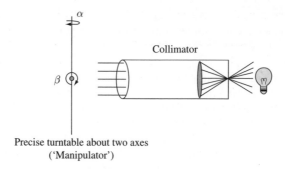

Precise turntable about two axes
('Manipulator')

Figure 5.3 Functional principle (sketch) of a collimator with a manipulator.

Offset, PRNU and DSNU Correction

A sensor line is normally composed of more than one *electronic channel* (e.g., RGB data define three channels, near-infrared imaging may define one more channel, and it is also possible to have sensor lines where odd or even detector elements define separate channels) for reading out its data. These channels differ in terms of offset, noise – such as dark signal non-uniformity (DSNU) and photo response non-uniformity (PRNU) – and possibly other properties, and these properties are assumed to be precalibrated by generating various functions in j and available (in digital form) for the given sensor line, where the variable j denotes a sensor cell (also called a *detector element*). Examples of such functions are the *offset* $O(j)$ and the *sensor cell signal* $S_j(t)$, where t denotes time.

Pixel values are per clock cycle in a read-out register, and read via available channels (including digital–analog conversion). These channels differ, and they have their own offset (even if only one channel). In the case of two channels, a typical effect might be a streaky image of odd and even pixels. The offset also creates a positive response for a pixel even if it was not hit by any photon; theoretically there should be no electron generated in this case, but the offset creates a false response. Furthermore, there is dark current. The initial aim is to generate a black image at pixels which did not receive any energy. The measurement splits now into offset and DSNU.

For measuring offset, we are reading more pixels than actually physically available. The mean of physically unavailable pixels defines the offset. All measurement operations have to be performed for individual channels.

To measure DSNU, we make use of the so-called *dark pixels* of a digital sensor. These are (a few) defined pixels which are (normally) covered with silver. Now we capture an image with the optics closed (i.e., cover on). The resulting image is nearly black due to dark current. Values are now divided by the mean of the values at dark pixels. The expectation is that this results in a normalized image with values equal to 1. Values not equal to 1 define the non-uniformity. When taking an image under "normal" conditions (not having the cover on), then we multiply the values at dark pixels with those of the normalized image, and we obtain the given dark current. This needs to be subtracted from the captured image.

In general, when processing images, DSNU and PRNU need to be eliminated. These effects are caused by changes in temperature or integration time, and by different sensitivities of the

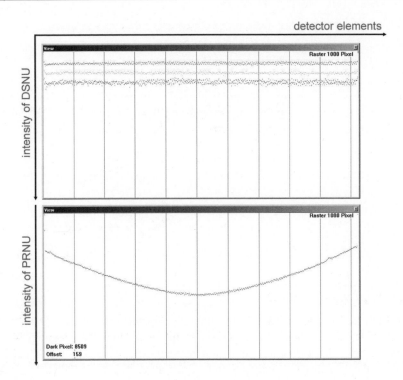

Figure 5.4 DSNU along a sensor line (top) PRNU, showing an optic gradient of the precalibrated CCD sensor line (bottom).

detector elements with regard to at-sensor radiances, respectively. To measure these effects, the camera has to be radiometrically and geometrically gauged (using a collimator).

The DSNU is measured in a precalibration laboratory for total absence of light and a long integration interval. Note that each detector element has its own dark signal noise depending on the integration time and temperature. Ideally, after the integration interval, each pixel should have a digital value of zero (i.e., black). Because of the dark signal noise, a pixel will have a digital value greater than zero. Therefore, a normalized detector factor D_j is first estimated during calibration, and then saved as a profile for the given electronic channel. Figure 5.4 shows measurements when calibrating DSNU and PRNU (along a sensor line).

During image acquisition, a sliding mean $\overline{D}_j(t)$ of DSNU may also be calculated (e.g., by analyzing an interval of $2k+1$ detector elements $j-k, \ldots, j+k$ of the sensor line, for example, sensor cells 10 to 30, for $k=10$). These elements are covered such that they are not illuminated, and then the calibrated DSNU level D_j is multiplied with a normalized detector factor and subtracted from the current signal (measured at all illuminated sensor cells). The DSNU-corrected signal is then calculated as follows:

$$S'_j(t) = S_j(t) - \overline{D}(t) \cdot D_j$$

with the estimated calibration profile

$$D_j = \frac{1}{n} \cdot \sum_{i=0}^{n-1} \frac{S_j(t_i)}{\overline{D}_j(t_i)}$$

in the case of n iterations.

For PRNU correction, the given electronic channel is illuminated with homogeneous light (e.g., using an *Ulbricht sphere*). For each detector element, a factor P_j is determined which allows each detector element to be normalized (e.g., by the maximum) to the same sensitivity:

$$P_j = \frac{1}{n} \cdot \sum_{i=0}^{n-1} \frac{\max(\{S_j(t_i)\})}{S_j(t_i)}.$$

Finally, the actual signal is noise-reduced and shifted to the origin, combining all three basic corrections as follows:

$$S'_j(t) = [S_j(t) - O(j) - \overline{D}_j(t) \cdot D_j] \cdot P_j.$$

5.2.2 Correction of Color Shift

A color shift is due to a spatial misalignment (i.e., a translational distance $\Delta_x > 0$ between corresponding detector elements on red, green or blue sensor lines; see Figure 5.5), and also defined by the pixel size τ and the acquisition geometry (i.e., the off-axis parameter R, and the attitude of the principal axis to the rotation axis, defined by the matrix $\mathbf{R}(\xi)$).

Single Projection Center

Assume that the rotation is about the projection center of the camera (i.e., $R = 0$), that the optical axis is perpendicular to the rotation axis (i.e., $\xi = 0$), and we consider a (visible) surface point at distance $h > 0$ from the projection center of the camera. In this case, the horizontal color shift between two color channels is the (small) angle $i \cdot \Delta\varphi$ (rounded to the nearest integer) by which the camera needs to be rotated to see the same surface point (at distance h) again in the second color channel.

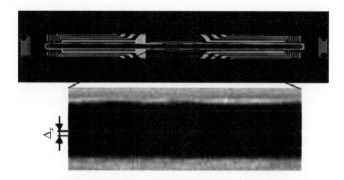

Figure 5.5 Kodak CCD line (top). Enlarged view illustrating the spacing between sensor lines for green and red channels (bottom).

The surface point is seen by the first color line, say, at (relative) angle $\varphi = 0$:

$$\begin{pmatrix} 0 \\ h \\ z \end{pmatrix} = \lambda \begin{pmatrix} 0 \\ f \\ j \cdot \tau \end{pmatrix}.$$

The same surface point is then seen (shortly afterward) by a second color line after the camera was rotated by $\mathbf{R}_{i \cdot \Delta\varphi}$; we have that

$$\begin{pmatrix} 0 \\ h \\ z \end{pmatrix} = \lambda \cdot \mathbf{R}_{i \cdot \Delta\varphi} \begin{pmatrix} \Delta_x \\ f \\ j \cdot \tau \end{pmatrix}. \tag{5.1}$$

The first of these three linear equations, which is

$$0 = \Delta_x \cdot \cos(i \cdot \Delta\varphi) - f \cdot \sin(i \cdot \Delta\varphi), \tag{5.2}$$

allows the horizontal color shift i (in pixels) to be calculated as follows:

$$i = \arctan\left(\frac{\Delta_x}{f}\right) \cdot \frac{1}{\Delta\varphi}. \tag{5.3}$$

Consequently, for RGB lines with spacing Δ_x between adjacent color lines, because of equations (2.2) and (5.3), the horizontal color shift is given as follows:

$$i = \frac{\Delta_x}{\tau} \qquad \text{for} \quad f \gg \tau.$$

Obviously, the horizontal color shift is independent of object distance h, and is constant for the whole panoramic image.

For the determination of vertical color shift, the system of equations (5.1) is solved for λ. The addition of the squares of the first two equations eliminates $\mathbf{R}_{\varphi(i)}$, and λ is determined as follows:

$$\lambda^2 = \frac{h^2}{\Delta_x^2 + f^2}.$$

Now, substitute λ into the third equation of (5.1) and solve it for j; this gives

$$j = \frac{z \cdot f}{\tau \cdot h} \sqrt{1 + \frac{\Delta_x^2}{f^2}}.$$

For simplicity, assume that $f \gg \Delta_x$; j can then be estimated as follows:

$$j = \frac{z \cdot f}{\tau \cdot h} \left(1 + \frac{1}{2} \frac{\Delta_x^2}{f^2}\right).$$

Assume that j_1 is specified by $\Delta_x = 0$ (thus defining a *nadir sensor line*); we obtain

$$j_1 = \frac{z \cdot f}{\tau \cdot h}.$$

This allows us to conclude that the vertical color shift (in pixels) is given by

$$|j_1 - j| = \frac{1}{2} \frac{z \cdot \Delta_x^2}{\tau \cdot f \cdot h}.$$

Obviously, in this case (vertical color shift) we have a dependency on the ratio z/h. The vertical color shift equals less than one pixel if the following condition is fulfilled:

$$\frac{z \cdot \Delta_x^2}{h \cdot \tau \cdot f} < 2.$$

This means that a vertical color shift can be practically excluded by using "reasonably accurate" RGB line sensors with $(f \cdot \tau) \gg \Delta_x$. We can also assume that $z/h \leq 1$. However, Chapter 10 on data fusion also covers vertical color shifts without much of an increase in the overall complexity of the method described: each of the three color lines is mapped separately using the correction formulas for horizontal or vertical color shift, which generates a color-distortion-free panoramic image in this preprocessing step.

Multiple Projection Centers

If the camera rotates with some eccentricity $R \neq 0$, a horizontal or vertical color shift (defining stereo lines with a small distance in-between) depends on a relation between object distance h and rotation radius R. For a start, in this case we have that

$$\begin{pmatrix} 0 \\ h \\ 0 \end{pmatrix} = \mathbf{R}_{i \cdot \Delta \varphi} \left[\lambda \begin{pmatrix} \Delta_x \\ f \\ j \cdot \tau \end{pmatrix} + \begin{pmatrix} 0 \\ R \\ 0 \end{pmatrix} \right].$$

Multiplication by $\mathbf{R}_{i \cdot \Delta \varphi}^{-1}$ leads to the following system of equations:

$$-\sin(i \cdot \Delta \varphi) \cdot h = \lambda \cdot \Delta_x \qquad (5.4)$$

$$\cos(i \cdot \Delta \varphi) \cdot h = \lambda \cdot f + R \qquad (5.5)$$

$$0 = \lambda \cdot j \cdot \tau. \qquad (5.6)$$

λ can be determined by adding the squares of equations (5.4) and (5.5), resulting in

$$h^2 = \lambda^2 \cdot \Delta_x^2 + \lambda^2 \cdot f^2 + 2\lambda \cdot f \cdot R + R^2. \qquad (5.7)$$

We derive and solve a polynomial in quadratic normal form:

$$0 = \lambda^2 + 2\lambda \frac{f \cdot R}{\Delta_x^2 + f^2} + \frac{R^2 - h^2}{\Delta_x^2 + f^2}$$

$$\lambda_{1,2} = \frac{1}{\Delta_x^2 + f^2} \left(-f \cdot R \pm \sqrt{h^2(\Delta_x^2 + f^2) - R^2 \cdot f^2} \right).$$

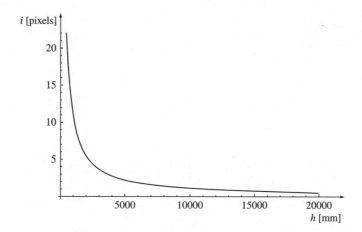

Figure 5.6 Color shift of RGB lines in off-axis mode for objects at distance h using a typical camera setup (see text for details).

We can assume that λ is positive (because of the forward-looking viewing direction of the camera); thus we have only one valid solution, namely λ_1. The color shift is now determined by dividing equations (5.4) and (5.5):

$$i = \arctan\left(\frac{\lambda \cdot \Delta_x}{\lambda \cdot f + R}\right)\frac{1}{\Delta\varphi}. \tag{5.8}$$

Finally, we use the known λ-value λ_1.

Figure 5.6 shows the color shift for objects in close range when the camera rotates off-axis. The figure sketches the color shift for an example of a typical setup, with $\Delta_x = 154\ \mu$m, $R = 500$ mm, and $\tau = 7\ \mu$m, using equation (5.8) to calculate i. The calculated values show that the color shift is less than one pixel for objects at distance $h > 11$ m, with $R = 0.5$ m.

Figure 5.7 illustrates the relation between R and h and the resulting color shift i. It follows that there is no relevant color shift if $\frac{R}{h} < \frac{1}{25}$.

Tilted principal axis ($\xi \neq 0$)

Figure 3.10 illustrates the "conical acquisition mode" for $\xi \neq 0$. Having a tilted principal axis, the color shift also depends on pixel position j in the CCD line and angle ξ. In this case, equation (5.2) is expanded by this rotation into the following:

$$0 = \Delta_x \cdot \cos(i \cdot \Delta\varphi) - \sin(i \cdot \Delta\varphi)[\cos(\xi) \cdot f - \sin(\xi) \cdot j \cdot \tau].$$

We solve this for i and obtain that

$$i = \arctan\left(\frac{\Delta_x}{\cos(\xi) \cdot f - \sin(\xi) \cdot j \cdot \tau}\right)\frac{1}{\Delta\varphi}.$$

This shows that the color shift (as a real) is different for each pixel position j on the CCD line.

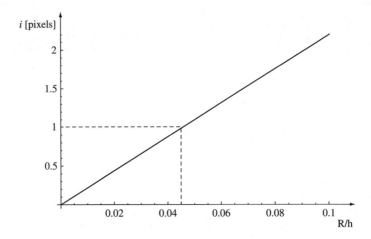

Figure 5.7 Color shift for RGB color lines in off-axis mode for objects in relation to ratio R/h.

5.2.3 Radiometric Corrections

Radiometric adjustments of different images are implemented as alignments of their cumulative histograms C_c, where subscript c denotes the channel (i.e., red, green, or blue). The cumulative histogram is defined by partial sums of the histogram h_c, and given as follows:

$$C_c(j) = \sum_{i=0}^{j} \frac{h_c(i)}{W \cdot H} \qquad \text{with } 0 \leq j < 2^n.$$

Here n is the radiometric resolution (in bits), W and H are the image dimensions (in pixels).

After determining a reference cumulative histogram from the reference image, each image can be transformed with respect to the reference image by using the defined transfer function (i.e., given by C_c). This function is implemented as a *lookup table*.

Figure 5.8 shows a reference image with its histogram, and cumulative histogram, and also a second image of the same scene (but under different lighting conditions) and its histograms. The second image is transformed as described, and the resulting image (with its histogram and lookup table) is shown in Figure 5.9.

5.2.4 Geometric Corrections

Before correcting errors caused by lens distortions or any other filters (e.g., an ultraviolet or infrared filter), or by a tilt of the CCD sensor line, geometric errors of positions of detector elements (on a given sensor line) need to be measured. For this purpose, the camera is mounted on a *manipulator* (e.g., at DLR Berlin), which is basically a high-precision turntable that can be rotated with an accuracy of 0.001°. A sufficient number of detector elements of the CCD line is then illuminated by a *collimator ray*. A collimator is a device that renders divergent or convergent light rays such that they are nearly parallel, thus illuminating an object as if from infinity. This ensures translation invariance.

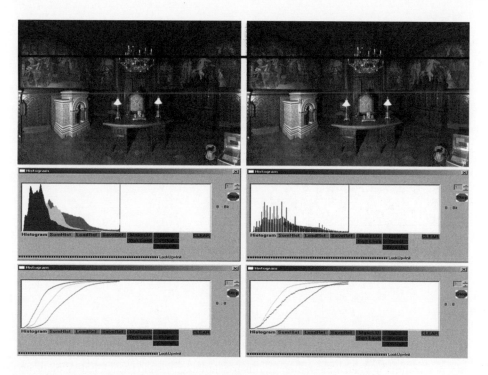

Figure 5.8 Reference image and its histogram and cumulative histogram (left), and a second image (right) with its histograms, taken under different lighting conditions (see Plate 27).

Histogram

Lookup table

Figure 5.9 Transformed image with its histogram and lookup table (see Plate 28).

Figure 5.10 shows (in a simplified scheme) the setup. After measuring two angles α and β for each detector element, defined with respect to the horizontal and vertical axis of the manipulator, the spatial attitude of the CCD line sensor is mapped into a virtual (ideal) focal plane. Coordinates

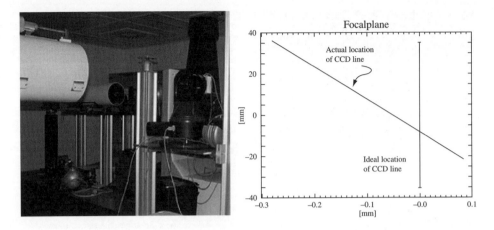

Figure 5.10 Camera mounted on a manipulator to measure geometric properties (the white tube is the collimator), and calculated virtual position of each detector element in an (ideal) focal plane.

$$x' = f \cdot \frac{\tan(\alpha)}{\cos(\beta)},$$

$$y' = f \cdot \tan(\beta)$$

define the position of a detector element in this focal plane; f is the focal length which should be used for subsequent calculations.

5.2.5 Correction of Mechanical Vibrations

A very simple image-based approach for correcting image distortion, caused by vibrations during image acquisition, is given by applying a *normalized cross-correlation* (NCC) algorithm. Here, the best correlation between two subsequent lines, or intervals on these lines, at times t and $t + 1$, will be found by shifting one line by σ pixel such that corresponding pixels (i.e., which continue the same object from t to $t + 1$) will have identical j-values. The correlation coefficient is given by

$$C_\sigma = \frac{\sum_j [S_j(t) - \overline{S}_j(t)] \cdot [S_{j+\sigma}(t+1) - \overline{S}_j(t+1)]}{\sqrt{\sum_j [S_j(t) - \overline{S}_j(t)]^2 \cdot \sum_j [S_{j+\sigma}(t+1) - \overline{S}_j(t+1)]^2}}$$

Summation is either over all pixels in a line, or over an interval $[j_{\min}, j_{\max}]$.

A best correlation is found for a maximum value of C_σ, with $-1 < C_\sigma < 1$. Lines are identical if $C_\sigma = 1$, and inverse if $C_\sigma = -1$. The use of the NCC algorithm is certainly a simple approach.

The common generalization is to compare image regions. For example, a set of templates of straight or curved arcs of defined length could be matched against an image region, but this may result in cases where curved edges are "corrected", for example, into straight segments.

Comparing edge templates with regions around (actually) straight object edges allows reasonable correction of image vibrations, but first these edges have to be estimated if they exist.

Figure 5.11 Corrected airport tower by using the NCC algorithm. The arrow points to a location where the restriction of using only relatively small regions for matching (see text) caused a false correlation (see Plate 29).

In brief, such a simple approach is insufficient in general; it works only partially. Figure 5.11 shows the corrected airport tower from Figure 2.23 when using the NCC algorithm for image regions.

5.3 A Least-Square Error Optimization Calibration Procedure

The common (and straightforward) point-based approach is to minimize the difference between ideal and actual projections of known 3D points, such as calibration marks on a calibration object, or localized points in the 3D scene. By taking many images of calibration marks, we are able to apply bundle adjustment and a least-square error (LSE) optimization procedure.

A pinhole-type camera is often used to explain calibration of a sensor-matrix camera, and we follow a similar path in this section. For this camera model, all rays pass through a small aperture (i.e., a very small hole); the smaller the aperture the better the approximation of the ideal pinhole case. (Of course, in practice we have apertures bigger than those of a pinhole, describable by Gaussian optics for a small paraxial region for those rays.)

5.3.1 Collinearity Equations

Photogrammetry often specifies image capture (for a pinhole-type camera) by image vectors, describing the position of pixels, all having the pinhole as origin (in the camera coordinate system). Typically we then have that

$$\begin{pmatrix} x \\ y \\ z \end{pmatrix} = \begin{pmatrix} i \cdot \tau \\ j \cdot \tau \\ 0 \end{pmatrix} + \begin{pmatrix} -x_0 \\ -y_0 \\ f \end{pmatrix}.$$

Therefore, a 3D surface point is "seen" by such an image vector, and transformed into a pixel position as follows:

$$\begin{pmatrix} X \\ Y \\ Z \end{pmatrix} = (X_0, Y_0, Z_0)^{\mathrm{T}} + \mathbf{R}\lambda \begin{pmatrix} x \\ y \\ f \end{pmatrix}.$$

Taking the inverse transform, we obtain that

$$\begin{pmatrix} x \\ y \\ f \end{pmatrix} = \frac{1}{\lambda} \mathbf{R}^{-1} \begin{pmatrix} X - X_0 \\ Y - Y_0 \\ Z - Z_0 \end{pmatrix}.$$

The derived collinearity equation is expanded by a correction term. For camera coordinate x we have that

$$x = f \cdot \frac{r_{11}(X - X_0) + r_{21}(Y - Y_0) + r_{31}(Z - Z_0)}{r_{13}(X - X_0) + r_{23}(Y - Y_0) + r_{33}(Z - Z_0)} + x_0 + \Delta x\,(x_0, y_0)\,.$$

Analogously, for camera coordinate y we have that

$$y = f \cdot \frac{r_{12}(X - X_0) + r_{22}(Y - Y_0) + r_{32}(Z - Z_0)}{r_{13}(X - X_0) + r_{23}(Y - Y_0) + r_{33}(Z - Z_0)} + y_0 + \Delta y\,(x_0, y_0)\,.$$

These two *collinearity equations* are the most frequently referenced equations in photogrammetry. Alternatively, computer vision uses so-called *fundamental* and *essential* matrices; this allows calibration and projective imaging to be described together within only one matrix, for the benefit of modern graphic card adapters.

We take a more detailed look at these collinearity equations. Of course, they are derived by dividing the general equations (with λ) either by x/f or y/f, respectively, to remove the unknown λ. Obviously, this is the central projection as discussed in Chapter 3, expanded by the terms $r_{11}, \ldots, r_{33}$ which are the elements of a rotation matrix $\mathbf{R}$, or of a matrix combining X_0, Y_0, Z_0 (which identifies the location of a projection center in a reference coordinate system, and thus the origin of the camera coordinate system), x_0, y_0 (the sensor's principal point), and $\Delta x, \Delta y$ as functions of the principal point. Both equations are solved for discrete image coordinates i, j, and are shown here in camera coordinates x, y. (Note that if we apply the rotation matrix $\mathbf{R}$ (a unit matrix), and have the camera's projection center at the origin of the reference coordinate system, then we obtain the basic equations

$$x = f\ X/Z \quad \text{and} \quad y = f\ Y/Z$$

of central projection.)

Why the name *collinearity equations*? The terms Δx and Δy describe a possible variation of the linear imaging process. Often this represents a radial symmetric distortion which appears when using an optics with slightly different entrance and exit angles over the whole aperture, thus causing the well-known *barrel* or *cushion distortions*. This function is rotation symmetric regarding the optical axis, and therefore a function of its intersection point with the sensor's principal point (x_0, y_0). Sometimes, these functions also compensate other deviations from the ideal linear Gaussian optics, or a tilted sensor. Often these functions also use further parameters when applying some higher-order terms.

Suppose we ignore the terms Δx and Δy, thus assuming distortion-free imaging. What are the consequences regarding the linear algebraic model? Collinearity is geometrically defined by a transformation of one vector space into another, where lines are mapped into lines. In other words, two vectors are collinear if they are linearly dependent; three or more points are collinear if they are located on a line. Affine transformations and projective imaging are special variants of collinear mappings. A central perspective transform is a common example of a collinear mapping.

However, the task of calibration is to determine the unknowns $f, r_{11}, \ldots, r_{33}$, and x_0, y_0, and this may be based on prior knowledge (by registration) of the extrinsic parameters, which is not urgently required. Parameters $r_{11}, \ldots, r_{33}$ allow the unknown intrinsic parameters α, β, δ (i.e., the tilting of a sensor combined in a common rotation matrix R_i; see the sensor model the in previous chapter) and the extrinsic parameters ψ, ϕ, κ to be derived. Furthermore, the matrix **R** can also include additional transformations (e.g., scaling, sheering) – in fact all permitted collinear transformations. (Obviously, it does not make sense to use many of them at once because we are already in 3D vector space.) Mixed terms such as $r_{11} = \cos \beta \sin \delta$ cause us "to leave linear algebra", and, typically, the collinearity equations are then first linearized before being solved.

5.3.2 Difference between Planar Capturing Surface and Panoramic Cylinder

We describe a difference between the planar capturing surface (of an ideal pinhole-type, a physically existing pinhole, or a sensor-matrix camera) and the panoramic cylindrical cases.

Assume a single sensor-line model. (The general case is described in detail later in this section.) An ideal panoramic scan with single projection center defines the following sensor model:

$$\begin{pmatrix} X \\ Y \\ Z \end{pmatrix} = (X_0, Y_0, Z_0)^{\mathrm{T}} + \mathbf{R}\mathbf{R}_{\varphi(i)}\mathbf{R}_i \lambda \begin{pmatrix} x \\ y \\ f \end{pmatrix},$$

here with $x = -x_0$ and $y = j\tau - y_0$. What has changed compared to the case of a planar capturing surface? We added a rotation $\mathbf{R}_{\varphi(i)}$ to the sensor model. This is the rotation $i \cdot \varphi$, dependent on the image coordinate i.

A separate calibration of intrinsic parameters α, β, and δ in the matrix $\mathbf{R}_i$ make sense for a rotating-sensor camera, because it is important for the imaging of the CCD line into the capturing surface. Remember the capturing surface is composed by stitching the image line by line; in the ideal case $\alpha = \beta = \delta = 0$ we have a cylindrical one. There is a need for an additional function $\Delta x, y (x_0, y_0)$ in order also to model any distortion parameter. Note that, in the shown example above this function also contains any affine transformation (i.e., a tilting of a matrix sensor).

We recall the collinearity equations for this camera model:

$$x = f \cdot \frac{r_{11}(X - X_0) + r_{21}(Y - Y_0) + r_{31}(Z - Z_0)}{r_{13}(X - X_0) + r_{23}(Y - Y_0) + r_{33}(Z - Z_0)} + x_0 + \Delta x (x_0, y_0),$$

$$y = f \cdot \frac{r_{12}(X - X_0) + r_{22}(Y - Y_0) + r_{32}(Z - Z_0)}{r_{13}(X - X_0) + r_{23}(Y - Y_0) + r_{33}(Z - Z_0)} + y_0 + \Delta y (x_0, y_0).$$

Obviously, these equations are similar to the usual approach. But there is of course a major difference, namely that these equations cannot be used for mapping a point from XYZ space into an xy capturing surface, because $r_{11}, \ldots, r_{33}$ also contain the rotation $\mathbf{R}_{\varphi(i)}$ which is a function of the image coordinate i. However, this additional feature is not important for calibration, because both the XYZ coordinates and the i, j coordinates are known. Only the collinearity has to be fulfilled.

5.3.3 Parameters and Objective Functions

In the sequel we describe a standard least-squares approach, as known from photogrammetry, but adapted to a rotating sensor-line camera. This approach determines the unknowns of the extrinsic parameter of the sensor, which are the matrices $\mathbf{R}(\psi, \phi, \kappa)$, $\mathbf{R}_\xi$, the coordinates X_0, Y_0, Z_0, off-axis distance R, and principal angle ω.

It also determines the intrinsic parameters, which are matrix $\mathbf{R}_i(\alpha, \beta, \delta)$ describing the tilt of the sensor, the "focal length" $f + z_0$, and the sensor's principal point (x_0, y_0), the latter also written as vector $\mathbf{\Delta}$.

The rotation angle φ of the rotating sensor (sensor-line camera or range-finder) may be measured using an internal measuring system of the turntable. Modern technology allows the angle (for each vertical scan line of the sensor) to be determined with an accuracy of $1/1000$ degree at least.

Note that it is frequently necessary to recalculate a "focal length" (i.e., of the camera constant) in order to obtain an exact determination of the (typically unknown) virtual projection center of a pinhole-type model, namely the distance between the entrance pupil and a virtual sensor plane which fulfills the linear imaging assumption (discussed in detail in Section 2.2). The origin of the sensor coordinate system (X_0, Y_0, Z_0) is denoted, as in previous chapters, by the vector $\mathbf{t}_0$.

An *observation* is a recorded calibration mark (with physically measured coordinates, identified with a point (X, Y, Z), such as the centroid of the mark) at corresponding image coordinates i and j (i.e., pixel (i, j) for the rotating sensor-line camera, when projecting point (X, Y, Z) into the cylindrical panorama). Note that two observations are derivable for one calibration mark by virtue of the two collinearity equations (i.e., one observation is given by two collinearity equations and its corresponding residues).

We have a linear system of n equations in m unknowns; the ith observation is given by l_i. The sum of all the observation can be written as

$$\sum_{i=0}^{n} l_i = a_{11} \cdot x_1 + a_{12} \cdot x_2 + \ldots + a_{im} \cdot x_m.$$

Observations are considered to be the residues of an iterative Taylor approximation of kth order (which defines a Newton method):

$$l = F(u) - \nabla F^k(\hat{u}) \Delta u.$$

For the determination of extrinsic parameters and the calibration of intrinsic parameters of a sensor, we place various calibration marks "around the sensor" in the scene. Some of them are projected into image data (depending on visibility), and we assume that all projected calibration marks can be uniquely identified in the resulting image data (e.g., in the panoramic image).

Assume that we have m unknowns in total (i.e., elements in matrices, vectors, and parameters), and that there are n observations, with $n \geq m$.

5.3.4 General Error Criterion

We use the general equation (3.16), recalled here for ease of reference:

$$
\mathbf{P}_w = \mathbf{t}_0 + \mathbf{RR}_{\varphi(i)} \left[\lambda \mathbf{R}_\xi \mathbf{R}_\omega \left[\mathbf{R}_i \left(\begin{array}{c} \Delta_x - x_0 \\ j\tau + \Delta_y - y_0 \\ \Delta_z \end{array} \right) + \left(\begin{array}{c} 0 \\ 0 \\ f \end{array} \right) \right] + \left(\begin{array}{c} 0 \\ 0 \\ R \end{array} \right) \right].
$$

We recall that the position vector $\mathbf{t}_0$ is defined by a translation of point $\mathbf{O}$ (the intersection of the rotation axis with the base plane, see Figure 2.11) in the world coordinate system.

Substituting $\mathbf{A} = \mathbf{RR}_{\varphi(i)}$, $\mathbf{B} = \mathbf{R}_\xi \mathbf{R}_\omega$, and $\mathbf{C} = \mathbf{BR}_i$ (with matrix elements $\mathbf{A} = a_{11}, \ldots, a_{33}$, $\mathbf{B} = b_{11}, \ldots, b_{33}$, and so on), the general equation is now given by

$$
\mathbf{P}_w = \mathbf{t}_0 + \mathbf{A}(\lambda \mathbf{B}(\mathbf{R}_i \mathbf{v}_{j,\Delta} + f\mathbf{z}^\circ) + R\mathbf{z}^\circ),
$$

$$
\mathbf{A}^{-1}(\mathbf{P}_w - \mathbf{t}_0) - R\mathbf{z}^\circ = \lambda \mathbf{C} \mathbf{v}_{j,\Delta} + \mathbf{B}f\mathbf{z}^\circ,
$$

where $\mathbf{v}_{j,\Delta}$ is the image vector

$$
\mathbf{v}_{j,\Delta} = (\mathbf{v}_x, \mathbf{v}_y, \mathbf{v}_z)^{\mathrm{T}} = \left(\begin{array}{c} \Delta_x - x_0 \\ j\tau + \Delta_y - y_0 \\ \Delta_z \end{array} \right).
$$

For all three components of this equation, and writing $\tilde{\mathbf{P}} = \mathbf{P}_w - \mathbf{t}_0$, we have

$$
a_{11}\tilde{\mathbf{P}}_x + a_{21}\tilde{\mathbf{P}}_y + a_{31}\tilde{\mathbf{P}}_z = \lambda(c_{11}\mathbf{v}_x + c_{12}\mathbf{v}_y + c_{13}\mathbf{v}_z + b_{13}f),
$$

$$
a_{12}\tilde{\mathbf{P}}_x + a_{22}\tilde{\mathbf{P}}_y + a_{32}\tilde{\mathbf{P}}_z = \lambda(c_{23}\mathbf{v}_x + c_{22}\mathbf{v}_y + c_{23}\mathbf{v}_z + b_{23}f),
$$

$$
a_{13}\tilde{\mathbf{P}}_x + a_{23}\tilde{\mathbf{P}}_y + a_{33}\tilde{\mathbf{P}}_z - R = \lambda(c_{31}\mathbf{v}_x + c_{32}\mathbf{v}_y + c_{33}\mathbf{v}_z + b_{33}f).
$$

The matrix of coefficients $a_{11}, \ldots, a_{33}$ is finally transposed because of the inversion of matrix $\mathbf{A}$. (For a rotation matrix we have that $\mathbf{I} = \mathbf{R} \cdot \mathbf{R}^{\mathrm{T}}$ is the unit matrix, and, consequently, $\mathbf{R}^{-1} = \mathbf{R}^{\mathrm{T}}$.)

By dividing these equations we may eliminate the scaling factor λ, and we obtain from the left-hand sides of those three equations the following two equations:

$$
F_{x/z} := \frac{a_{11}(\tilde{\mathbf{P}}_x) + a_{21}(\tilde{\mathbf{P}}_y) + a_{31}(\tilde{\mathbf{P}}_z)}{a_{13}(\tilde{\mathbf{P}}_x) + a_{23}(\tilde{\mathbf{P}}_y) + a_{33}(\tilde{\mathbf{P}}_z) - R}
$$

and

$$
F_{y/z} := \frac{a_{12}(\tilde{\mathbf{P}}_x) + a_{22}(\tilde{\mathbf{P}}_y) + a_{32}(\tilde{\mathbf{P}}_z)}{a_{13}(\tilde{\mathbf{P}}_x) + a_{23}(\tilde{\mathbf{P}}_y) + a_{33}(\tilde{\mathbf{P}}_z) - R}.
$$

For the right-hand sides we obtain

$$G_{x/z} := \frac{c_{11}\mathbf{v}_x + c_{12}\mathbf{v}_y + c_{13}\mathbf{v}_z + b_{13}f}{c_{31}\mathbf{v}_x + c_{32}\mathbf{v}_y + c_{33}\mathbf{v}_z + b_{33}f}$$

and

$$G_{y/z} := \frac{c_{23}\mathbf{v}_x + c_{22}\mathbf{v}_y + c_{23}\mathbf{v}_z + b_{23}f}{c_{31}\mathbf{v}_x + c_{32}\mathbf{v}_y + c_{33}\mathbf{v}_z + b_{33}f}.$$

These are the general collinearities given by $F_{x/z} = G_{x/z}$ and $F_{y/z} = G_{y/z}$ in short form.

By linearization of these equations it is now possible to estimate iteratively the unknown parameters for the left-hand side $F_{x/z}$ and $F_{y/z}$,

$$\mathbf{u} = (t_{x0}, t_{y0}, t_{z0}, \psi, \phi, \kappa, R),$$

or for the right-hand side $G_{x/z}$ and $G_{y/z}$,

$$\mathbf{u} = (\xi, \alpha, \beta, \delta, \omega, f, y_0, x_0).$$

The upper index k is the number of the iteration step. The linearization is given as follows:

$$\nabla(G_{x/z} - F_{x/z}) = \left(\frac{\partial G_{x/z}}{\partial u_1} - \frac{\partial F_{x/z}}{\partial u_1}, \frac{\partial G_{x/z}}{\partial u_2} - \frac{\partial F_{x/z}}{\partial u_2}, ..., \frac{\partial G_{x/z}}{\partial u_m} - \frac{\partial F_{x/z}}{\partial u_m} \right),$$

$$F_{x,z}^k - G_{x,z}^k = \nabla(G_{x,z} - F_{x,z})^k \cdot \Delta\mathbf{u},$$

$$\mathbf{l} = \mathbf{M} \cdot \Delta\mathbf{u}.$$

For $n = m$, the solution is uniquely given by

$$\Delta\mathbf{u} = \mathbf{M}^{-1} \cdot \mathbf{l},$$

assuming linear independence between equations.

For $n > m$ observations (i.e., a typical adjustment problem), we now apply the method of least-square error minimization. The error is given by

$$\mathbf{v} = \mathbf{M} \cdot \Delta\hat{\mathbf{u}} - \mathbf{l}.$$

The error function (which needs to be minimized) is defined as follows:

$$\min = \mathbf{v}^{\mathsf{T}}\mathbf{v}$$
$$= \left(\mathbf{M} \cdot \Delta\hat{\mathbf{u}} - \mathbf{l} \right)^{\mathsf{T}} \left(\mathbf{M} \cdot \Delta\hat{\mathbf{u}} - \mathbf{l} \right)$$
$$= \Delta\hat{\mathbf{u}}^{\mathsf{T}}\mathbf{M}^{\mathsf{T}}\mathbf{M} \cdot \Delta\hat{\mathbf{u}} - 2\mathbf{l}^{\mathsf{T}}\mathbf{M} \cdot \Delta\hat{\mathbf{u}} + \mathbf{l}^{\mathsf{T}}\mathbf{l}.$$

For identifying the minimum, we differentiate and have the resulting function equal to zero:

$$\frac{\partial(\mathbf{v}^{\mathsf{T}}\mathbf{v})}{\partial\Delta\hat{\mathbf{u}}} = 2\Delta\hat{\mathbf{u}}^{\mathsf{T}}\mathbf{M}^{\mathsf{T}}\mathbf{M} - 2\mathbf{l}^{\mathsf{T}}\mathbf{M} = 0.$$

This leads to the solution

$$\Delta\hat{\mathbf{u}} = \left(\mathbf{M}^{\mathrm{T}}\mathbf{M}\right)^{-1}\mathbf{M}^{\mathrm{T}}\mathbf{l}. \tag{5.9}$$

The Jacobian matrix M contains all first-order partial derivatives, and $\mathbf{l}$ are the residues as defined above. This is solved by means of iterations; the vector $\Delta\mathbf{u}$ contains the corrections of of each unknown. A minimum is found if the unknowns do not change significantly in an iteration step (e.g., $\sum_{i=0}^{m}|\Delta\mathbf{u}_i| < \varepsilon$, with $\varepsilon = 10^{-9}$).

5.3.5 Discussion

Human intervention is generally required for this calibration approach for identifying the projections of those 3D points in a real scene (e.g., the projected points) used as calibration marks. If a specially designed calibration object is used, this process can be supported by an automatic calibration mark detection algorithm, where marks are located with subpixel accuracy (using, for example centroid calculation within a mark's region, or intersection points of approximated straight lines when using a checkerboard). Special feature detectors may be used for solving this task automatically.

The least-squares approach described has been used in many applications of panoramic sensors, and is so far our recommended way for calibrating all the parameters mentioned, possibly also including a tilt of the rotation axis of the sensor.

Figure 5.1 shows the calibration courtyard at the Institute for Photogrammetry at TFH Berlin. The locations of the control points along the buildings can be measured with a theodolite and are photogrammetrically balanced. The deviation of our calculated points from the given control points is shown in Table 5.1. Further experiments confirmed that the estimation of the parameters of the intrinsic attitude is stable, which means that it does not change significantly by changing the parameters of the extrinsic attitude, and therefore these parameters are properly separated from each other.

Control point	Error in pixel coordinate i	Error in pixel coordinate j
7	0.38	0.03
9	0.09	−0.44
10	0.09	0.08
11	−0.23	0.54
12	−0.17	0.28
14	−0.16	0.05
15	−0.14	−0.37
16	−0.27	−0.35
17	0.26	−0.01
19	0.22	0.30
50	0.01	−0.12
51	−0.09	0.11

Table 5.1 Example of a typical panoramic adjustment, here for calibration marks shown in Figure 5.1. The table lists deviations between calculated image coordinates and their actual reference coordinates. All listed values are in subpixel scale.

To ensure a separated calibration of intrinsic and extrinsic parameters, a balanced calculation, called *bundle adjustment*, is commonly applied in photogrammetry. The method is based on using a set (or "bundle") of images, taken from different positions. The extrinsic parameters change but not the intrinsic parameters, thus the dependency between these parameters is minimized. The point-based calibration approach introduced actually does allow such bundle adjustment. For such adjustment, we capture panoramas at different positions (for $R > 0$), and the use of wide-angle (say, full 360°) panoramas stabilizes the calibration.

5.4 Geometric Dependencies of R and ω

This section aims to present a more theoretical comparison of different options for calibrating these two important parameters (i.e., R and ω) of rotating sensor-line cameras, and in particular to discuss a difference between linear and non-linear parameters. The following three methods have been implemented and tested, but not at the same scale as the least-square error method described above.[4]

5.4.1 Three Methods

In this discussion we address the general issue of how the existing concepts, methods developed for pinhole-type cameras, can be applied to the calibration problem described. In particular, how far can we restrict ourselves to using only the linear geometric features in the scene, such as calibration points on a calibration plane, or sets of line segments? Compared to the previous discussions in this chapter, we now address a more theoretical performance analysis issue than a practical comparative test (based on simplification, but illustrating possible alternatives).

The least-squares method presented above is an example of a point-based approach whose basic idea is to find the minimum difference between actual and ideal projections of 3D points (i.e., control points or calibration marks) where relative coordinates are known based on measurements. This proved to be robust and accurate in extensive applications. We now simplify this to make it comparable with the other two techniques. This simplified point-based approach (i.e., ideal off-axis case without distortions, additional axis alignments, or tilting of the focal plane) is our first method in this section, and it allows a discussion on unstable parameter estimation for panoramic cameras due to non-linearity and high dimensionality.

The second method is an *image correspondence approach*, where we have a set of pairs of corresponding points (in different images). Structure from motion is based on using (just) image correspondence information for the registration of pinhole-type cameras (applying epipolar geometry), but it is well known that this cannot be used for the calibration of intrinsic parameters. This approach requires neither scene measures nor any calibration object. Therefore it almost avoids the influence of extrinsic camera parameters on the calibration of intrinsic

---

[4] These methods require some knowledge of the epipolar geometry of multi-center panoramas, as discussed in the previous chapter.

parameters using the concentric case (i.e., only R and ω are changed in the second image). (This approach is potentially possible, but requires further research to be to make it a practical alternative to the least-square error optimization approach discussed above.)

Finally, we present a *parallel line approach* that uses geometric properties of parallel line segments (calibration lines) available in the scene. In this case, a panoramic camera is well posed if the axis of the image cylinder is parallel to at least three of those straight line segments. This approach theoretically allows the most accurate and numerically stable calibration (compared to the other two methods). However, more work is needed on numerical optimization in order to develop this technique into a practical alternative to a point-based calibration approach.

In order to discuss these three general techniques, we first split the calibration process into two steps, one for linear and one for non-linear components. The first step is to calibrate the effective focal length f and the central row j_c[5] (Note, that these parameters are calculated as approximate values.). The second step is to calibrate the off-axis distance R and principal angle ω. This split allows exploration of linear features independently of non-linear features, to reduction of dimensionality, and reduction of computational complexity. This two-step calibration process is theoretically applicable regardless of the panoramic camera or progressive scanning approach used. (In practice, they actually depend on each other in iterative or error-minimization approaches.)

Section 5.4.2 discusses the "linear" step of panoramic camera calibration for f and j_c. The rest of the section presents the three approaches (listed above) for the "non-linear" calibration step for R and ω. The standard approach for pinhole-type cameras (i.e., the point-based approach) is still preferred in applications, but also serves as a motivation for discussing two more approaches for potentially further improvements. Indeed, this section indicates that the use of parallel lines (note: the exploitation of linear features) may lead to most stable solutions among these three methods.

5.4.2 Focal Length and Central Row

We estimate the focal length f (measured in pixels) and the *central row* j_c. Given a calibration object or singular calibration marks distributed within the scene, we estimate f and j_c by minimizing differences between actual and ideal projections of known 3D points defined by calibration marks.

In this case a 2D space is sufficient to describe necessary geometric relations; see Figure 5.12. **C** is the associated camera projection center. The 2D coordinate system shows projections of 3D calibration points (positioned on a 3D calibration object). The coordinates are denoted by (Y_c, Z_c). (We use Y and Z instead of X and Y in correspondence with the 3D coordinate system used in this book.) The Y-axis of the camera coordinate system is defined parallel to the sensor column; the Z-axis is perpendicular to the sensor column and passes through it.

Two image coordinate systems are used for describing an image. One is discrete with integer coordinate j, and the other one is the Euclidean image coordinate system with real coordinate y; see Figure 5.12. The coordinate value j could appear to be a real due to the arithmetic calculation, but is then rounded into the nearest integer when considering it as the position of a pixel. The coordinates j and y are related by the expression

---

[5] The image row that intersects the base plane.

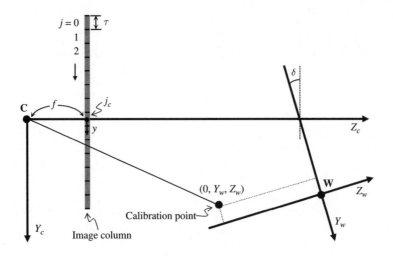

Figure 5.12 The first ("linear") step of panoramic camera calibration.

$$j = j_c + \frac{y}{\tau}$$

where j_c is the coordinate of the image point where the Z-axis of the camera coordinate system intersects with the image, and τ is the height of a pixel (i.e., in the cell model) in the image. The image point with coordinate j_c is referred to as the *image center* of the sensor or image.

The direction of the Y-axis of the camera coordinate system is defined to be the same as that of the y-axis in the Euclidean image coordinate system. A point (Y_c, Z_c) is projected onto an image column at position y as follows:

$$y = \frac{f Y_c}{Z_c}$$

where f is the effective focal length of the panoramic camera. The relation between pixel j (in the point model) and (Y_c, Z_c) can be described in matrix form as

$$\begin{bmatrix} sj \\ s \end{bmatrix} = \begin{bmatrix} \frac{f}{\tau} & j_c \\ 0 & 1 \end{bmatrix} \begin{bmatrix} Y_c \\ Z_c \end{bmatrix},$$

where s is any scalar. Let $f_\tau = \frac{f}{\tau}$ denote the camera's effective focal length measured in pixels.

The calibration object in Figure 5.12 is (theoretically) a planar region. In the plane of this calibration object we consider a 2D world coordinate system, with its origin at **W**. Any point in the plane of Figure 5.12 thus has world coordinates (Y_w, Z_w).

The acute angle between the Y_w-axis and the Z-axis (of the camera coordinate system) is denoted by δ. The origin **C** of the camera coordinate system has world coordinates (t_y, t_z).

A calibration point (Y_w, Z_w) is transformed from world coordinates into camera coordinates as follows:

$$\begin{bmatrix} Y_c \\ Z_c \end{bmatrix} = \begin{bmatrix} \cos(\delta) & -\sin(\delta) & t_y \\ \sin(\delta) & \cos(\delta) & t_z \end{bmatrix} \begin{bmatrix} Y_w \\ Z_w \\ 1 \end{bmatrix}.$$

Thus, the relation between a calibration point (Y_w, Z_w) and its projection j in the image can be expressed as

$$\begin{bmatrix} sj \\ s \end{bmatrix} = \begin{bmatrix} f_\tau & j_c \\ 0 & 1 \end{bmatrix} \begin{bmatrix} \cos(\delta) & -\sin(\delta) & t_y \\ \sin(\delta) & \cos(\delta) & t_z \end{bmatrix} \begin{bmatrix} Y_w \\ Z_w \\ 1 \end{bmatrix}$$

$$= \begin{bmatrix} f_\tau c(\delta) + j_c s(\delta) & -f_\tau s(\delta) + j_c c(\delta) & f_\tau t_y + j_c t_z \\ s(\delta) & c(\delta) & t_z \end{bmatrix} \begin{bmatrix} Y_w \\ Z_w \\ 1 \end{bmatrix}.$$

The value of j is given by

$$\frac{Y_w(f_\tau c(\delta) + j_c s(\delta)) + Z_w(j_c c(\delta) - f_\tau s(\delta)) + (f_\tau t_y + j_c t_z)}{Y_w s(\delta) + Z_w c(\delta) + t_z}. \tag{5.10}$$

The values of f_τ and j_c can therefore be estimated by a given set of calibration points and their corresponding projections. Equation (5.10) can be rearranged into a linear equation of five unknowns X_k, $k = 1, 2, \ldots, 5$:

$$Y_w X_1 + Z_w X_2 - j Y_w X_3 - j Z_w X_4 + X_5 = j$$

where

$$X_1 = \frac{f_\tau \cos(\delta) + j_c \sin(\delta)}{t_z},$$

$$X_2 = \frac{j_c \cos(\delta) - f_\tau \sin(\delta)}{t_z},$$

$$X_3 = \frac{\sin(\delta)}{t_z},$$

$$X_4 = \frac{\cos(\delta)}{t_z},$$

$$X_5 = \frac{(f_\tau t_y + j_c t_z)}{t_z}.$$

It follows that at least five pairs of calibration points and their projections are necessary to determine f_τ and j_c.

Now assume that N such pairs $\{(Y_{wk}, Z_{wk}), j_k\}$ are given, where $N > 5$; (Y_{wk}, Z_{wk}) denotes the kth calibration point, and j_k denotes the corresponding projection. This allows the following overdetermined system of linear equations to be solved:

$$\begin{bmatrix} Y_{w1} & Z_{w1} & -j_1 Y_{w1} & -j_1 Z_{w1} & 1 \\ Y_{w2} & Z_{w2} & -j_2 Y_{w2} & -j_2 Z_{w2} & 1 \\ \vdots & \vdots & \vdots & \vdots & \vdots \\ Y_{ws} & Z_{wN} & -j_N Y_{wN} & -j_N Z_{wN} & 1 \end{bmatrix} \begin{bmatrix} X_1 \\ X_2 \\ X_3 \\ X_4 \\ X_5 \end{bmatrix} = \begin{bmatrix} j_1 \\ j_2 \\ \vdots \\ j_N \end{bmatrix}.$$

Having obtained a least-squares solution of $(X_1, X_2, X_3, X_4, X_5)^{\mathrm{T}}$, the values of f_τ and j_c can be calculated as follows:

$$\begin{bmatrix} f_\tau \\ j_c \end{bmatrix} = \begin{bmatrix} X_4 & X_3 \\ -X_3 & X_4 \end{bmatrix}^{-1} \begin{bmatrix} X_1 \\ X_2 \end{bmatrix}.$$

This completes the first step of the calibration of the panoramic camera, which provides the effective focal length f (measured in pixels) and the central row j_c.

The following sections present and characterize different options for a second camera calibration step. Henceforth we can always assume that the effective focal length f_τ and central row j_c are known.

5.4.3 Point-based Approach

The point-based approach (i.e., "point" = calibration mark) is our first option for calibrating off-axis distance R and principal angle ω. (Such a point-based approach assumes pre-calibrated images; it was also used in Section 5.4.2 for the first step of the calibration process.) Assume N pairs of known 3D points (X_{wk}, Y_{wk}, Z_{wk}) (in world coordinates) and their actual projections $(\hat{i}_k, \hat{j}_k)$ (in image coordinates). The use of a hat "ˆ" indicates that this parameter may be inaccurate.

In contrast, let (i_k, j_k) be the correct projection of the 3D calibration mark (X_{wk}, Y_{wk}, Z_{wk}). We wish to minimize the error

$$\sum_{k=1}^{N} \left(\hat{i}_k - i_k \right)^2 + \left(\hat{j}_k - j_k \right)^2, \tag{5.11}$$

where the value of i_k can be obtained from equations (3.6) and (3.7) and the value of j_k can be obtained from equations (3.8) and (3.9). After some minor algebraic transformations, this problem proves to be equivalent to minimizing

$$\sum_{k=1}^{N} \left(\sin\left(\frac{2\hat{i}_k \pi}{W} + \omega \right) - \frac{X_{ok} A + Z_{ok} R \sin\omega}{X_{ok}^2 + Z_{ok}^2} \right)^2 + \left(\hat{j}_k - \frac{f_\tau t_{ok}}{A - R\cos\omega} + j_c \right)^2,$$

where $A = \sqrt{X_{ok}^2 + Z_{ok}^2 - R^2 \sin^2\omega}$ and

$$\begin{bmatrix} X_{ok} \\ Y_{ok} \\ Z_{ok} \end{bmatrix} = \begin{bmatrix} X_{wk}t_{11} + Y_{wk}t_{12} + Z_{wk}t_{13} + t_{14} \\ X_{wk}t_{21} + Y_{wk}t_{22} + Z_{wk}t_{23} + t_{24} \\ X_{wk}t_{31} + Y_{wk}t_{32} + Z_{wk}t_{33} + t_{34} \end{bmatrix}.$$

Parameters f_τ, j_c, derivations from ideal Gaussian imaging, and intrinsic parameters are assumed to be pre-calibrated and (thus) known (and not estimated as in the section before). Therefore, there are 14 parameters in total here to be estimated using a non-linear least-square optimization method. These 14 parameters consist of the parameters R, ω as well as the 12 (intermediate) unknowns of the transformation matrix.

This approach does not cover the full complexity of a calibration scenario of all the remaining intrinsic parameters: the rotation axis may also be tilted with respect to three coordinate axis,

Approaches	Point-based	Image correspondence	Parallel-line-based
Number of images required	One	Two	One
Pre-process	Identifying projections of 3D points	Searching image corresponding points	Identifying projections of lines
Complexity	Non-linear form e.g., sine, products of unknowns, and square roots	Linear form	Linear form
Dimensionality	14 unknowns, $[\mathbf{R}|\mathbf{t}]$ unavoidable	Four unknowns, $[\mathbf{R}|\mathbf{t}]$ not required	Three unknowns, $[\mathbf{R}|\mathbf{t}]$ avoidable
Initial values dependence	High	Ignorable	Ignorable
Growth of error sensitivity	Exponential	Exponential	Linear
Estimated R	Actual value	Ratio	Actual value
Estimated ω	Actual value	Actual value	Actual value

Table 5.2 Summary of the performance of the three camera calibration approaches discussed for non-linear parameters.

and so forth (as discussed for the general calibration procedure above). However, even for this simplified case, the error function of equation (5.11) is already rather complicated. The parameters to be estimated are enclosed in sine functions, and square roots are involved in both numerator and denominator of the fractions. The dimensionality is high due to the fact that the extrinsic parameters in $\mathbf{R}_{wo}$ and $\mathbf{t}_{wo}$ are unavoidable in this approach. Thus, a "large" set of 3D points is needed for reasonably accurate estimation.

Following this approach, the accuracy of the calibration result critically depends on the accuracy of initial values (i.e., 3D points and identified projections). Error sensitivity analysis showed exponential growth of resulting errors.

The high dimensionality of the error function is a critical issue of this point-based approach, and a substantial reduction of the exponential growth of errors in relation to errors of initial values is an interesting problem.

Assessments are summarized in Table 5.2, including those for two more approaches still to be presented in this section; these two approaches will not require the camera's extrinsic parameters for calibrating R and ω.

5.4.4 Image Correspondence Approach

This section discusses the possibility of calibrating R and ω by using corresponding image points in two uncalibrated concentric panoramas. This approach requires neither scene measures nor a calibration object and is (thus) independent of the camera's extrinsic parameters.

The basic idea of this approach is similar to a recovery of epipolar geometry: first use equations of epipolar curves (see Chapter 4) to link the corresponding points; then calibrate parameters by optimization. The first step is known in computer vision as fundamental (or essential) matrix estimation in the case of sensor-matrix cameras.

Chapter 4 provides the analytic representation of epipolar curves for the general case of multi-view panoramas as well as for a few specific categories of multiple panoramas (defined by some geometric constraints on their poses). This section makes use of a result reported for concentric panoramic images.

The goal is to estimate values of R and ω, and thus it is essential to keep all other sensor parameters (such as effective focal length, the angular unit and all other extrinsic and intrinsic parameters, except R and ω) constant for the two panoramas used. A concentric panoramic pair can be acquired in various ways, by using different or identical off-axis distances, or different or identical principal angles. The three possible options are as follows: same off-axis distance plus different principal angles; same principal angle plus different off-axis distances; or both different off-axis distances and principal angles.

For the first option, two different values ω_1 and ω_2 can be obtained, but it is impossible to estimate the value of R. For the second option, the value of ω and the ratio of R_1 and R_2 can be estimated. Estimation results obtained by the first two options are of acceptable accuracy if we have a 'good' guess for the initial values. The third option leads to relatively poor results for estimating ω_1 and ω_2 in comparison to the first option, and the true values of R_1 and R_2 cannot be calibrated in this case either. This subsection describes the estimation method for the second option only, since the optimization methods used for these three options are basically identical.

Let (φ_1, L_1) and (φ_2, L_2) be a pair of corresponding image points in a concentric pair of panoramas $E_{\mathcal{P}_1}(R_1, f, \omega, \gamma)$ and $E_{\mathcal{P}_2}(R_2, f, \omega, \gamma)$. Using the epipolar curve equation provided in equation (4.13), these two image points have to satisfy the equation

$$L_2 = L_1 \cdot \frac{R_2 \sin \omega - R_1 \sin(\varphi_2 + \omega - \varphi_1)}{R_2 \sin(\varphi_1 + \omega - \varphi_2) - R_1 \sin \omega}. \tag{5.12}$$

The equation can be rearranged as follows:

$$L_2 R_2 \sin((\varphi_1 - \varphi_2) + \omega) - L_2 R_1 \sin \omega + L_1 R_1 \sin((\varphi_2 - \varphi_1) + \omega) - L_1 R_2 \sin \omega = 0.$$

Let $(\varphi_1 - \varphi_2) = \sigma$. We have

$$L_2 \sin \sigma R_2 \cos \omega + L_2 \cos \sigma R_2 \sin \omega - L_1 \sin \sigma R_1 \cos \omega$$

$$+ L_1 \cos \sigma R_1 \sin \omega - L_2 R_1 \sin \omega - L_1 R_2 \sin \omega = 0.$$

Furthermore, we obtain

$$L_2 \sin \sigma R_2 \cos \omega + (L_2 \cos \sigma - L_1) R_2 \sin \omega$$

$$- L_1 \sin \sigma R_1 \cos \omega + (L_1 \cos \sigma - L_2) R_1 \sin \omega = 0.$$

We observe from these equations that the ratio $R_1 : R_2$ and the value of ω can be calibrated. The values of R_1 and R_2 are not computable when using this approach alone.

Now assume N pairs of corresponding image points (φ_{1k}, L_{1k}) and (φ_{2k}, L_{2k}) in both panoramas, where $k = 1, \ldots, N$. The values of corresponding indices σ_k are assumed to have been calculated in advance. We use an optimization method of sequential quadratic programming to estimate $\frac{R_1}{R_2}$ and ω. The error function to be minimized is

$$\sum_{k=1}^{N}[L_{2k}\sin\sigma_k X_1 + (L_{2k}\cos\sigma_k - L_{1k})X_2 - L_{1k}\sin\sigma_k X_3 + (L_{1k}\cos\sigma_k - L_{2k})X_4]^2,$$

subject to the equality constraint $X_1 X_4 = X_2 X_3$, where $X_1 = R_2\cos\omega$, $X_2 = R_2\sin\omega$, $X_3 = R_1\cos\omega$, and $X_4 = R_1\sin\omega$.

Once the values of X_1, X_2, X_3, and X_4 are obtained, $\frac{R_1}{R_2}$ and ω can be calculated by

$$\frac{R_1}{R_2} = \frac{\sqrt{X_3^2 + X_4^2}}{\sqrt{X_1^2 + X_2^2}}$$

and

$$\omega = \arccos\left(\frac{X_1}{\sqrt{X_1^2 + X_2^2}}\right).$$

The error function of corresponding point approach is basically in linear form for all three options, and there are only four unknowns to be estimated for the first and second options. This means that at least four pairs of corresponding image points (in general positions) are required for a unique solution to this minimization problem.

However, the experimental evidence indicates that the method gives rise to inaccuracies; estimated values for concentric panoramas of real scenes appeared to be erroneous compared to the known (i.e., pre-calibrated) parameter values. Such an experiment using real scene data is illustrated in Figure 5.13. In this example there are 35 pairs of corresponding image points

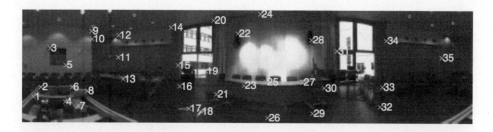

Figure 5.13 Two concentric panoramas (showing a seminar room at DLR in Berlin) with 35 pairs of corresponding image points, used for the calibration of off-axis distance R and principal angle ω.

Input data error (pixels)	Estimated ω error (%)
0.0	0.00
0.5	0.17
1.0	0.52
1.5	2.78
2.0	10.34
3.0	28.14
4.0	68.25
5.0	155.62

Table 5.3 Error sensitivity results of calibration by image correspondence.

identified manually, marked by crosses and numbered. We omit the details because results also depend on the accuracy of the pre-calibration method.

The error sensitivity of this approach can be analyzed in an objective way by a simulation using synthetic data. Ground-truth data[6] can be generated by attempting a simulation of real scenes (e.g., as shown in Figure 5.13), and errors are then simulated by additive random noise with a normal distribution, perturbating the coordinates of ideal pairs of corresponding image points. A few calibration results for ω as a function of additive errors are shown in Table 5.3 and illustrate that the estimated result is rather sensitive to these errors. In general, errors of estimated parameters increased exponentially with respect to errors in input data.

A reason why this image correspondence approach is sensitive to input error is that the values of the coefficients in the error function are likely to be numerically very close to one another, and these values depend on the selected pairs of corresponding points. Possible ways of suppressing the error sensitivity (without relying on the "robustness" of numerical methods) include:

(1) increasing the number of pairs of corresponding points;
(2) placing calibration object closer to the camera (i.e., larger disparities);
(3) designing a special calibration object that allows all coded points on the calibration object to appear twice in a single panorama (e.g., inward looking case).

Proposal (1) seemed (in experiments) to allow only minor improvements.

Despite the error problem, this approach is unable to recover the absolute value of R, only relative values as stated above. Assessments for this approach are summarized in Table 5.2. The drawbacks of the two approaches discussed so far provide the motivation to search for an approach where linear geometric relations between 3D scenes panoramic images can be

---

[6] The term "ground truth" was coined in photogrammetry, when results calculated based on recorded aerial views were compared against values obtained by measuring 'on the ground', such as distances between selected landmarks. Today, the term is commonly used in general when defining true values to be used for performance evaluation of algorithms.

utilized. The next section investigates a parallel-line-based approach which allows a further reduction of the dimensionality of the error function, simplifies the computational complexity by using linear geometric features only, and proved to be less sensitive to errors compared to the first two approaches (only as specified above; further refinements might be able to change these relations).

5.4.5 Parallel-line-based Approach

This section calibrates the off-axis distance R and principal angle ω using an approach that has been widely used for sensor-matrix camera calibration: we call it the parallel-line-based approach.

The general intention is to find a single linear equation that links 3D geometric scene features to the image cylinder such that (by providing sufficient scene measurements) we are able to calibrate R and ω with high accuracy. Suitable features are, for example, distances, lengths, or angular configurations of straight line segments. This section describes two alternatives for such an approach, one based only on parallel line segments, and another which also takes orthogonal configurations of straight lines into consideration.

We assume there are at least three straight line segments in the real scene captured (e.g., straight edges of doors or windows), which are parallel to the axis of the associated image cylinder. (The latter assumption can normally be satisfied by using a "bulls-eye" or a more advanced leveling device.)

For each straight line segment we assume that both end points are visible (from the camera) and identifiable in the panoramic image, and that we have an accurate measurement of the distance between these two end points. The projected line segment (in the panoramic image) should ideally be in a single image column, and we will assume this.

Furthermore, for each of the straight line segments selected we assume either (for the first alternative) that there is a second usable straight line segment in the scene which is parallel to the first, and where the distance between both lines is also accurately measurable, or (for the second alternative) that there exist two more usable parallel straight line segments such that all three segments form an orthogonal configuration.

The focal length f (in pixels) and central row j_c are again assumed to be already known (from the first step of the calibration procedure).

First Alternative: Distance Constraint

Any usable straight line segment in the 3D scene is denoted by $\mathcal{L}$ and indexed where needed for the distinction of multiple lines. The (Euclidean) distance of two visible points on a line $\mathcal{L}$ is denoted by H (as in "height"). The length of a projection of a line segment on an image column i is denoted by h and measured in pixels. Examples of H_k and corresponding h_k values are illustrated in Figure 5.14(a), where $k = [1, \ldots, 5]$.

The distance D_{kl} between two parallel lines $\mathcal{L}_k$ and $\mathcal{L}_l$ is the length of a line segment that connects both and is perpendicular to them. If the distance between two straight line segments is available then we say that both lines form a *pair of lines*. A line segment may be paired up

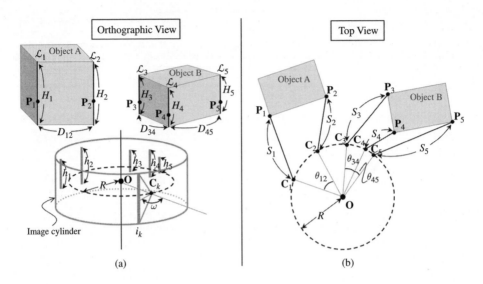

Figure 5.14 Configurations of parallel straight lines in the 3D scene and on the image cylinder: (a) orthographic side view; (b) top view.

with more than just one other line segment. Figure 5.14(a), left, shows three pairs of lines, namely $(\mathcal{L}_1, \mathcal{L}_2)$, $(\mathcal{L}_3, \mathcal{L}_4)$, and $(\mathcal{L}_4, \mathcal{L}_5)$.

Consider two straight segments $\mathcal{L}_k$ and $\mathcal{L}_l$ in 3D space and the image columns of their projections, denoted by i_k and i_l, respectively. The optical centers associated to image columns i_k and i_l are denoted by $\mathbf{C}_k$ and $\mathbf{C}_l$, respectively. The distance between the two associated image columns is $d_{kl} = |u_k - u_l|$ (in pixels). The angular distance of two image columns, associated to line segments $\mathcal{L}_k$ and $\mathcal{L}_l$, is the angle between line segments $\overline{\mathbf{C}_k\mathbf{O}}$ and $\overline{\mathbf{C}_l\mathbf{O}}$, where $\mathbf{O}$ is the center of the base circle. We denote the angular distance between a pair $(\mathcal{L}_k, \mathcal{L}_l)$ of lines by θ_{kl}. Examples of angular distances for two pairs of lines are given in Figure 5.14(b). The angular distance θ_{kl} and d_{kl} are related by

$$\theta_{kl} = \frac{2\pi d_{kl}}{W}$$

where W is the width of the panorama in pixels.

The distance S between a line segment $\mathcal{L}$ and the associated optical center (which "sees" this line segment) is defined by the length of the line segment starting at the optical center, ending on $\mathcal{L}$ and perpendicular to $\mathcal{L}$. We have that

$$S = \frac{f_\tau H}{h},$$

where f_τ is the pre-calibrated effective focal length of the camera.

Geometric Relation
We are now ready to formulate a distance constraint by combining all the geometric information previously provided. A 2D coordinate system is defined on the base plane for every pair of lines

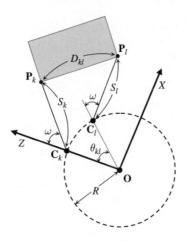

Figure 5.15   Coordinate system of a pair of lines.

$(\mathcal{L}_k, \mathcal{L}_l)$; see Figure 5.15. Note that even though all the measurements are defined in 3D space, the geometric relation of interest can be described in a 2D space since all the straight segments are assumed to be parallel to the axis of the image cylinder. The origin of the coordinate system is **O**, and the Z-axis is incident with the camera focal point $\mathbf{C}_k$. The X-axis is orthogonal to the Z-axis and is incident with the base plane. (This coordinate system coincides with the camera coordinate system previously defined, but without the Y-axis.) Such a coordinate system is defined for each pair of lines.

The position of $\mathbf{C}_k$ can now be described by coordinates $(0, R)$, and the position $\mathbf{C}_l$ can be described by coordinates $(R \sin \theta_{kl}, R \cos \theta_{kl})$. The intersection point of the line $\mathcal{L}_k$ with the base plane, denoted by $\mathbf{P}_k$, can be expressed by a sum of the vector $\overrightarrow{OC_k}$ and the vector $\overrightarrow{C_k P_k}$. Thus, we have

$$\mathbf{P}_k = \left[\begin{array}{c} S_k \sin \omega \\ R + S_k \cos \omega \end{array} \right].$$

Analogously, the intersection point of the line $\mathcal{L}_l$ with the base plane, denoted by $\mathbf{P}_l$, can be described by a sum of vectors $\overrightarrow{OC_l}$ and $\overrightarrow{C_l P_l}$. We have

$$\mathbf{P}_l = \left[\begin{array}{c} R \sin \theta_{kl} + S_l \sin(\theta_{kl} + \omega) \\ R \cos \theta_{kl} + S_l \cos(\theta_{kl} + \omega) \end{array} \right].$$

The distance D_{kl} between $\mathbf{P}_k$ and $\mathbf{P}_l$ has been measured. We have the following equation:

$$D_{kl}^2 = (S_k \sin \omega + R \sin \theta_{kl} - S_l \sin(\omega + \theta_{kl}))^2$$
$$+ (R + S_k \cos \omega - R \cos \theta_{kl} - S_l \cos(\omega + \theta_{kl}))^2.$$

This equation can be expanded and rearranged as follows:

$$
\begin{aligned}
D_{kl}^2 &= S_k^2 \sin^2 \omega + R^2 \sin^2 \theta_{kl} + S_l^2 \sin^2(\omega + \theta_{kl}) - 2S_k R \sin \omega \sin \theta_{kl} \\
&\quad - 2S_k S_l \sin \omega \sin(\omega + \theta_{kl}) + 2RS_l \sin \theta_{kl} \sin(\omega + \theta_{kl}) \\
&\quad + R^2 + S_k^2 \cos^2 \omega + R^2 \cos^2 \theta_{kl} + S_l^2 \cos^2(\omega + \theta_{kl}) \\
&\quad + 2RS_k \cos \omega - 2R^2 \cos \theta_{kl} - 2RS_l \cos(\omega + \theta_{kl}) - 2S_k R \cos \omega \cos \theta_{kl} \\
&\quad - 2S_k S_l \cos \omega \cos(\omega + \theta_{kl}) + 2RS_l \cos \theta_{kl} \cos(\omega + \theta_{kl}) \\
&= S_k^2 + 2R^2 + S_l^2 + 2RS_k \cos \omega - 2R^2 \cos \theta_{kl} \\
&\quad - 2S_l R \cos(\omega + \theta_{kl}) - 2S_k R(\sin \omega \sin \theta_{kl} + \cos \omega \cos \theta_{kl}) \\
&\quad - 2S_k S_l(\sin \omega \sin(\omega + \theta_{kl}) + \cos \omega \cos(\omega + \theta_{kl})) \\
&\quad + 2S_l R(\sin \theta_{kl} \sin(\omega + \theta_{kl}) + \cos \theta_{kl} \cos(\omega + \theta_{kl})) \\
&= S_k^2 + S_l^2 + 2R^2(1 - \cos \theta_{kl}) + 2S_k R \cos \omega \\
&\quad - 2S_l R(\cos \omega \cos \theta_{kl} - \sin \omega \sin \theta_{kl}) - 2S_k R(\sin \omega \sin \theta_{kl} + \cos \omega \cos \theta_{kl}) \\
&\quad - 2S_k S_l \cos \theta_{kl} + 2S_l R \cos \omega \\
&= S_k^2 + S_l^2 + 2R^2(1 - \cos \theta_{kl}) + 2(S_k + S_l) R \cos \omega \\
&\quad - 2(S_k + S_l) R \cos \omega \cos \theta_{kl} - 2(S_k - S_l) R \sin \omega \sin \theta_{kl} \\
&\quad - 2S_k S_l \cos \theta_{kl}.
\end{aligned}
$$

Finally, we obtain that

$$
\begin{aligned}
0 = &\ (1 - \cos \theta_{kl}) R^2 \\
&+ (S_k + S_l)(1 - \cos \theta_{kl}) R \cos \omega \\
&- (S_k - S_l) \sin \theta_{kl} R \sin \omega \\
&+ \frac{S_k^2 + S_l^2 - D_{kl}^2}{2} - S_k S_l \cos \theta_{kl}.
\end{aligned}
\tag{5.13}
$$

Error Function
Basically we use equation (5.13) as an error function. The values of S_k, S_l, D_{kl}, and θ_{kl} are known. Thus, equation (5.13) can be arranged into the linear form

$$
A_1 X_1 + A_2 X_2 + A_3 X_3 + A_4 = 0
$$

with coefficients A_n, $n = 1, 2, 3, 4$, defined as follows:

$$
\begin{aligned}
A_1 &= 1 - \cos \theta_{kl}, \\
A_2 &= (S_k + S_l)(1 - \cos \theta_{kl}), \\
A_3 &= -(S_k - S_l) \sin \theta_{kl}, \\
A_4 &= \frac{S_k^2 + S_l^2 - D_{kl}^2}{2} - S_k S_l \cos \theta_{kl}.
\end{aligned}
$$

For the three linearly independent variables X_n, $n = 1, 2, 3$, we have

$$X_1 = R^2,$$
$$X_2 = R \cos \omega,$$
$$X_3 = R \sin \omega.$$

In this case we can solve for absolute (not just relative) values R and ω by using all three equations. (If more than three equations are provided then it is possible to apply a linear least-squares technique.) The values of R and ω may be calculated by

$$R = \sqrt{X_1} = \sqrt{X_2^2 + X_3^2}$$

and

$$\omega = \arccos \left(\frac{X_2}{\sqrt{X_1}} \right) = \arcsin \left(\frac{X_3}{\sqrt{X_1}} \right) = \arccos \left(\frac{X_2}{\sqrt{X_2^2 + X_3^2}} \right).$$

The given dependencies among variables X_1, X_2, and X_3 define multiple solutions of R and ω. To tackle this multiple-solution problem, we constrain the parameter estimation process further by

$$X_1 = X_2^2 + X_3^2$$

which is valid because

$$R^2 = (R \cos \omega)^2 + (R \sin \omega)^2.$$

Assume that N copies of equation (5.13) are given. We wish to minimize the expression

$$\sum_{n=1}^{N} (A_{1n} X_1 + A_{2n} X_2 + A_{3n} X_3 + A_{4n})^2 \tag{5.14}$$

subject to the equality constraint $X_1 = X_2^2 + X_3^2$, where the values of A_{1n}, A_{2n}, A_{3n}, and A_{4n} are calculated based on measurements in the real scene and in the image. We also known that $X_1 = R^2$, $X_2 = R \cos \omega$, and $X_3 = R \sin \omega$. Now, the values of R and ω can be uniquely (!) calculated as

$$R = \sqrt{X_1}$$

and

$$\omega = \arccos \left(\frac{X_2}{\sqrt{X_1}} \right).$$

Note that even though the additional constraint forces the use of a non-linear optimization method, the accuracy of the method remains comparable to that of a linear parameter estimation procedure.

Second Alternative: Orthogonality Constraint

We say that three parallel line segments $\mathcal{L}_k$, $\mathcal{L}_l$, and $\mathcal{L}_m$ are *orthogonal* if and only if the plane incident with $\mathcal{L}_k$ and $\mathcal{L}_l$ and the plane incident with $\mathcal{L}_l$ and $\mathcal{L}_m$ are orthogonal. (It follows that $\mathcal{L}_l$

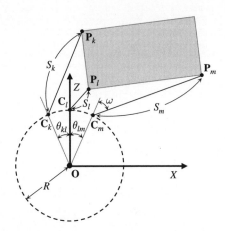

Figure 5.16   Coordinate system for three orthogonal line segments.

is the intersection of both planes.) For example, line segments $\mathcal{L}_3$, $\mathcal{L}_4$, and $\mathcal{L}_5$ in Figure 5.14(a) are orthogonal.

Consider three orthogonal line segments $\mathcal{L}_k$, $\mathcal{L}_l$, and $\mathcal{L}_m$ in 3D space. We have the measured values of S_k, S_l, S_m, θ_{kl}, and θ_{lm}, obtained the same way as in the case of the distance constraint. We define a 2D coordinate system for each group of three orthogonal line segments; see Figure 5.16 for segments $(\mathcal{L}_k, \mathcal{L}_l, \mathcal{L}_m)$.

The position of $\mathbf{C}_l$ is given by coordinates $(0, R)$, the position of $\mathbf{C}_k$ by coordinates $(-R \sin \theta_{kl}, R \cos \theta_{kl})$, and the position of $\mathbf{C}_m$ by coordinates $(R \sin \theta_{lm}, R \cos \theta_{lm})$. Intersection points of line segments $\mathcal{L}_k$, $\mathcal{L}_l$, and $\mathcal{L}_m$ (or their linear extensions) with the base plane are denoted by $\mathbf{P}_k$, $\mathbf{P}_l$, and $\mathbf{P}_m$, respectively. We have that

$$\mathbf{P}_k = \begin{bmatrix} -R \sin \theta_{kl} + S_k \sin(\omega - \theta_{kl}) \\ R \cos \theta_{kl} + S_l \cos(\omega - \theta_{kl}) \end{bmatrix},$$

$$\mathbf{P}_l = \begin{bmatrix} S_l \sin \omega \\ R + S_l \cos \omega \end{bmatrix},$$

and

$$\mathbf{P}_m = \begin{bmatrix} R \sin \theta_{lm} + S_m \sin(\theta_{lm} + \omega) \\ R \cos \theta_{lm} + S_m \cos(\theta_{lm} + \omega) \end{bmatrix}.$$

The vectors $\overrightarrow{\mathbf{P}_k \mathbf{P}_l}$ and $\overrightarrow{\mathbf{P}_l \mathbf{P}_m}$ are orthogonal; thus we have the following equation:

$$\begin{aligned} 0 = & (-R \sin \theta_{kl} + S_k \sin(\omega - \theta_{kl}) - S_l \sin \omega) \\ & \times (R \sin \theta_{lm} + S_m \sin(\omega + \theta_{lm}) - S_l \sin \omega) \\ & + (R \cos \theta_{kl} + S_l \cos(\omega - \theta_{kl}) - R - S_l \cos \omega) \\ & \times (R \cos \theta_{lm} + S_m \cos(\omega + \theta_{lm}) - R - S_l \cos \omega). \end{aligned}$$

This equation can be transformed as follows:

$$
\begin{aligned}
0 = & (1 - \cos\theta_{kl} - \cos\theta_{lm} + \cos(\theta_{kl} + \theta_{lm}))R^2 + (2S_l - (S_l + S_m)\cos\theta_{kl} \\
& - (S_k + S_l)\cos\theta_{lm} + (S_k + S_m)\cos(\theta_{kl} + \theta_{lm}))R\cos\omega \\
& + ((S_m - S_l)\sin\theta_{kl} + (S_l - S_k)\sin\theta_{lm} + (S_k - S_m)\sin(\theta_{kl} + \theta_{lm}))R\sin\omega \\
& + S_l^2 + S_k S_m \cos(\theta_{kl} + \theta_{lm}) - S_k S_l \cos\theta_{kl} - S_l S_m \cos\theta_{lm}.
\end{aligned}
\tag{5.15}
$$

The values of S_k, S_l, S_m, θ_{kl}, and θ_{lm} in equation (5.15) are known. The equation can be expressed in the following linear form

$$
B_1 X_1 + B_2 X_2 + B_3 X_3 + B_4 = 0
$$

where B_n, $n = 1, 2, 3, 4$, are the following coefficients:

$$
\begin{aligned}
B_1 &= 1 - \cos\theta_{kl} - \cos\theta_{lm} + \cos(\theta_{kl} + \theta_{lm}), \\
B_2 &= 2S_l - (S_l + S_m)\cos\theta_{kl} - (S_k + S_l)\cos\theta_{lm} + (S_k + S_m)\cos(\theta_{kl} + \theta_{lm}), \\
B_3 &= (S_m - S_l)\sin\theta_{kl} + (S_l S_k)\sin\theta_{lm} + (S_k - S_m)\sin(\theta_{kl} + \theta_{lm}), \\
B_4 &= S_l^2 + S_k S_m \cos(\theta_{kl} + \theta_{lm}) - S_k S_l \cos\theta_{kl} - S_l S_m \cos\theta_{lm}.
\end{aligned}
$$

Furthermore, X_n, $n = 1, 2, 3$, are three linearly independent variables, with

$$
\begin{aligned}
X_1 &= R^2, \\
X_2 &= R\cos\omega, \\
X_3 &= R\sin\omega.
\end{aligned}
$$

Note that this linear form is the same as for the distance constraint except that the coefficients are different. This allows us to reuse the minimization approach as defined by equation (5.14) and the way of calculating R and ω as specified for the first alternative.

5.4.6 Experimental Results

We briefly report on experiments with the WAAC line camera (Wide-Angle Air- (see Figure 5.17). The specifications of the model used are as follows. There are three (monochromatic) sensor lines on the focal plate of the camera, defining angles $\omega = 335°$, $\omega = 0°$ (the *center line*), and $\omega = 25°$. (For airborne missions these lines can be identified as the forward-looking line, nadir or center line, and backward-looking line.) Rotating a WAAC on a tripod or turntable would actually allow the capture of three panoramic images during one rotation. In the following experiments only image data captured via the backward line have been used. Each sensor line (image line) has 5,184 CCD cells (pixels). The effective focal length of the camera is 21.7 mm for the center line, and 23.94 mm for the backward or forward line. The size (edge length) of one of WAAC's CCD cells is 0.007 mm. Thus, the value of f_τ (for the backward line) is equal to 3,420 pixels. The camera was mounted on a turntable which supported an extension arm allowing R to be up to 1.0 m.

Figure 5.17   WAAC built by DLR in the 1990*s*.

Figure 5.18   Test panorama with eight highlighted and indexed pairs of lines.

Figure 5.18 shows a panoramic image taken in 2001 in a seminar room of the Institute for Space Sensor Technology and Planetary Exploration, at the DLR in Berlin. This seminar room is about 120 m^2 in size. The image has a resolution of 5,184 × 21,388 pixels. Eight pairs of lines are highlighted in the figure and indexed. They are used for estimating R and ω based on the distance constraint. The value of R was set to be 100 mm. The principal angle ω is equal to 155° in the sense of the definition of our general panoramic sensor model.[7] The lengths of line segments used were measured, with an expected error of no more than 0.5% of the length readings. The data for these eight pairs of lines (shown in Figure 5.18) are summarized in Table 5.4.

The calibration proceeds as described above using an optimization method of sequential quadratic programming for estimating R and ω, in particular for minimizing equation (5.14). Results for the specified example are as follows: the use of all pairs of lines produces $R = 103.2$ mm and $\omega = 161.68°$; for pairs {2,3,4,7,8}, $R = 108.7$ mm and $\omega = 151.88°$; and for pairs {2,4,8}, $R = 108.3$ mm and $\omega = 157.21°$. In general, such experiments showed that calibration accuracy is influenced by sample selection (e.g., aim at a uniform distribution of segments in the scene pictured) and quality of sample data (e.g., 0.5% error in length readings as in the given example is insufficient for high-accuracy calibration).

The error sensitivity of both alternatives of the parallel-line based approach can be evaluated by experiments using synthetic data. For example, we use (in an assumed scene geometry) one pair of lines for the first alternative, and one triple of orthogonal lines for the second.

---

[7] This would be 25° in terms of the WAAC specification as given in (Reulke and Scheele, 1998).

Index	$H_k = H_l$ (m)	h_k (pixels)	h_l (pixels)	D_{kl} (m)	d_{kl} (pixels)
1	0.0690	91.2	133.8	1.4000	1003.1
2	0.6320	600.8	683.0	1.0000	447.3
3	0.5725	351.4	367.4	1.5500	490.5
4	1.0860	1269.0	1337.6	0.6000	360.9
5	0.2180	273.0	273.6	0.2870	180.1
6	0.0690	81.8	104.2	1.4000	910.5
7	0.5725	318.0	292.0	1.5500	398.2
8	1.3300	831.2	859.4	1.3400	422.5

Table 5.4 Data for the eight pairs of lines shown in Figure 5.18.

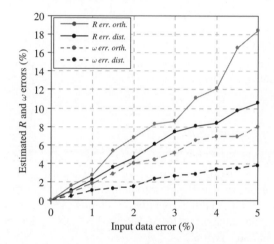

Figure 5.19 Error sensitivities for R and ω when using either the distance constraint (plotted in black) or the orthogonality constraint (plotted in gray): the distance constraint performs better in both cases. The error of R is depicted in solid lines and the error of ω in dashed lines.

The simulated ground-truth data are $R = 100$ mm and $\omega = 155°$. We introduce errors to the ground-truth data S_k, D_{kl}, and θ_{kl}, independently and with a maximum of 5% additive random (normally distributed) noise. The range of S_k is from 1 m to 8 m, and the range of θ_{kl} is from 4° to 35°. The sample size is 8. Mean errors (for 100 trials in each case) are shown in Figure 5.19. Results suggest that estimated parameters using the orthogonality constraint are more sensitive to error than when using the distance constraint. The errors of the estimated parameters increase (only) linearly with respect to input errors for both cases. A combination of both constraints might be useful.

Table 5.2 provided a performance comparison of the point-based, image correspondence, and parallel-line-based calibration approaches for the non-linear part of intrinsic parameter calibration. The parallel-line-based approach performs best among these three (for their given state of specification). The parallel-line-based approach may also be based on a designed calibration object. It could be considered for inclusion in a general least-square error calibration procedure.

5.5 Error Components in LRF Data

Note that 3D points, calculated at a single LRF attitude, are inaccurate, and we assume an error $\varepsilon > 0$ describing a sphere $U_\varepsilon(\mathbf{p})$ around a measured 3D point $\mathbf{p}$. The correct 3D point is in this sphere. In practice, we can assume that ε is specified by several components such as:

- a distance error (caused by the LRF measurement unit),
- the eccentricity of the scan center,
- the incident angle (angle between the laser ray and the surface),
- different surface material properties, such as of wood or metal,
- a collimation axis error (error of the principal axis),
- a vertical or horizontal axis error,
- the trunnion axis error (i.e., the oscillation around axes),
- and a scale factor.

Calibration of an LRF also involves an understanding of inaccuracies at various distances, for various surfaces, or under specific conditions.

5.5.1 LRF Used in Experiments

Figure 5.20 shows an IMAGER 5003 laser scanner. This system is based on the phase-shift measurement principle. Scans of an LRF are generated point by point, and are uniformly defined in two dimensions, vertically by a rotating a deflecting mirror (defining a "sensor line" in the previous section), and horizontally by rotating the whole measuring system. For example, the vertical scan range of the IMAGER 5003 is 310° (which leaves 50° uncovered), and the horizontal scan range is 360°.

Actually, a horizontal scan range of 180° is sufficient to measure each visible 3D point once, because an LRF typically scans "overhead" (180°). However, it is of benefit to scan the full 360° horizontally, thus having each visible 3D point scanned twice. The redundancy can be used for calibration or enhanced measurement accuracy. Figure 5.21 shows a raw data set without such redundancy: both images combined show a 310° horizontal angle times a 180° vertical angle.

5.5.2 Error Measurement

We measured errors of distances on a dark gray, planar surface (with 20% of maximum intensity), where scans hit this surface approximately normally (i.e., perpendicular to the surface's tangential plane). In related work, Schulz and Ingensand (2004) provided a detailed analysis of accuracy of LRF distances by using an interferometer.[8] This allows measurements with an accuracy of 1 μm, a thousand times more accurate than the measured distances of the laser

---

[8] This is an instrument that uses the principle of interference of electromagnetic waves for purposes of measurement.

Figure 5.20 Left: Zoller and Fröhlich IMAGER 5003 with (from left to right) electronic module, optics (basically a golden mirror), and mechanical module (for motion control and transmission). Right: visualization of both scan directions (vertical and horizontal).

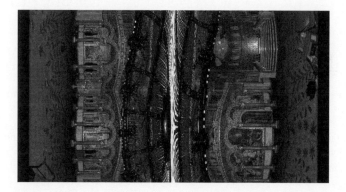

Figure 5.21 Raw data (without redundancy) of an uncalibrated LRF scan. The uncalibrated image geometry is caused by motions of the mirror or the LRF (one row corresponds to one mirror rotation).

scanner. Thus, they can be treated as *nominal distances*. LRF distance errors are illustrated in Figure 5.22. The standard deviation and the differences between nominal distances and measured distances (mean distance of 1,000 measurements) are shown in this figure; it can be seen that absolute errors are about 5 mm or less, increase towards a range of about 20 meters and then decrease again slightly. Some systematic corrections (e.g., addition of a constant of

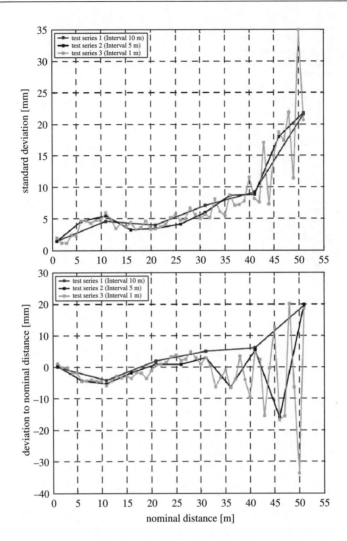

Figure 5.22 Standard deviation (top) and deviations between minimum and maximum of measured distances (mean values of 1,000 measurements) and nominal distances (bottom).

approximately 5 mm) or a negative scaling factor (for distances greater than 20 m), can be used to be specified by calibration. The distance accuracy of the laser scanner only influences the longitudinal direction.

Photogrammetry specifies ways of calibrating rotational measuring devices (e.g., theodolite systems) and how to measure errors along rotation axes. These errors are classified into *vertical, horizontal,* or *principal collimation errors,* and *scale factors.* Figure 5.23 illustrates the principal Z-axis as an axis orthogonal both to the corresponding vertical rotation axis V and the horizontal rotation axis (*tilt axis*) K.

Note that we have to consider the elimination order of the errors because of interdependencies between errors. To measure a 3D point twice at one LRF position means typically that a 3D

Plate 1 Fan Kuan's *Travelers among Streams and Mountains*, painted around the year 1000 and over 2 metres tall. Courtesy of the National Palace Museum, Taipei, Taiwan, Republic of China.

Plate 2 *Along the River during the Ch'ing-ming Festival*, painted during the Ch'ing dynasty, following a design created by Chang Tse-tuan in the early 12th century (National Palace Museum, Taipei). Bottom: copy of painting (original length: 1,152.8 cm). Middle row: a part of the painting. Top row: further magnification of a part of the painting. Courtesy of the National Palace Museum, Taipei, Taiwan, Republic of China.

Plate 3   A segment of the Bayeux Tapestry, which is over 70 m long. It is an embroidery, stitched, not woven, in woollen yarn on linen.

Plate 4   Top: adjacent images are registered. Below: pixel values are unified to hide lighting variations. From Chen and Klette (1999).

Plate 5   Interface of an 1998 interactive anaglyph panorama stereo player: the globe is an example of an independent object which can be rotated independent of the global view. Anaglyphic eyeglasses are required for stereo viewing. From Wei et al. (1998).

Plate 6   A panoramic image captured by a rotating line camera in 2001. The vertical resolution of the panoramic image is more than 10,000 color pixels. A reproduction of it could be close to 4 meters in height at a resolution of 72 pixels per inch.

Plate 7 Part of an anaglyphic panorama (Tamaki Campus, University of Auckland, 2002), generated using a symmetric panorama. Proper 3D viewing requires anaglyphic eyeglasses.

Plate 8 Illustration of projecting a sphere onto a tangential plane.

Plate 9 Top: Panoramic image of the Gendarmenmarkt in Berlin; the unrolled cylinder shows straight lines of the tile pattern of this square as curves. Bottom: Rectified, central-perspective view, by mapping onto a tangential plane.

Plate 10 Correct stereo projection of the same hall as shown in Plate 12; the anaglyph uses red for the left eye.

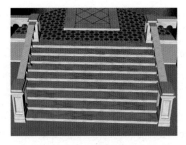

Plate 11   Corrected stairs: image-based and using planarity tests in LRF data, with some minor final manual touch-ups.

Plate 12   Panoramic image data have been fused (see colored subwindow) with the range-scan shown (throne room at Neuschwanstein castle).

Plate 13   Left: 3D model with mapped texture (intensity channel of the LRF scan). Right: mapped texture based on recorded panoramic images.

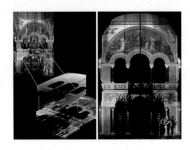

Plate 14   Left: panoramic image projected into a defined orthoplane (based on a DSM). Right: unfinished orthoimage (still with shadows, etc.).

Plate 15 Anaglyphic 360° panorama (Neuschwanstein, Germany). 3D viewing requires anaglyphic eyeglasses.

Plate 16 Orthophoto from HRSC data (Nymphenburg, Munich, Germany). Courtesy by H. Hirschmüller.

Plate 17 3D model extracted from HRSC data, with a (partial) texture map; texture captured by a terrestrial panoramic camera.

Plate 18 Untextured (top) and textured (bottom) 3D close-up on the Berlin Reichstag. Courtesy by H. Hirschmüller.

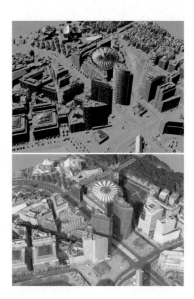

Plate 19 Untextured (top) and textured (bottom) 3D aerial view of Berlin around the Sony Center. Courtesy by H. Hirschmüller.

Plate 20 A panoramic camera (*périphote*), built in 1901, with a lens rotating 360° (manufacturer: Lumière Frères, Paris). Reproduced by permission of George Eastman House, Rochester, NY (see Figure 1.3).

Plate 21 A high-accuracy panoramic image (view towards Rangitoto Island, Bastion Point, Auckland, 2002). Each column of this image consists of 10,200 color pixels; the size of the panorama is about 4 GB (see Figure 1.12).

Plate 22 A view into the 3D reconstructed (using multiple laser range-scans) throne room of Neuschwanstein castle, with color texture generated by multiple, very-high resolution panoramic scans. Spatial resolution on surfaces is of the order of 10 mm, and texture resolution is about 1 mm (see Figure 1.18).

Plate 23 This panoramic 380° image was taken in 2001 at Auckland's Northcote Point, and shows a view near the Harbour Bridge towards Auckland's CBD. The labeled rectangular window of this image is shown again in Plate 24(a), (see Figure 2.21).

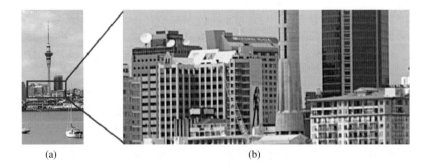

(a) (b)

Plate 24 (a) The rectangular window in Plate 23 showing a part of Auckland's CBD. (b) Zooming in further reveals this detail of 500 × 200 pixels (see Figure 2.22).

Plate 25 Two windows within one panoramic image captured for surveillance of Braunschweig airport: airport tower with distortions caused by vibrations (left), and people ("ghosts") move while the rotating line sensor is scanning in their direction (right) (see Figure 2.23).

Plate 26 Anaglyph of a symmetric panoramic pair. It requires a presentation in color and anaglyphic eyeglasses for 3D viewing (see Figure 4.9).

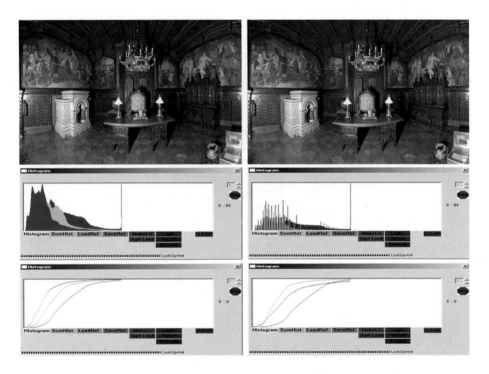

Plate 27 Reference image and its histogram and cumulative histogram (left), and a second image (right) with its histograms, taken under different lighting conditions (see Figure 5.8).

Plate 28   Transformed image with its histogram and lookup table (see Figure 5.9).

Plate 29   Corrected airport tower by using the NCC algorithm. The arrow points to a location where the restriction of using only relatively small regions for matching (see text) caused a false correlation (see Figure 5.11).

Plate 30   Huang et al. (2001a) report on the acquisition of images with a rotating line camera in New Zealand, and give several examples. The figure shows a resolution-reduced part of one of these (single-line) 7 GB panoramas (view from the roof of the Auckland Museum) (see Figure 7.1).

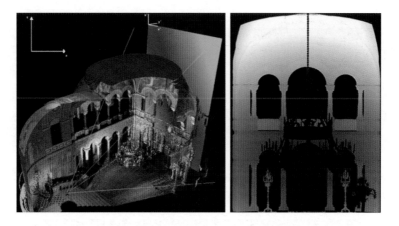

Plate 31   Creating a digital surface model (depth map). A defined orthoplane "behind" the generated 3D data (left). Gray-value encoded and orthogonally projected range data for those surface points which are at a distance of 2 metres from the defined plane (right) (see Figure 9.5).

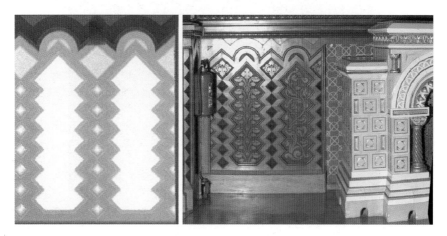

Plate 32   A digital surface model (left) used for bump mapping for the timber panel (right) around the fireplace (see Figure 9.6).

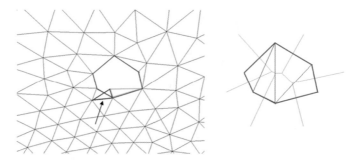

Plate 33   Left: triangulated point cloud (Delaunay) with a hole in the mesh at the center; the arrow shows an edge which remains to be filtered (see text). Right: Delaunay triangulated hole; the green lines are the Voronoi diagram (see Figure 9.10).

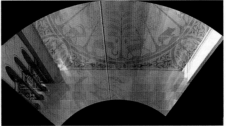

Plate 34 Panoramic input image: perspective visualization (left); orthogonal visualization (right) (see Figure 9.16).

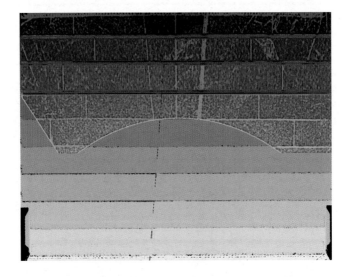

Plate 35 DSM (gray) with an overlayed rectified panoramic image (blue) (see Figure 9.17).

Plate 36 A low-resolution overview of individualy triangulated surfaces generated for Neuschwanstein castle within a project of Illustrated Architecture, Berlin (see Figure 9.18).

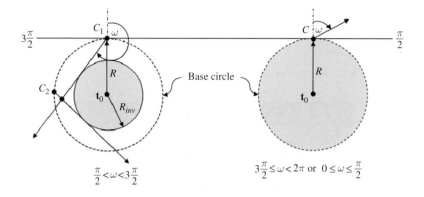

$$\frac{\pi}{2} < \omega < 3\frac{\pi}{2}$$

$$3\frac{\pi}{2} \le \omega < 2\pi \text{ or } 0 \le \omega \le \frac{\pi}{2}$$

Plate 37 Inward (left) and outward (right) cases (see Figure 10.1).

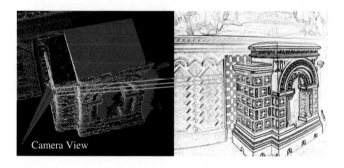

Camera View

Plate 38 Illustration of the raytracing approach using a stencil buffer. Right: Triangles closest to the camera are masked in the stencil buffer. Left: Lines from right to left indicate such a masked triangle in the left image; triangles which are behind this triangle do not obtain valid texture coordinates using the stencil test (see Figure 10.3).

Plate 39 A selection of airborne cameras (from left to right): Lawrence's camera, Leica's ADS40, Vexel's UltraCam, and two DLR cameras, the MFC and HRSC – the latter is currently in use on a Mars mission (see Figure 10.9).

Plate 40   Textured 2.5D top view of the scene shown in Figure 10.10 (see Figure 10.14).

Plate 41   Textured 3D view of the scene shown in Figures 10.10 and 10.14 (see Figure 10.15).

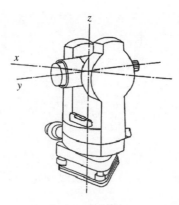

Figure 5.23   Theodolite with three axes: the German terms *Zielachse* (roll axis), *Kippachse* (pitch axis) and *Stehachse* (yaw axis) specify in photogrammetry the principal y-axis and an orthogonal x-axis and z-axis. A rangefinder measures along a variable y-axis, which may be affected by horizontal (i.e., along the x-axis) or vertical (i.e., along the z-axis) errors.

point on a surface is first measured, then both rotation axes are turned by about 180°, and the same surface point is measured again. We use this redundancy (each visible point is scanned twice) for calibration.

To determine the *Zielachsfehler* (i.e., errors caused by the principal axis), 3D points are measured twice near the *equator* (i.e., the "horizon" of the LRF data set). The Zielachsfehler describe the deviation of the principal axis with respect to an ideal line, orthogonally to the horizontal axis.

For the determination of the *Kippachsfehler* (i.e., errors caused by the horizontal axis) the Zielachsfehler has to be known first. The Kippachsfehler describes an error between the vertical axis and the horizontal axis. Ideally, both axes have to be orthogonal. To determine the Kippachsfehler, a point near the *pole* has to be measured twice. The pole column indicates the "north pole" of a theodolite system. If the LRF is aligned horizontally then the pole column indicates the correct *zenith*, which is the point on the scanned surface exactly "on top" of the LRF (i.e., $\vartheta = 0$), and this point will be scanned repeatedly for any rotation of the LRF at this position. It is important to confirm that the zenith is actually uniquely defined in 3D space for the whole combined scan of 360°.

The scale factor needs to be determined for each scan. The scale factor provides the largest or most critical error for the LRF used (which is about 1 mm per 1 m distance). The scale factor depends on temperature changes and is determined via control points or corresponding points in different scans.

The collimation axis error is about 0.003° whereas the horizontal axis error is about 0.03°. The larger value of the horizontal axis error is due to limited accuracy when leveling the laser scanner (thus, it is recommended to use very precise levels and compensators). Figure 5.24 depicts a calibrated LRF scan (i.e., after removing errors as discussed). The inaccuracies of the LRF as described can be due to errors of the distance measurement unit, or to mechanical errors. LRF measurements can be basically either reflectorless, or reflector-based (e.g., using a prism). Measuring with reflectors is much more accurate because of a unique returning ray.

Figure 5.24   Calibrated LRF image in spherical coordinates.

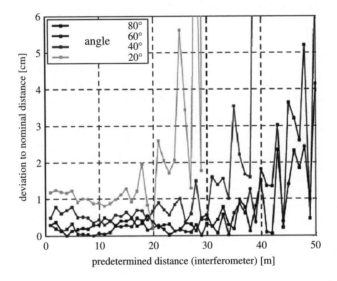

Figure 5.25   Dependencies between incident angle of the laser ray and the surface.

For measurements on natural or synthetic surfaces without reflectors or prisms, the intensity E of the returning ray is crucial for the achievable accuracy of distance measuring. Note that the intensity E is inversely proportional to the square of the distance s, formally expressed by $E \sim \frac{1}{s^2}$.

The intensity is influenced by different parameters (e.g., the angle between laser ray and surface tangential; we call this the *incident angle*). Figure 5.25 illustrates dependencies between the incident angle and the accuracy. An optimal incident angle is orthogonal to the tangential plane at the surface point. Furthermore, the intensity is influenced by properties of surface materials. Resulting errors are illustrated in Figure 5.26. In the case of styrofoam materials, the laser ray "sinks" into the surface. The intensity returned is too large for metallic surfaces due to specularities.

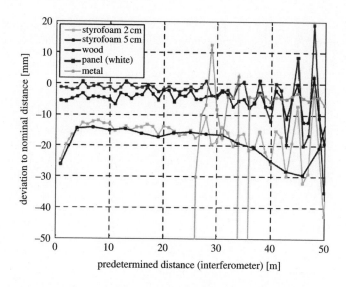

Figure 5.26 Analysis with respect to different surface materials.

5.6 Exercises

5.1. Assume an RGB CCD line camera with 10,000 pixels and a spacing of $\Delta x = 0.154$ mm between color lines; also assume a pixel size of $\tau = 0.007$ mm and an optics with $f = 35$ mm. Calculate the color shift in pixels between RGB lines for a center and border pixel of the line, for different object distances ($h = 5,000$ mm, $h = 15,000$ mm, $h = 25,000$ mm, and $h = 50,000$ mm) for the following cases:

1. ideal cylindrical single-projection-center panorama,
2. ideal cylindrical multi-projection-center panorama, with $R = 500$ mm, and
3. conic single-projection-center panorama with $\xi = 45°$.

5.2. Suppose that the the pixel positions of a multi-sensor-line camera ($\omega = \pm 20°$) are measured with a collimator. The static axis of the manipulator (i.e., a turntable rotates the camera to allow illumination of a selected pixel) is vertical in 3D space, and called the α-axis. The second vertical axis is the β-axis (see Figure 5.3).

Angles α and β measurements are given here for some pixels:

Pixel no.	CCD1	α	β	CCD2	α	β	CCD3	α	β
first		-20	-45		0	-45		$+20$	-45
center		-20	0		0	0		$+20$	0
last		-20	$+45$		0	$+45$		$+20$	$+45$

1. How are these three lines positioned on the focal plane? How is the focal plane mounted on the manipulator? Draw a simple sketch of this three-CCD-line focal plane.
2. See the data in the above table; can we assume distortion-free imaging?

3. Calculate the metric pixel position (x, y) for an ideal pinhole-type model (with $f = 35$ mm) using the given viewing direction defined by α and β. Check your result against your answer to part (2).

4. What happens to the measured viewing direction if the projection center of the camera is translated towards the collimator beam?

5.3. [Possible lab project] You may use any matrix-type image (e.g., which you used for one of the previous exercises). Implement a program which estimates the normalized cross-correlation factor between two adjacent image columns. Shift these lines pixelwise up or down, such that a maximum correlation between resulting lines is achieved. Do this for the whole image and compare your results with the statement in Section 5.2.4 about a "best correlation".

5.4. [Possible lab project] Implement a program which radiometrically equalizes one image to another. Input images are easily created for a scene, for example, by varying illumination (use lights of different color temperatures), or by using an image tool that allows colors to be changed in some non-linear way.

5.5. [Possible lab project] Implement a program which estimates the camera's attitude by using calibration marks distributed within your scene. Now make a slight change to the intrinsic parameters of your camera, and recalculate the extrinsic parameters. Do this for a common matrix-type camera and a panoramic camera model (if the latter is not available, then discuss this case theoretically using synthetic 3D data).

5.7 Further Reading

For close-range focus, see, for example, the *bb.Fokal* project (Scheele et al., 2004) which used holographic optical elements for calibrating cameras, which is an alternative method to the least-squares approach. For the phase-shift measurement principle, see Heinz et al. (2001). The accuracy of LRF scans, in particular with respect to volume measurements, is discussed in Zhang et al. (2005). The WAAC line camera is described in Reulke and Scheele (1998).

On estimating fundamental and essential matrices for sensor-matrix cameras, see, among others, Liu et al. (2005) and Sugaya and Kanatani (2007) for recent studies.

A nonlinear least-squares optimization method is described by Gill et al. (1981). They discuss the influence of the "robustness" of numerical methods on the possible suppression of error-sensitivity; they also specify the sequential quadratic programming optimization method.

The length of a projected line segment in an arbitrary position in the panoramic image, forming a curved line, can be accurately measured with methods as described, for example, in Klette and Rosenfeld (2004). Schulz and Ingensand (2004) provided a detailed analysis of the accuracy of LRF distances by using an interferometer. Däumlich and Steiger (2002) suggested that both rotation axes are turned by about 180°, and the same surface point is measured again. Figures 5.22, 5.25, and 5.26 are from Schulz and Ingensand (2004).

6

Spatial Sampling

Spatial sampling describes how a 3D space is sampled by stereo pairs of images, without considering geometric or photometric complexities of 3D scenes. This chapter presents studies on spatial sampling for stereo panoramas which are a pair of panoramic images which only differ in the chosen value of the principal angle ω. Symmetric panoramas are a special case of stereo panoramas. Recall from our previous definition of symmetric panoramas that this is a pair of stereo panoramas whose associated principal angles sum to 360°.

6.1 Stereo Panoramas

Consider a stereo panorama $E_{\mathcal{P}_R}(R, f, \omega_R, \gamma)$ and $E_{\mathcal{P}_L}(R, f, \omega_L, \gamma)$, with $0° \leq \omega_R < \omega_L \leq 360°$. Without loss of generality, we call the panorama that is associated with the smaller (larger) value of the principal angle the right (left) image.

In the symmetric case (i.e., $\omega_R = \omega$ and $\omega_L = 360° - \omega$), we only consider ω between 0° and 180°. In order to reduce repetition, only the right image parameters will be referred to in the analysis and calculations throughout this chapter. The chapter will mostly consider stereo panoramas which are symmetric; non-symmetric cases are explicitly identified as such.

Section 6.2 is about sampling structures of stereo panoramic cameras, in which mainly the spatial distribution of samples will be discussed. Section 6.3 turns to sampling resolution and addresses the question how many samples can be acquired by a pair of stereo panoramas with image dimensions $W \times H$. Section 6.4 computes different sample distances and discusses spatial dependencies of sample density.

6.2 Sampling Structure

A *spatial sample* is defined by an intersection of two projection rays, one normally defined by the right and the other by the left image pixel. However, exceptions may occur in the panoramic image case based on our panoramic camera model: two projection rays forming a spatial sample may be defined by two image pixels on the same panoramic image (see Section 6.2.2). This interesting phenomenon implies that two corresponding points may actually lie on the

Panoramic Imaging: Sensor-Line Cameras and Laser Range-Finders   F. Huang, R. Klette, and K. Scheibe
© 2008 John Wiley & Sons, Ltd

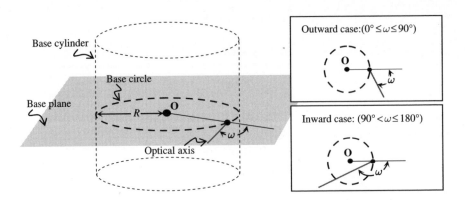

Figure 6.1 A base cylinder is defined orthogonal to the base plane, centered at **O**, and has radius R. Inward and outward cases are specified according the different ranges of principal angle value.

same image, which is different from the case of planar images where image correspondences occur only between pixels in two different images of a stereo pair. Thus, the *structure of spatial samples* that describes the distribution of sampling points in 3D space has been a new challenge for stereo panoramas.

To generalize sampling structure for a single projection direction (say, tangential, $\omega = 90°$), we investigate panoramic sampling structures for the omnidirectional ($\omega \in [0°, 180°]$) case. A cylinder, referred to as a *base cylinder*, is introduced here in order to classify the sampling structure in different cases. The base cylinder, illustrated in Figure 6.1, is orthogonal to the base plane, centered at **O**, and has radius R. Moreover, the base cylinder is defined to be of infinite height so that it partitions the 3D space into two portions, namely inside and outside the cylinder. According to the basic differences in sampling structure characteristics, this section is divided into two subsections, addressing the *outward case* defined by $0° \leq \omega \leq 90°$, and the *inward case* defined by $90° < \omega \leq 180°$, as shown in Figure 6.1 (the two top-view drawings on the right).

6.2.1 Outward Case

In the outward case, no projection ray passes through the interior of the base cylinder; see Figure 6.2. This figure presents "top views" of spatial sample formations of a pair of symmetric panoramas, where (a) and (b) show projection rays emitting from projection centers on the base circles, (a) for the right and (b) for the left stereo panorama. Note that the number of projection rays should be identical to the width of the panorama. In this figure, an extremely low-resolution panorama pair is presented for clarity. Figure 6.2(c) is just the superimposition of (a) and (b). Figure 6.2(d) sketches the basic structure of stereo samples (i.e., of all intersection points in (c)), which form *supporting concentric circles*, as shown in (e), (actually *supporting concentric cylinders* in 3D). The innermost (black) circle in Figure 6.2(e) is the base circle; this should not be confused with a supporting concentric circle. The radii of concentric circles increase non-linearly.

Motivated by the fact that samples are layered in depth (= distance from **O**), we also refer to these supporting concentric circles (cylinders, in 3D) as *depth layers*.

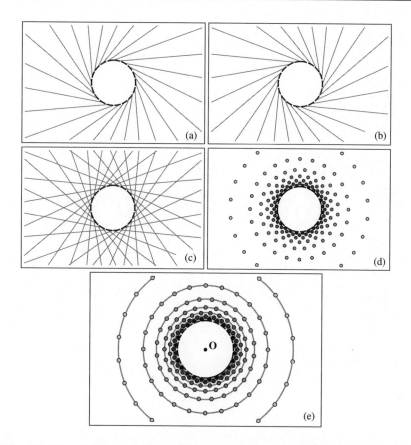

Figure 6.2 "Top views" of the sampling structure for a symmetric outward case. Projection rays of the (a) right and (b) left panoramic images, respectively. (c) Superimposition of (a) and (b). (d) Intersection points of (c). (e) The supporting concentric circles of those intersection points.

Figure 6.3 is a 3D visualization of the same sampling structure as in Figure 6.2(e). Here only samples within a 240° view are shown for clarity. Samples are depicted by spheres in different sizes (the larger they are, the further away from the base cylinder). From this 3D visualization, we may also observe that the vertical distances between any two adjacent samples increase with distance from the base cylinder. The depth layers are shaded in different gray levels for clarity. The structure of all depth layers can be described as a set of straight concentric cylinders whose top and bottom frontiers are incident with a pair of co-axial hyperboloids which are symmetric to the base plane. This observation can also be confirmed by the fact that the epipolar surface for stereo panoramas is a half-hyperboloid.

The density of depth layers increases inversely with distance from the base cylinder. (This will be different for the inward case.) A change in the off-axis distance R has the effect of increasing (reducing) density of depth layers towards (away from) **O**. For example, a large value of R allows more depth samples for a distant object in a scene, compared to a small value of R. This effect is analogous to changes in the base distance between two cameras in

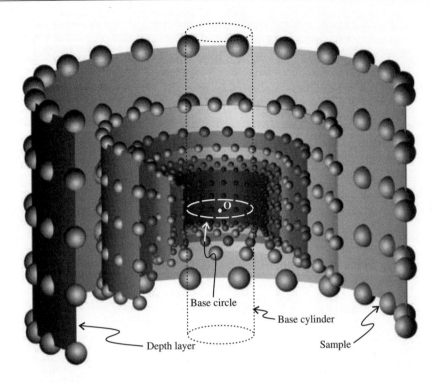

Figure 6.3 Three-dimensional view of sampling structure for a symmetric outward case. Samples are depicted by spheres of different sizes and supported by a set of concentric cylinders in different gray levels.

pinhole binocular stereo. Note that a change in R does not alter the number of depth layers, just their spatial distribution.

A change in the principal angle ω alters the total number of depth layers in 3D space (an exact analysis will be provided later). Figure 6.4 shows three examples of sampling structures for different angles ω (30°, 60°, and 90°), where R is assumed to remain constant. The number of depth layers increases as ω increases in the interval defining the outward case. A similar effect is known for pinhole-type converged stereo pairs, where an increase in vergence angle increases the number of (potential) depth levels in 3D space, and thus also the (potential) accuracy in stereo reconstruction. However, pinhole converged stereo pairs suffer from the change in epipolar geometry compared to *standard stereo geometry* (i.e., parallel optical axes and coplanar image planes, where we have image row y in the left image collinear to image row y in the right image, for all image rows). Stereo panoramas in the symmetric case still have all epipolar lines on their image rows, for any angle ω. This is a remarkable advantage of stereo panoramas compared to pinhole stereo pairs.

We conclude that the cardinality and density distribution of spatial samples are influenced by the chosen values of ω and R, without leaving (for any chosen pair of values!) the space of stereo-viewable panoramas, which also allows the use of computational stereo as designed for standard (pinhole) stereo geometry. (This advantage also allows control of spatial sampling dependent on the spatial complexity of the scene; see Chapter 7.)

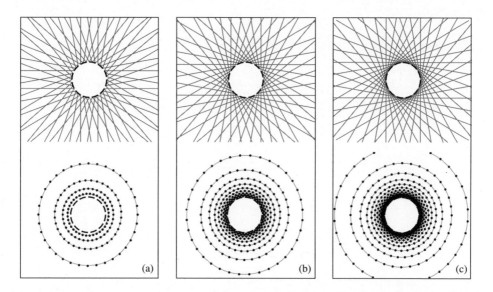

Figure 6.4 The "top views" of sampling structure for a symmetric outward panoramas with different principal angles. The upper row shows the intersecting projection rays; lower row illustrates their intersecting points (i.e., samples) with supporting concentric circles. (a) $\omega = 30°$; (b) $\omega = 60°$; and (c) $\omega = 90°$.

6.2.2 Inward Case

We now assume a principal angle ω greater than 90° and less than or equal to 180°. All projection rays intersect the interior of the base cylinder. The inward case has a sampling structure which is more complex than that of the outward case. It also has various geometric features which characterize it as being of potential value for applications including scene reconstruction for close-range (indoor) scenes, design of calibration objects, or shape reconstruction of 3D objects.

Figure 6.5 presents a top view of the sampling structure of a symmetric panorama in the inward case. Figure 6.5(a) shows a single projection ray of the right panorama. The projection ray starts at the base circle, first intersects the interior of the base cylinder shown as a bold dashed circle, and then continues into its exterior towards infinity. Figure 6.5(b) shows multiple projection rays of both the left and the right panoramas, and (c) illustrates the supporting concentric circles (i.e., cylinders in 3D) for this case.

There are three zones of sampling structures (listed from the center outwards): (1) a *sample-free zone* enclosed by the *inner surface* tangential to all projection rays, followed by (2) an *in-between zone* enclosed by the base cylinder, and then (3) the zone outside the base cylinder.

In this subsection we discuss first the sampling structure of the in-between zone, then that of the sample-free zone, but not that of the zone outside of the base cylinder because it follows similar (general) observations as discussed for the outward case. (A formal analysis be given in the next section.)

Figure 6.5(c) is an enlarged representation of Figure 6.5(b). The dashed circle is the base circle, not to be confused with a depth layer (i.e., a supporting concentric circle). There are three kinds of depth layers in the in-between zone. One is formed by intersections of projection

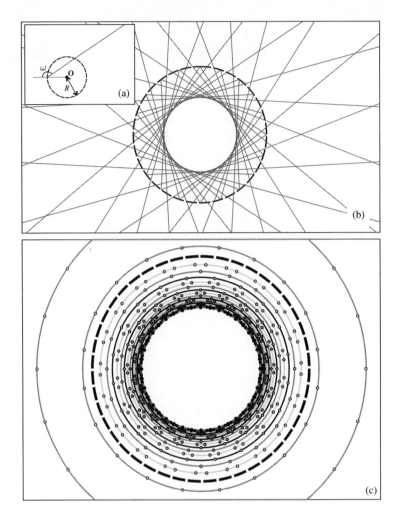

Figure 6.5 The "top views" of sampling structure of a symmetric panorama in the inward case. (a) A single projection ray of the right panorama. (b) Multiple projection rays of both the left and the right panoramas. (c) The supporting concentric circles for this case.

rays before the rays intersect the tangential inner surface, and those layers are shown in gray. Another kind is formed by intersections of projection rays after the rays intersect the tangential inner surface, shown in dark gray. Finally, there are depth layers shown in light gray; these are created by two rays where one is not yet incident with the inner surface, but the other is. The first two kinds are *identical-type depth layers*.

Potential depth samples are evenly distributed on identical-type depth layers. The third kind, *opposite-type depth layers*, are characterized by uneven distributions except under a special condition (to be discussed below).

More importantly, only identical-type depth layers of inward stereo panoramas follow the symmetric principle of epipolar geometry which ensures that epipolar lines coincide with image rows. Opposite-type depth layers contain potential samples which are defined by only one image of the stereo pair. In these cases an image point has a corresponding point in the same image, which means that the epipolar curve is also defined in the same image, and not a line anymore which coincides with an image row.

In general, we would assume that a sample in the in-between zone is uniquely defined by two rays which intersect at this point (the same as in the pinhole planar geometry case). However, when the angular unit γ is, for example, a factor of $2\omega - 180°$, then a sample is defined by two pairs of intersecting rays, where one is of identical type and the other is of opposite type. Figure 6.6 shows an example of this special case, where $\omega = 160°$ and $\gamma = 10°$. Note that under this special condition, *all* samples are uniformly distributed on their depth layers, including opposite-type depth layers. In Figure 6.6, a point **P** is the intersection of rays emitting from $\mathbf{C}_i$ and $\mathbf{C}_j$, as well as of rays emitting from $\mathbf{C}_k$ and $\mathbf{C}_l$, both being identical-type. In addition, **P** is the intersection of rays emitting from $\mathbf{C}_i$ and $\mathbf{C}_l$, and rays emitting from $\mathbf{C}_j$ and $\mathbf{C}_k$, both of opposite-type. Thus, ideally, a point in 3D can be "stereo-sampled" up to four times. However, the occurrence of this ideal case is constrained by occlusions in the scenes captured. Only one of the identical-type (opposite-type) can be valid at one time.

Whether opposite-type samples just define a problem for stereo applications of inward stereo panoramas, or also allow a particular utilization, has not yet been studied. A possible application is the design of calibration objects for panoramic camera calibration. A special calibration object can be designed such that calibration marks on the calibration object are at positions sampled twice in a single (!) panorama. This may contribute to calibration accuracy. Moreover, accurate camera calibration requires the projection of a calibration object that covers most of the area of the captured image. Obviously, compared to the outward case, the inward case allows much smaller calibration objects (with a diameter less than R).

The density of depth layers increases towards the inner surface of the sample-free zone (rather than towards the base cylinder as in the outward case). A change in the off-axis distance R has a similar effect to the outward case (i.e., no alteration of numbers of samples, but a dynamic change of the density pattern). A change in the principal angle ω, however, allows different observations compared to the outward case.

Figure 6.7 illustrates three sampling structures for ω equal to 130°, 155°, and 175°; R remains constant for all three angles. A larger value of ω increases the number of samples. But note

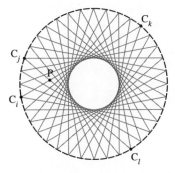

Figure 6.6 Special case of an inward sampling structure where samples can be defined by multiple intersection points of rays.

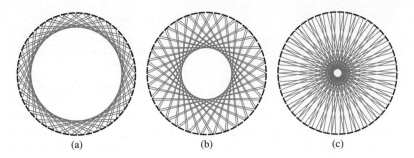

Figure 6.7 "Top views" of the sampling structure for a symmetric inward panorama with different principal angles: (a) $\omega = 130°$, (b) $\omega = 155°$, and (c) $\omega = 175°$.

that a less vergenced angle reduces accuracy in stereo reconstruction. The value of ω has an effect on the size of the sample-free zone; see Figure 6.7. This can be expressed formally.

The sample-free zone is enclosed by the inner surface (a cylinder) which is tangential to all projection rays. Let r be the radius of this cylinder. It satisfies the following relation:

$$r = R \cdot \sin(|180° - \omega|)$$

The off-axis distance R is (just) a scaling factor for r. The principal angle ω, however, offers more flexibility.

This is illustrated in Figure 6.8 by (symmetric) stereo panoramas. A synthetic rhino is placed in the base circle of the panoramic camera, and a sequence of stereo panoramas is acquired using different values of ω. The figure shows stereo pairs where (ω_R, ω_L) is equal to $(155°, 205°)$, $(160°, 200°)$, $(165°, 195°)$, $(172°, 188°)$, and $(179°, 181°)$.

A stereo reconstruction (e.g., of surface "slices" of the rhino) may be based on individual stereo panoramas, and reconstruction accuracy may be enhanced by utilizing interrelations among reconstructions obtained from individual pairs. This finally supports accumulated stereo reconstruction of the whole (visible) surface.

6.3 Spatial Resolution

The previous section discussed the general geometric structure of potential samples in stereo panoramas. This section focuses on a more specific issue, namely the *spatial resolution*, defined by the number of samples per volume element.

Since sampling density is not uniform in 3D space, spatial resolution depends upon position. This section provides a first answer, and a more detailed analysis (characterizing distances between samples) follows in the next section.

6.3.1 Indexing System

The following (unified) indexing system for the outward and inward cases is such that calculations of sampling resolution or sample distance will not depend anymore on ω, instead just on discrete indices (i, j, k) of samples.

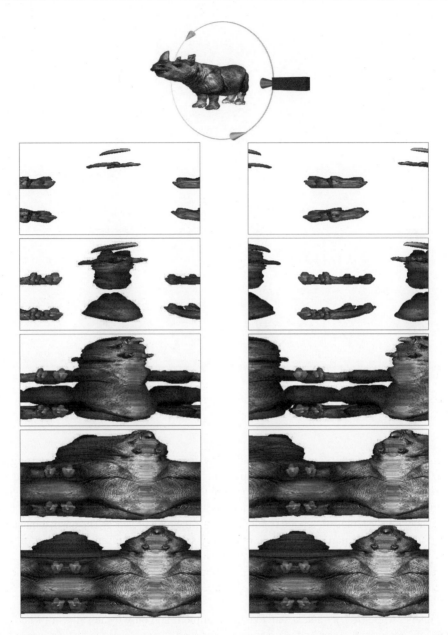

Figure 6.8 Top: a synthetic rhino. Row by row: five stereo panoramic pairs, with (from top to bottom) (ω_R, ω_L) equal to $(155°, 205°)$, $(160°, 200°)$, $(165°, 195°)$, $(172°, 188°)$, and $(179°, 181°)$, respectively.

Every sample in 3D space is indexed by (i, j, k), where i uniquely identifies the direction (i.e., one of finitely many intervals of directions of rays emerging at **O**) to the sample, j uniquely represents the elevation of the sample (in its depth layer) with respect to the base plane, and k specifies the depth layer of the sample. Index i starts with value 0 at a predefined direction and

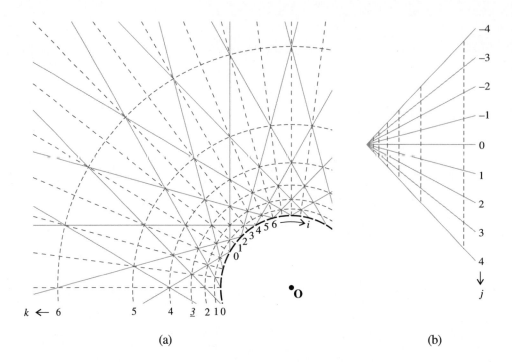

Figure 6.9 Indexing system for sampling structures. (a) A partial top view shows indices i (directions, with respect to **O**) and k (depth layers). (b) A partial side view of a single (perspective) projection that shows indices j; we have $j = 0$ at the base plane and positive j downwards.

increases clockwise. Index j is not equal to the elevation; it just numbers samples starting with value 0 at the base plane, and decreasing upward or increasing downward. Index k starts with value 0 at the base cylinder and increases for samples with their distance to the associated projection center (note that this increase in distance is also uniquely specified for the inward case).

The indexing system is illustrated in Figure 6.9 for the outward case. The dashed black bold arc in Figure 6.9(a) is part of the base circle. Gray solid lines represent projection rays into the 3D space. Gray dashed straight lines represent directions indexed by i. Gray concentric dashed arcs indicate different indices of k (i.e., of depth layers). Figure 6.9(b) illustrates index j. Vertical gray dashed lines show different depth layers. The value of j is zero at the base plane; positive values are downwards.

An integer triple (i, j, k) indexes an existing sample if and only if the values of i and k are either both odd or both even (with an exception for $k = 0$; in this case we also assume $j = 0$). Index $(i, 0, 0)$, with i even, refers to a projection center on the base circle.

The same indexing system also applies for the inward case, as shown in Figure 6.10. The only apparent difference between outward and inward cases is the index k, which looks more complicated in the inward case. In this figure, a pair of very low-resolution stereo panoramas is illustrated for the sake of clarity. Consider a particular set of rays (thicker gray line) emitting from **C**; the samples on these rays are depicted by gray circles. The index k increases with the corresponding sample distance from the projection center **C**.

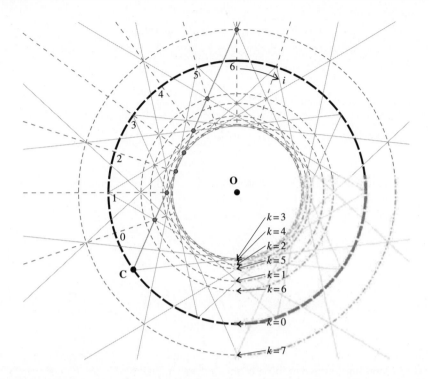

Figure 6.10 This illustrates the same indexing system of sampling structures for the inward case. The indices k increases as the samples (e.g., gray circles) is further away from the associated projection center (e.g., **C**).

6.3.2 Computation of Resolution

Recall that W is the number of image columns in a panorama, and H the number of image rows. Following the above definitions, index i starts at zero and goes to $2W - 1$. If i exceeds $2W - 1$ then we consider it modulo $2W$. Index j goes from $-(H - 1)/2$ to $(H - 1)/2$ if H is odd, or from $-(H - 2)/2$ to $H/2$ if H is even.

The interval of possible k-values is more difficult to specify. Obviously, index k goes from zero to a value k_{max} which specifies the maximum number of depth layers. This number can be calculated from the principal angle ω and unit angle γ as follows:

$$k_{max} = \left\lfloor \frac{2\omega}{\gamma} \right\rfloor .$$

Note that 2ω is the angular (i.e., angle with center point at **O**) distance between two projection centers whose associated projection rays are parallel. For a symmetric panorama we only have to consider ω between $0°$ and $180°$. (If $\omega = 180°$, then $2\omega = 360° \equiv 0°$ and $k_{max} = 0$.) Thus we can rewrite the above equation for k_{max} as follows:

$$k_{max} = \left\lfloor \frac{2(\omega \bmod 180°)}{\gamma} \right\rfloor .$$

Alternatively, k_{max} can also be determined by ω and W as follows:

$$k_{max} = \left\lfloor \frac{(\omega \bmod 180°)W}{180°} \right\rfloor. \tag{6.1}$$

The discussion above has identified the following extreme cases:

6.1. PROPOSITION. *A symmetric panorama with ω equal to $0°$ or $180°$ does not allow any spatial sample.*

The combination of both representations of k_{max} leads to the following theorem for symmetric panoramas. The total number of spatial samples of a stereo panorama is also called its *spatial sampling resolution*.

6.1. THEOREM. *For of a symmetric panorama of image resolution $W \times H$, where the principal angle ω (of the right camera) is in the interval $(0°, 180°)$, the spatial sampling resolution is equal to*

$$W \times H \times \left\lfloor \frac{\omega W}{180°} \right\rfloor. \tag{6.2}$$

Proof There are $2W$ different values of i (from zero to $2W - 1$). There are H different values of j according to the definition of this index. The value of index k goes from zero to $\lfloor \omega W/180° \rfloor$, thus there are $\lfloor \omega W/180° \rfloor + 1$ different values of k.

The total number of spatial samples is thus upper-bounded by the product of these three values $2W$, H, and $\lfloor \omega W/180° \rfloor + 1$. However, there are only W spatial samples for each pair of values of j and k (see Figure 6.10), and there are no samples for $k = 0$. Therefore, the total number of spatial samples is equal to $W \times H \times \lfloor \omega W/180° \rfloor$. $\square$

6.1. COROLLARY. *The off-axis distance R of a symmetric panorama has no impact on the spatial sampling resolution.*

6.2. COROLLARY. *The spatial sampling resolution of a symmetric panorama increases as ω increases towards $180°$.*

In the inward case, the spatial sampling resolution increases with ω following Corollary 6.2, but the samples are clustered more and more towards the center of the base circle (see Figure 6.7); the number of samples outside the base cylinder decreases with increasing ω.

6.3. COROLLARY. *In the outward case, the spatial sampling resolution of a symmetric panorama reaches a maximum at $\omega = 90°$.*

Proof By definition of the outward case (i.e., $\omega \in [0°, 90°]$). $\square$

6.4 Distances between Spatial Samples

Samples are non-uniformly distributed in 3D space. Figure 6.11 illustrates a local neighborhood of samples (shown as shapes of different shades) around a sample (i, j, k). As stated in Section 6.3.2, not every triple (i, j, k) has a corresponding sample. Positions of samples can be identified with a regular grid where each grid point (i.e., sample) has exactly six adjacent samples. The definition of adjacency follows below (which does not correspond to edges shown in Figure 6.11).

When adjusting an image acquisition system to a particular scene it might be of value to know the exact distance between adjacent samples at a specified *depth* (i.e., distance from the camera).

Sample distance denotes the distance between two adjacent samples; it is undefined for pairs of non-adjacent samples. This section defines sample distances for stereo panoramas, derives related formulas, and discusses changes in sample distances depending on depth or depth layer.

6.4.1 Basic Definitions

Figure 6.11 shows projection rays as solid gray lines. As always, samples are at intersections of those rays. We use five different shapes to indicate different i values, five different levels of shading to indicate different depth layers (i.e., different k values), and three 3×3 grids (i.e., top, middle, and bottom) with lines of different gray levels to indicate the three height values of all shown samples (i.e., different j values).

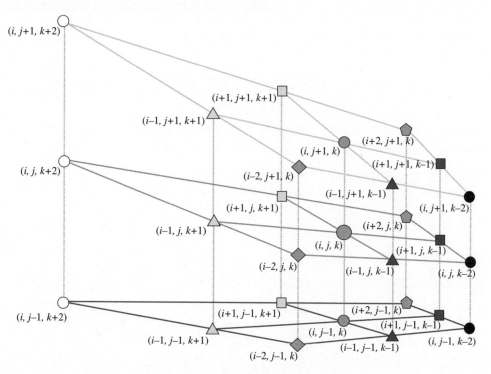

Figure 6.11 Indexed samples in the neighborhood of sample (i, j, k).

DEFINITION **6.1.** *The sample* (i, j, k) *is* horizontally adjacent *to samples* $(i \pm 2, j, k)$, vertically adjacent *to samples* $(i, j \pm 1, k)$, *and* depth-adjacent *to samples* $(i, j, k \pm 2)$. *Two samples are* adjacent *if they are either horizontally, vertically, or depth-adjacent.*

Adjacency is thus an irreflexive and symmetric relation. Each sample is adjacent to six other samples. For further analysis on the distance between any pair of adjacent samples, we define the following sample distances.

DEFINITION **6.2.** *The* horizontal sample distance, *denoted as* G_k, *is the distance between sample* (i, j, k) *and one of the two horizontally adjacent samples.*

DEFINITION **6.3.** *The* vertical sample distance, *denoted as* H_k, *is the distance between sample* (i, j, k) *and one of the two vertically adjacent samples.*

DEFINITION **6.4.** *The* depth sample distance, *denoted as* U_k, *is the distance between sample* (i, j, k) *and its depth-adjacent sample* $(i, j, k + 2)$ *after projecting both into the base plane* *(i.e., distance between* $(i, 0, k)$ *and* $(i, 0, k + 2)$*).*

The distances G_k, H_k and U_k are constant for all samples in the same depth layer; thus we do not have to add variables i or j. For an illustration of those defined sample distances, see Figure 6.12 (which follows the graphical conventions of Figure 6.11).

In order to find sample distances to a reference sample at a particular distance, we need one more definition:

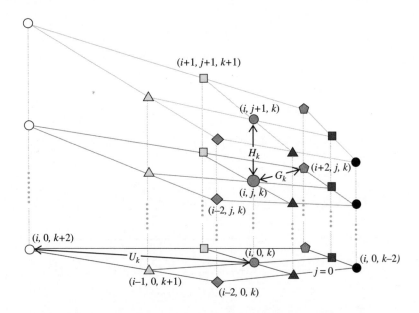

Figure 6.12   Illustration of horizontal, vertical, and depth sample distances.

DEFINITION **6.5.** *The depth D_k of a sample (i, j, k) is the distance between sample $(i, 0, k)$ and the center* **O** *of the base circle.*

Depth is defined on the base plane with $j = 0$. Index i is not needed in order to specify D_k because all samples in one depth layer have a constant depth.

Figure 6.13 illustrates the depth of a sample (i, j, k), which is the distance between **P** and **O**, where point **P** denotes sample $(i, 0, k)$. Consider the triangle $\triangle \mathbf{P}\mathbf{C}_m\mathbf{O}$. We know that $\overline{\mathbf{C}_m\mathbf{O}} = R$. Furthermore, we have $\angle \mathbf{P}\mathbf{C}_m\mathbf{O} = 180° - \omega$ and $\angle \mathbf{C}_m\mathbf{O}\mathbf{P} = k\gamma/2$. Thus, by the sine theorem, the value D_k is calculated as follows:

$$D_k = \frac{R \sin \omega}{\sin\left(\omega - \dfrac{k\gamma}{2}\right)}. \tag{6.3}$$

6.4.2 Horizontal Sample Distance

In order to calculate a horizontal sample distance, we define two auxiliary distances B_k and S_k; see Figure 6.13. B_k is the distance between two projection centers indexed as $(i, 0, 0)$ and $(i + 2k, 0, 0)$. It is defined on the base plane, in the interior of the base circle. Based on this definition, B_{k+1} will be the distance between $(i, 0, 0)$ and $(i + 2k + 2, 0, 0)$, and it can also

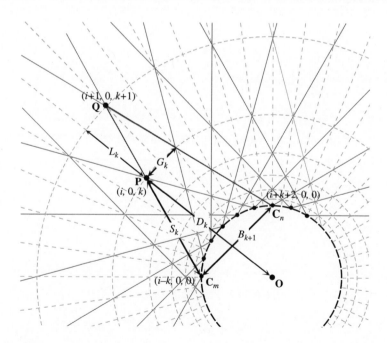

Figure 6.13 Auxiliary distances B_k and S_k, used for calculating sample distances G_k, H_k, and U_k of a reference sample (i, j, k) at distance D_k.

represent the distance between $(i - k, 0, 0)$ and $(i + k + 2, 0, 0)$, due to to the fact that index i is uniformly distributed (in an angular sense) on the base circle.

Distances B_k are simply the lengths of chords of the base circle, defined by k, and from elementary geometry it is known that

$$B_k = 2R \sin \left(\frac{k\gamma}{2} \right).$$

S_k is the distance between the projection center with index $(i, 0, 0)$ and sample $(i + k, 0, k)$, where $k \neq 0$. It is also defined on the base plane and independent of i. In Figure 6.13, S_k represents the distance between $(i - k, 0, 0)$ and $(i, 0, k)$, indicated by $\mathbf{C}_m$ and $\mathbf{P}$, respectively.

Similar to the calculation of the depth of a sample, we consider the triangle $\triangle \mathbf{PC}_m \mathbf{O}$. Applying the sine theorem, the value S_k is calculated as follows:

$$S_k = \frac{R \sin \left(\dfrac{k\gamma}{2} \right)}{\sin \left(\omega - \dfrac{k\gamma}{2} \right)}.$$

To calculate the horizontal sample distance at sample (i, j, k), we consider the triangle $\triangle \mathbf{QC}_m \mathbf{C}_n$ with vertices $(i + 1, j, k + 1)$, $(i - k, 0, 0)$, and $(i + k + 2, 0, 0)$; see Figure 6.13. (Note that in this example, $k = 5$, so $n = m + k + 1 = m + 6$.) We have

$$\frac{G_k}{B_{k+1}} = \frac{S_{k+1} - S_k}{S_{k+1}}.$$

Thus, the horizontal sample distance G_k can be calculated as follows:

$$
\begin{aligned}
G_k &= B_{k+1} - \frac{B_{k+1} S_k}{S_{k+1}} \\[2mm]
&= 2R \sin \left(\frac{(k+1)\gamma}{2} \right) - \frac{2R \sin \left(\dfrac{k\gamma}{2} \right) \sin \left(\omega - \dfrac{(k+1)\gamma}{2} \right)}{\sin \left(\omega - \dfrac{k\gamma}{2} \right)} \\[2mm]
&= \frac{2R}{\sin \left(\omega - \dfrac{k\gamma}{2} \right)} \left(\sin \left(\frac{(k+1)\gamma}{2} \right) \left(\sin \omega \cos \left(\frac{k\gamma}{2} \right) - \cos \omega \sin \left(\frac{k\gamma}{2} \right) \right) \right. \\[2mm]
&\qquad \left. - \sin \left(\frac{k\gamma}{2} \right) \left(\sin \omega \cos \left(\frac{(k+1)\gamma}{2} \right) - \cos \omega \sin \left(\frac{(k+1)\gamma}{2} \right) \right) \right) \\[2mm]
&= \frac{2R \sin \omega}{\sin \left(\omega - \dfrac{k\gamma}{2} \right)} \sin \left(\frac{(k+1)\gamma}{2} - \frac{k\gamma}{2} \right) \\[2mm]
&= \frac{2R \sin \omega \sin \left(\dfrac{\gamma}{2} \right)}{\sin \left(\omega - \dfrac{k\gamma}{2} \right)}.
\end{aligned}
$$

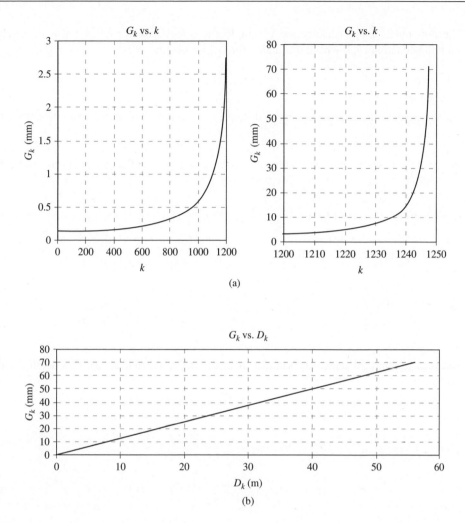

Figure 6.14 Horizontal sample distance G_k versus (a) depth layer index k (note the different scales) and (b) D_k.

Figure 6.14(a) illustrates general relations between horizontal sample distances and depth layers. We assume $W = 5,000$ pixels and let $R = 0.1$ m and $\omega = 45°$. Therefore, $\gamma = 360°/5,000 = 0.072°$. The number of depth layers is equal to 1,250 in this case.

The values of G_k are very small for $k < 1,200$, and much larger otherwise. Therefore we split the diagram of G_k versus k into two parts: the one on the left of Figure 6.14(a) shows values of k from 0 to 1,200, and the one on the right shows the values of k from 1,200 to 1,248. We do not include G_k for $k = 1,249$ or 1,250 because G_{1249} and G_{1250} are extremely large numbers (and this would distract from the basic pattern shown in those diagrams). Both diagrams show an exponential increase for horizontal sample distances, for increasing depth layers.

The horizontal sample distance at a sample can also be expressed in terms of depth. Combining equation (6.3) with the equation for G_k, we obtain

$$G_k = 2D_k \sin\left(\frac{\gamma}{2}\right).$$

We conclude that the horizontal sample distance of a sample is linearly proportional to its depth.

A diagram of G_k versus D_k is shown in Figure 6.14(b). The maximum value of G_k is less than 70 mm, and there is no sample anymore if the distance D_k is greater than 60 m (approximately).

6.4.3 Vertical Sample Distance

We have $H_k : S_k = \tau : f$, where τ is the size of a pixel (i.e., length of edge in the cell model). Thus, the value of H_k can be calculated as follows:

$$H_k = \frac{\tau R \sin\left(\dfrac{k\gamma}{2}\right)}{f \sin\left(\omega - \dfrac{k\gamma}{2}\right)}.$$

The vertical sample distance at a sample can also be represented in terms of its depth value. From equation (6.3), we obtain

$$\sin\left(\omega - \frac{k\gamma}{2}\right) = \frac{R \sin \omega}{D_k}$$

and

$$\sin\left(\frac{k\gamma}{2}\right) = \sin\left(\omega - \arcsin\left(\frac{R \sin \omega}{D_k}\right)\right).$$

Substituting these two terms into the equation for H_k, we have

$$
\begin{aligned}
H_k &= \frac{\tau D_k \sin\left(\omega - \arcsin\left(\dfrac{R \sin \omega}{D_k}\right)\right)}{f \sin \omega} \\[2ex]
&= \frac{\tau D_k \left(\sin \omega \cos \arcsin\left(\dfrac{R \sin \omega}{D_k}\right) - \cos \omega \dfrac{R \sin \omega}{D_k}\right)}{f \sin \omega} \\[2ex]
&= \frac{\tau D_k \left(\sin \omega \sqrt{\dfrac{D_k^2 - R^2 \sin^2 \omega}{D_k^2}} - \dfrac{R \cos \omega \sin \omega}{D_k}\right)}{f \sin \omega} \\[2ex]
&= \frac{\tau}{f}\left(\sqrt{D_k^2 - R^2 \sin^2 \omega} - R \cos \omega\right).
\end{aligned}
$$

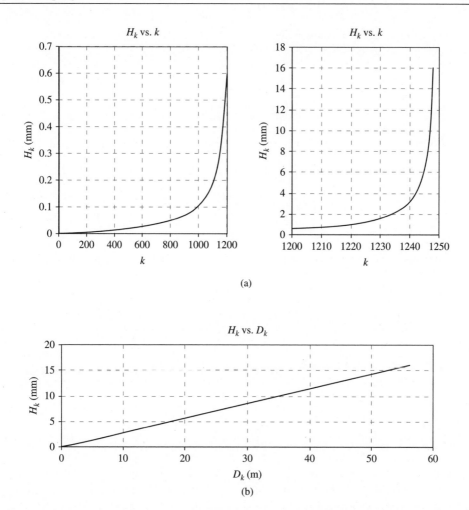

Figure 6.15 Vertical sample distance H_k versus (a) index k (note the different scales) and (b) depth D_k.

Figure 6.15 shows diagrams of vertical sample distances with respect to different layers k (again there are two different diagrams due to the large variation in values) and depth values D_k. The values of R, ω, and γ are as in the example for the horizontal case. Let $f = 35$ mm and $\tau = 0.01$ mm. For such a configuration we observe that vertical sample distances are always smaller than horizontal sample distances at the same depth. Although the relation between H_k and D_k is not linear (as we can see from the equation above), the diagram of H_k versus D_k also indicates that this relation is "almost linear".

6.4.4 Depth Sample Distance

DEFINITION **6.6.** *The depth layer distance L_k at depth layer k is the distance $D_{k+1} - D_k$ between depth values of layers k and $k + 1$.*

The depth layer distance is given by

$$L_k = R \sin \omega \left(\frac{1}{\sin\left(\omega - \frac{(k+1)\gamma}{2}\right)} - \frac{1}{\sin\left(\omega - \frac{k\gamma}{2}\right)} \right).$$

It follows that the depth sample distance U_k at sample (i, j, k) is the sum of L_k and L_{k+1}. We have

$$U_k = R \sin \omega \left(\frac{1}{\sin\left(\omega - \frac{(k+2)\gamma}{2}\right)} - \frac{1}{\sin\left(\omega - \frac{k\gamma}{2}\right)} \right).$$

The depth sample distance at a sample can also be represented in terms of its depth value. Combining equation (6.3) with the equation for U_k, we obtain

$$U_k = R \sin \omega \left(\frac{\sin\left(\omega - \frac{k\gamma}{2}\right) - \sin\left(\omega - \frac{(k+2)\gamma}{2}\right)}{\sin\left(\omega - \frac{k\gamma}{2}\right)\sin\left(\omega - \frac{(k+2)\gamma}{2}\right)} \right)$$

$$= D_k \left(\frac{\sin\left(\omega - \frac{k\gamma}{2}\right) - \sin\left(\omega - \frac{(k+2)\gamma}{2}\right)}{\sin\left(\omega - \frac{(k+2)\gamma}{2}\right)} \right)$$

$$= D_k \left(\frac{\sin\left(\omega - \frac{k\gamma}{2}\right)}{\sin\left(\omega - \frac{(k+2)\gamma}{2}\right)} - 1 \right)$$

$$= D_k \left(\frac{R \sin \omega}{D_k \sin\left(\omega - \frac{k\gamma}{2} - \gamma\right)} - 1 \right)$$

$$= \frac{R \sin \omega}{\sin\left(\omega - \frac{k\gamma}{2}\right)\cos\gamma - \cos\left(\omega - \frac{k\gamma}{2}\right)\sin\gamma} - D_k$$

$$= \frac{R \sin \omega}{\frac{R \sin \omega \cos\gamma}{D_k} - \sin\gamma\sqrt{\frac{D_k^2 - R^2 \sin^2 \omega}{D_k^2}}} - D_k$$

$$= \frac{D_k R \sin \omega}{R \sin \omega \cos\gamma - \sin\gamma\sqrt{D_k^2 - R^2 \sin^2 \omega}} - D_k.$$

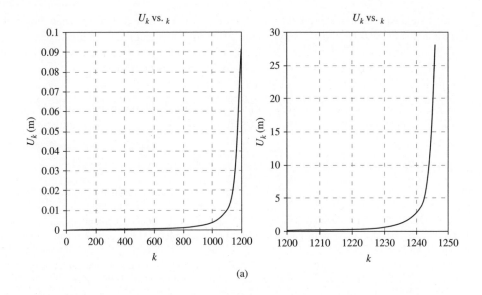

(a)

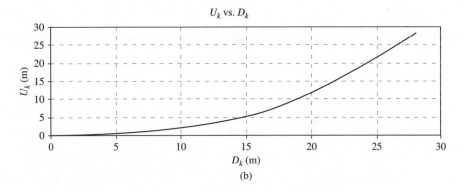

(b)

Figure 6.16 Depth sample distance U_k versus (a) index k (note the different scales) and (b) depth D_k.

Figure 6.16 shows diagrams of depth sample distances with respect to different layers k or depth values D_k. All parameter settings are as in the previous examples for horizontal or vertical sample distances. In contrast to those cases, the value of U_k increases exponentially with increases in D_k.

6.5 Exercises

6.1. Prove equation (6.3) by applying the sine theorem.

6.2. Calculate the spatial sampling resolution of a symmetric panorama $E_{\mathcal{P}_R}(100, 35, 60, 1)$ and $E_{\mathcal{P}_L}(100, 35, 300, 1)$; both images have an image height of 300 pixels. Off-axis distance and focal length are measured in millimeters, and the principal angle and angular unit are measured in degrees.

6.3. Calculate the horizontal sample distance G_5, the vertical sample distance V_5, and the depth sample distance U_5 for the same symmetric panorama as specified in Exercise 6.2.

6.4. [Possible lab project] Figure 6.8 illustrates how a synthetic 3D object (in this case, a rhino) is visualized by different inward panoramic pairs. For this lab project, you are requested to perform experiments as illustrated in this figure:

(i) Generate or reuse a textured 3D object of comparable shape complexity to the rhino.
(ii) Calculate five stereo panoramic pairs for pairs of angles as used in Figure 6.8.

6.6 Further Reading

Both stereo panoramas and symmetric panoramas have received attention in the context of various applications; see, for example, Ishiguro et al. (1992), Murray (1995), Huang and Hung (1998), Wei et al. (1998, 2002a, 2002b), Peleg and Ben-Ezra (1999), Shum et al. (1999), Huang et al. (2001c), Hung et al. (2002), Chan et al. (2005) and Chen et al. (2006). The epipolar surface for stereo panoramas is a half-hyperboloid (Seitz, 2001).

The approach by Shum et al. (1999) discusses only the sampling structure of a tangential projection direction (i.e., $\omega = 90°$). Corollary 6.2 coincides with results in (Shum et al., 1999).

Shi and Zheng (2003) study the spatial resolution of route panoramas (i.e., continuous 2D images extracted from a video sequence taken by a sensor-line camera mounted on a moving vehicle, showing scenes along a route). Here, "spatial resolution" basically means blurring (and depth estimation is based on the degree of blurring).

7

Image Quality Control

In this chapter, the camera is assumed to be a precisely defined parameterized model, and we do not apply any further constraints on camera parameters. Pinhole or multiple-projection-center panoramic cameras are examples of parameterized camera models. This chapter discusses image quality in the context of stereo data acquisition. For example, the number of potential stereo samples should be maximized for the 3D space of interest.

The chapter introduces four application-specific parameters, namely the scene range of interest enclosed by two coaxial cylinders, where we have radius D_1 for inner cylinder and radius D_2 for outer cylinder, the imaging distance H_1 (it also characterizes the vertical field of view), and the width θ_w of the angular disparity interval (θ_w specifies stereoacuity). Estimated values of these four parameters are required as inputs for the image quality control method which allows optimum sensor parameters R and ω to be calculated.

7.1 Two Requirements

Image quality has been addressed in camera developments since the first photo ever was taken by Joseph Nicéphore Niépce (1765–1833) in 1826, when an exposure time of eight hours produced (what would today be regarded as) a highly unsatisfactory image of parts of farm buildings. Image quality has considerable influence on the range of possible applications (e.g., in computer vision or photogrammetry); it is of continuous interest in camera design, production, and application. *Image quality* can be specified with respect to capabilities (e.g., speed), geometric or photometric accuracy, or circumstances of acquisition (e.g., against a light source, or on a vibrating bridge), and we refrain from attempting a formal definition of image quality.

Consequently, *image quality control* also remains vaguely defined in general terms as a method which identifies proper settings of camera parameters such that image quality meets the defined requirements or criteria. Those requirements or criteria (e.g., optimized for a particular 3D scene complexity, or a maximum of spatial stereo samples within a particular range of depth) are specified by the acquisition context.

Panoramic Imaging: Sensor-Line Cameras and Laser Range-Finders   F. Huang, R. Klette, and K. Scheibe
© 2008 John Wiley & Sons, Ltd

The design of an image quality control method requires the understanding of geometric or photometric relations between the given requirements or criteria (defined by the context of image acquisition) and possible parameter settings. Requirements or criteria vary with the architecture of the camera used. This chapter addresses image quality control for two geometric requirements when capturing a scene with a symmetric panoramic camera (for stereo viewing or stereo reconstruction).

First, an optimal *scene composition* (here basically defined only in terms of distances between camera and objects in the scene, and not by, for example, artistic considerations) ensures an optimum visualization of all the objects of interest (i.e., with respect to visibility and size) within the resulting symmetric panoramas.

Second, *stereoacuity* should be maximized for an assumed stereo-viewing situation (not, for example, modeling the stereoacuity of individual eyes, but the geometry of the viewing situation) for the symmetric panorama captured. This means that the number of spatial samples, as defined in Chapter 6, within the "range of interest" should be maximized.

The first requirement could possibly be achieved by the following individual or combined settings: firstly, different positions for acquiring images; secondly, different camera focal lengths; and finally, different camera off-axis distance R and principal angle ω. The last option offers different distances between a camera's optical center and the objects in the scene. Users will often decide on a spot for image shooting first, which leaves three camera parameters, namely f, R, and ω, to optimize scene composition. These parameter values can be traded off against each other to hopefully achieve less image distortion, large vertical field of view, and enough image resolution (with respect to the projected sizes on image) for the objects of interest in the scene.

The stereoacuity problem is to analyze the settings of camera parameters in relation to the resultant numbers of depth layers. *Proper stereoacuity* for scenes of interest is defined by optimized representations of geometric scene complexities. Insufficient stereoacuity produces a *cardboard effect*, where the 3D scene is perceived as a set of parallel pieces of cardboard, sorted in depth, one sitting in front of the other. Another case of improper stereoacuity is when the upper disparity limit for human vision is exceeded; this causes double images, called *dipodia*, which results in uncomfortable stereo viewing as well as eyestrain.

Both quality issues are often (merely) mentioned. However, image quality control is not discussed in these studies. In particular, some approaches restrict camera parameters to constant values (e.g., constant principal angle $\omega = 90°$ or $270°$) or to a specific interval (e.g., $0° \leq \omega \leq 90°$). Under such a priori restrictions, image quality control becomes difficult or even impossible with respect to both specified requirements.

Panorama acquisition is (still) a time-consuming, storage-consuming[1] and costly process. Without using a proper image quality control method, the quality of resultant images cannot be assured, and hence further acquisitions might be required, incurring more costs. On the other hand, we do not want to introduce unnecessary overheads (e.g., time, costs, or weights) to the original image acquisition process or device. An image quality control method has to be time-efficient and simple to handle.

---

[1] The panoramic image in Figure 7.1 was recorded in 2000 and it is 7 GB in raw format; the figure only shows it at low resolution. These images were acquired with a line consisting of about 10,000 pixel sensors. Recent developments have seen this number rise to about 14,000 pixels in a single sensor line, or to 24,000 if two 12,000 pixel lines are either staggered (as in case of the ADS40) or side by side, with a half-pixel shift between both (as in the case of Kompsat3).

Figure 7.1 Huang et al. (2001a) report on the acquisition of images with a rotating line camera in New Zealand, and give several examples. The figure shows a resolution-reduced part of one of these (single-line) 7 GB panoramas (view from the roof of the Auckland Museum) (see Plate 30).

This chapter describes an image quality control method for the two selected criteria. We provide a formula for optimizing camera parameters (such as off-axis distance R, principal angle ω, or angular increment γ) for the acquisition of symmetric panoramas. Section 7.2 describes basic geometric relationships between the panoramic camera (as specified by the parameters defined), the range of interest in the 3D scene, and further application requirements. This section leads to a more precise formulation of the image quality control problem. Section 7.3 presents a solution for this problem by computing optimized values for camera parameters. A few examples demonstrate the practical use of the formulas derived. Section 7.4 discusses how the measurement errors (which cannot be totally excluded) may (still) impact the image quality, assuming that our image quality control method has been used for parameter adjustment.

7.2 Terminology

7.2.1 Range of Interest in the Scene

A *range of interest* (RoI) is defined as a volume between two concentric cylinders; see Figure 7.2. The inner cylinder (radius D_1) is defined by the closest objects of interest, and the outer cylinder (radius D_2) by the furthest objects of interest. (We assume that $D_1 \neq D_2$.) In some applications, one might like to set $D_2 = \infty$ to reflect the fact that the furthest object of interest is not an issue and can be ignored. However, Theorem 6.1, which states that the total number of stereo samples is finite, also implies that there are no samples where the distance to **O** exceeds a certain upper bound. This upper bound turns out to be typically less than 200 m in practice due to equipment parameters, such as the smallest angular increment (say, 0.5°) or the size of the camera slider (e.g., limiting off-axis distance R to a maximum of 3 m). As a result, it does not matter whether we set, for example, $D_2 = 500$ or 1,000 m. Thus, if the value D_2 is assumed infinite in some applications, then it may simply be replaced by a value of 1,000 m in subsequent analysis or calculations.

This model of an RoI corresponds to the spatial sampling structure of symmetric panoramas. The goal is to have optimized stereo visualization within a given RoI, that is, to maximize the number of depth layers under the constraint of human stereo viewing (or binocular fusion) ability.

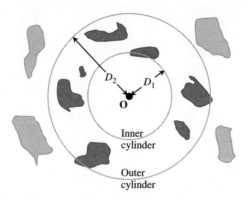

Figure 7.2 Top view of an RoI. Darker objects indicate locations which are of (major) interest while light gray areas are not, or are of minor interest. The figure indicates that both radii are in practice estimated based on 2.5D visibility (with respect to **O**) of objects of interest.

7.2.2 Distance to Target Range

The target range is defined by inner and outer cylinders as specified in the previous subsection. The distance to the target range, denoted by H_1, is defined to be the shortest distance between a camera's optical center and the object at the inner border of the RoI in the camera's viewing direction. Figure 7.3 (top left) illustrates the definition of H_1, where the camera's optical center

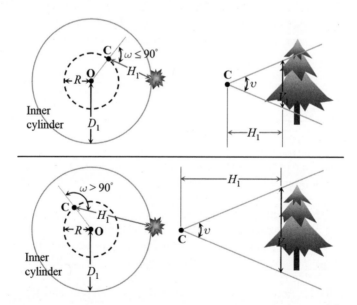

Figure 7.3 On the left are two top views of geometrical camera configurations including the closest object of interest (a tree in this example): the off-axis distance R remains constant, but ω is shown for an outward case (top) and an inward case (bottom); both cases define different values of H_1. On the right are the corresponding side views; both values of H_1 define different values of height V_1 at the inner cylinder.

is denoted by **C** and the object of interest is a tree. Note that this distance is defined on the base plane.

The *height* V_1 ("V" for vertical) of an RoI is estimated at the location of the inner cylinder. Let υ be the angle which specifies the vertical (angular) field of view, defined by the given or chosen lens of the camera. Then

$$V_1 = 2H_1 \tan\left(\frac{\upsilon}{2}\right) \tag{7.1}$$

where H_1 is the distance between a focal point **C** (of a line image) and the point where that projection ray which is incident with the base plane intersects the inner cylinder (of radius D_1). Figure 7.3 illustrates the distance H_1 for the outward case ($\omega \leq 90°$, top left) and inward case ($\omega > 90°$, bottom left), respectively. The value of distance H_1 depends on sensor parameters ω and R, but not on the choice of the camera location on the base circle.

A measurement of height V_1 is more difficult in practice than a measurement of H_1. Thus we use H_1 and υ to describe the height of an RoI. The measurement (or estimation) of distances to the nearest or furthest objects of interest defines the input for our quality control method, and various tools can be used (e.g., a laser range-finder or GPS for larger distances). The value of H_1 should be larger than the shortest focusable distance, and it will be smaller than $2D_1$. (Note that if $H_1 > 2D_1$, then the sensor will need a space larger than perimeter of the inner cylinder for a full 360° scan, which does not make sense.)

If D_1 is small, typically a wide-angle lens would be chosen. Note that H_1 (based on angle ω) allows the distance to the inner cylinder of the RoI to be increased, which means there is a reduced need for a wide-angle lense (which causes more lens distortion).

The positioning of the camera within the inner cylinder is typically very much constrained by the given scenery (architecture, landscape, etc.). Thus we do not discuss further optimization at this point, and simply take the camera position as a predefined input.

7.2.3 Depth, Disparity, and Angular Disparity

All spatial samples of a symmetric panorama (see Theorem 6.1) are in general partially inside a given RoI, and partially outside. The aim is to maximize the number of depth layers between D_1 and D_2.

We use two different representations of depth layers; one is specified by the *angular disparity* θ, which is defined by the angle (centered at **O**) between two corresponding points (on the image cylinder) in the left and right panorama of a symmetric panorama; see Figure 7.4. The figure illustrates a symmetric panorama and two points in 3D space, one on the inner and one on the outer cylinder of the RoI. Both points define angular disparities, θ_1 and θ_2, and the *width of the angular disparity interval* is defined by $\theta_w = \theta_2 - \theta_1$. Angular disparities enable the representation of depth layers.

A second representation is specified by the *disparity*, d, which is defined by the *horizontal parallax* (i.e., the difference in positions between two corresponding points in the left and right panorama, measured as the number of image columns between both on the image cylinder). Let d_1 and d_2 be disparities for 3D points on the inner and outer cylinder of an RoI, respectively. The *width of the disparity interval* is defined by $d_w = d_2 - d_1$. (Note that disparities are calculated

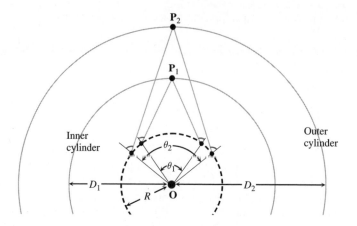

Figure 7.4 Two angular disparities, θ_1 and θ_2, defined by (arbitrary) 3D points on the inner and outer cylinder of an RoI.

along (one-dimensional) lines for symmetric panoramas, which means that corresponding points lie on identical image rows.)

Obviously, disparity and angular disparity are related as follows:

$$\theta = \frac{360d}{W} \tag{7.2}$$

where W is the width of the symmetric panorama in pixels. Valid ranges of θ and d are $0° < \theta < 180°$ and $0 < d < W/2$.

The depth D of a 3D point (i.e., the Euclidean distance between point and $\mathbf{O}$ on the base plane) is defined by either angular disparity or disparity, assuming R and ω are known:

$$D = \frac{R \sin \omega}{\sin \left(\omega - \frac{180d}{W} \right)} = \frac{R \sin \omega}{\sin \left(\omega - \frac{\theta}{2} \right)}.$$

This formula is of analogous simplicity to the depth formula of standard stereo geometry in the pinhole model case.

We now define *stereoacuity* in mathematical terms as the number of depth layers having a non-empty intersection (in terms of spatial samples) with a given RoI. Maximization of stereoacuity will provide more depth levels and thus a smoother depth transition for stereo visualization of the symmetric panorama. Moreover, it is also appropriate for computational stereo, because the reconstruction result can be more accurate.[2]

Stereo viewing also requires that disparities for an RoI are kept below the maximum disparity limit for human stereo fusion, which is approximately equal to

$$0.03 \times \text{viewing distance}.$$

---

[2] The accuracy of computational stereo is always limited by the stereoacuity available for a symmetric panorama within the RoI, regardless of the performance of the stereo-matching algorithm chosen.

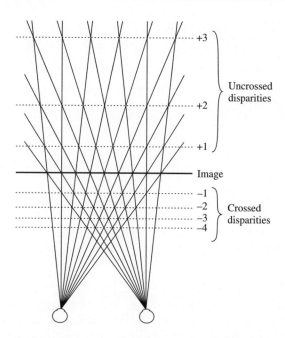

Figure 7.5 The human eye may fuse corresponding points as being in front of the fixation plane (crossed disparities; object appears to be in front of the image plane) or behind the fixation plane (uncrossed disparities; object appears to be behind the image).

This value characterizes the "fusibility" of corresponding points under particular viewing conditions, for example for viewing distances in relation to the size of a computer screen. Consider stereo viewing on a 17″ screen with a resolution of 1,024 × 768 pixels, and a frontal position at a distance of 0.4 m; the upper disparity limit is approximately $d \approx 70$ pixels for fusible and comfortable stereo viewing.

Calculations based on the upper disparity limit also lead to this value of $d \approx 70$ pixels. Interestingly, this upper disparity limit remains roughly constant for both crossed and uncrossed disparity fields. (Figure 7.5 illustrates the definition of crossed and uncrossed disparities.)

We are aware that there is still only partial understanding of human vision with respect to the enormous variety of binocular fusibility and viewing conditions, and there are still limited insights into related characteristics of computer displays. This implies that it is still difficult to ensure proper stereoacuity, for different viewers and different displays.

7.2.4 Image Resolution

In contrast to single-center panoramic acquisition, where the number W of image columns can be chosen independently of the given 3D scene complexity, the number of image columns of a symmetric panorama should be chosen depending on the geometric relationship between the camera (specified by defined parameters) and the given RoI.

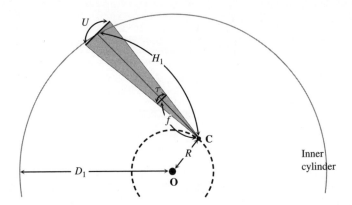

Figure 7.6 Geometric model for computing the number of image columns W of a symmetric panorama. U is the ground sampling distance defined by the effective focal length f and the pixel size τ.

Figure 7.6 illustrates our geometric model for computing the number of image columns W of a symmetric panorama. Without loss of generality assume a depth layer exactly at distance D_1; the inner cylinder of the RoI is selected to be the *uniform sampling target*, where U is the *ground sampling distance* (defined by the effective focal length f) on this sampling target. A first (and simple) observation is that

$$W = \frac{2\pi D_1}{U} \tag{7.3}$$

for a 360° panorama. Given an effective focal length f, pixel size τ, and the measurement of the corresponding value H_1, equation (7.3) can be rewritten as

$$W = \frac{2\pi D_1}{U} = \frac{2\pi f D_1}{H_1 \tau} = \frac{2\pi f}{\tau} \frac{D_1}{H_1} \tag{7.4}$$

where $2\pi f/\tau$ is the horizontal sampling rate for the single-center panorama case, and D_1/H_1 characterizes the potential difference to single-center (i.e., $R = 0$ with $D_1 = H_1$) panoramas.

For a single-center panorama, the number of projected points (at distance D_1, or wherever projection rays hit a surface) is simply defined by f and τ. On the other hand, for symmetric panoramas, with $R > 0$, if we have $H_1 > D_1$, then equation (7.4) suggests that a lens of larger focal length can be used without decreasing W. Therefore, image resolution can be preserved.

The image width W determines the total number of possible samples in 3D space, which is equal to

$$W \times H \times \left\lfloor \frac{\omega W}{180°} \right\rfloor.$$

For example, in experiments with the WAAC camera (e.g., with image height $H = 5,184$ pixels and pixel size $\tau = 0.007$ mm) we had a lens with $f = 21.7$ mm, and thus $W \le 19,478$ pixels. Larger values of W correspond to smaller values of ω; however, smaller values of ω reduce the total number of spatial samples. An indoor RoI could, for example, have $D_1 = 6\,\text{m}$, and thus $H_1 < 12\,\text{m}$ and $W > 9{,}739$ pixels, for $R < D_1$ and arbitrary ω. The value of W is reduced (for fixed f) when ω goes to $180°$, but this may still allow the total number of spatial samples to be maximized.

7.3 Parameter Optimization

The optimization problem is defined for computational stereo by the aim of maximizing the number of spatial samples in the given RoI; for stereo viewing we also have to stay below the upper (angular) disparity limit. Values of the four parameters D_1, D_2, H_1, and θ_w are assumed as input. For computational stereo, the upper angular disparity limit θ_w can be chosen, but cannot be larger than $180°$.

7.3.1 Off-Axis Distance R and Principal Angle ω

Figure 7.7 shows the notation and geometric relations as used in this subsection. Let $\mathbf{P}_1$ and $\mathbf{P}_2$ be two 3D points at depth D_1 and D_2, respectively, such that $\mathbf{P}_1$, $\mathbf{P}_2$, and $\mathbf{O}$ are collinear. To ensure that the desired scene objects captured at $\mathbf{P}_1$ meet the composition requirement on the resulting image, the camera has to be constantly positioned at a distance H_1 away from the object. A circle $\mathcal{H}$, centered at point $\mathbf{P}_1$ and of radius H_1, defines possible locations of projection centers of the camera when acquiring the object at $\mathbf{P}_1$.

Points $\mathbf{P}_1$ and $\mathbf{P}_2$ are projected onto image columns associated with the optical centers at $\mathbf{C}_1$ and $\mathbf{C}_2$, respectively. (The image cylinder is not shown in the figure.) In order to maximize the number of different depth perceptions within the specified RoI, the angle defined by line segments $\overline{\mathbf{C}_1\mathbf{O}}$ and $\overline{\mathbf{C}_2\mathbf{O}}$ has to equal to $\theta_{\max}/2$. We also have $\angle\mathbf{OC}_1\mathbf{P}_1 = \angle\mathbf{OC}_2\mathbf{P}_2$ (i.e., the same value of ω at both positions).

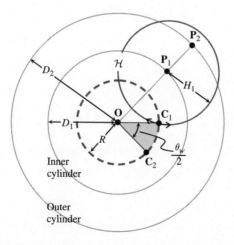

Figure 7.7 Geometric situation when optimizing R and ω.

7.1. THEOREM. *There exists a unique solution for R and ω which maximizes the number of spatial samples in the RoI and satisfies the upper disparity limit (i.e., angular disparity less than or equal or θ_w):*

$$R = \sqrt{D_1^2 + H_1^2 + 2D_1 H_1 \frac{D_1 - D_2 \cos\left(\frac{\theta_w}{2}\right)}{\sqrt{A}}},$$

$$\omega = \arccos\left(\frac{D_1 D_2 \cos\left(\frac{\theta_w}{2}\right) - D_1^2 - H_1\sqrt{A}}{R\sqrt{A}}\right),$$

where $A = D_1^2 + D_2^2 - 2D_1 D_2 \cos\left(\frac{\theta_w}{2}\right)$.

Proof Consider the triangles $\triangle OP_1 C_1$ and $\triangle OP_2 C_2$ in Figure 7.8(a). By rotating the first one clockwise around O, with rotation angle $\theta_w/2$, point C_1 coincides with C_2. The result is shown in Figure 7.8(b).

After the rotation, let angle $\angle OP_1 P_2$ be denoted by α, and angle $\angle OP_1 C_2$ be denoted by β. Obviously, $\beta = 180° - \alpha$. Consider $\triangle OP_1 P_2$ in Figure 7.8(b); the length L of side $\overline{P_1 P_2}$ can be calculated by the formula

$$L = \sqrt{D_1^2 + D_2^2 - 2D_1 D_2 \cos\left(\frac{\theta_w}{2}\right)}. \tag{7.5}$$

Still considering $\triangle OP_1 P_2$ in Figure 7.8(b); angle α has the following relationship with the three sides of this triangle:

$$D_2^2 = D_1^2 + L^2 - 2D_1 L \cos\alpha.$$

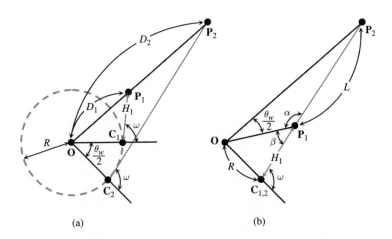

(a)   (b)

Figure 7.8 (a) Initial geometric situation for the calculation of R and ω. (b) Geometric situation after rotation.

This implies that

$$\cos\alpha = \frac{D_1^2 + L^2 - D_2^2}{2D_1 L}.$$

(7.6)

Now substitute the formula for L in equation (7.5) into equation (7.6):

$$\cos\alpha = \frac{D_1^2 + D_1^2 + D_2^2 - 2D_1 D_2 \cos\left(\frac{\theta_w}{2}\right) - D_2^2}{2D_1 \sqrt{D_1^2 + D_2^2 - 2D_1 D_2 \cos\left(\frac{\theta_w}{2}\right)}}$$

$$= \frac{D_1 - D_2 \cos\left(\frac{\theta_w}{2}\right)}{\sqrt{D_1^2 + D_2^2 - 2D_1 D_2 \cos\left(\frac{\theta_w}{2}\right)}}.$$

(7.7)

The angle β is defined to be equal to $180° - \alpha$, thus we have

$$\cos\beta = \cos(180° - \alpha)$$

$$= -\cos\alpha.$$

(7.8)

Next we consider $\triangle \mathbf{OP_1 C_2}$ in Figure 7.8(b). The value of R can be calculated using the formula

$$R = \sqrt{D_1^2 + H_1^2 - 2D_1 H_1 \cos\beta}.$$

Summarizing the results of equations (7.7) and (7.8), the value of R can be obtained as follows:

$$R = \sqrt{D_1^2 + H_1^2 + 2D_1 H_1 \frac{D_1 - D_2 \cos\left(\frac{\theta_w}{2}\right)}{\sqrt{D_1^2 + D_2^2 - 2D_1 D_2 \cos\left(\frac{\theta_w}{2}\right)}}}.$$

(7.9)

Angle ω satisfies in $\triangle \mathbf{OP_1 C_2}$ the following equation:

$$D_1^2 = R^2 + H_1^2 - 2RH_1 \cos(180° - \omega)$$

$$= R^2 + H_1^2 + 2RH_1 \cos\omega.$$

This implies that

$$\omega = \arccos\left(\frac{D_1^2 - H_1^2 - R^2}{2H_1 R}\right).$$

(7.10)

We substitute the formula for R in equation (7.9) into the equation for ω in equation (7.10), obtaining

$$
\omega = \arccos \left(\frac{D_1^2 - H_1^2 - D_1^2 - H_1^2 - 2D_1 H_1 \frac{D_1 - D_2 \cos\left(\frac{\theta_w}{2}\right)}{\sqrt{D_1^2 + D_2^2 - 2D_1 D_2 \cos\left(\frac{\theta_w}{2}\right)}}}{2H_1 \sqrt{D_1^2 + H_1^2 + 2D_1 H_1 \frac{D_1 - D_2 \cos\left(\frac{\theta_w}{2}\right)}{\sqrt{D_1^2 + D_2^2 - 2D_1 D_2 \cos\left(\frac{\theta_w}{2}\right)}}}} \right)
$$

$$
= \arccos \left(\frac{-H_1 + \frac{D_1 D_2 \cos\left(\frac{\theta_w}{2}\right) - D_1^2}{\sqrt{A}}}{\sqrt{\frac{(D_1^2 + H_1^2)\sqrt{A} + 2D_1^2 H_1 - 2D_1 D_2 H_1 \cos\left(\frac{\theta_w}{2}\right)}{\sqrt{A}}}} \right)
$$

$$
= \arccos \left(\frac{D_1 D_2 \cos\left(\frac{\theta_w}{2}\right) - D_1^2 - H_1 \sqrt{A}}{\sqrt{(D_1^2 + H_1^2)A + 2D_1 H_1 \left(D_1 - D_2 \cos\left(\frac{\theta_w}{2}\right)\right)\sqrt{A}}} \right) \tag{7.11}
$$

where $A = \left(D_1^2 + D_2^2 - 2D_1 D_2 \cos\left(\frac{\theta_w}{2}\right)\right)$.

The values of parameters D_1, D_2, H_1, and θ_w have to satisfy the physical constraints as specified in Section 7.2; equations (7.9) and (7.11) then lead to the conclusion that solutions for R and ω are unique.   □

7.1. COROLLARY. *In the computational stereo case of $\theta_w = 180°$, the unique solution for R and ω simplifies to*

$$
R = \sqrt{D_1^2 + H_1^2 + 2D_1^2 \frac{H_1}{\sqrt{A}}}, \qquad \omega = \arccos \left(\frac{-D_1^2 - H_1 \sqrt{A}}{R\sqrt{A}} \right),
$$

where $A = D_1^2 + D_2^2$.

The possible range of R is $0 < R < D_1$. However, in practice the value of R is constrained by the limitations of the equipment used. For the solution of ω, only the interval $0° < \omega < 180°$ needs to be considered (for the right camera of the symmetric panorama). If the value of ω is the solution (for the right camera), then $360° - \omega$ is the solution for the left camera. The extreme cases of $R = 0$, $\omega = 0°$ or $\omega = 180°$ do not support the capture of a stereoscopic panoramic pair (i.e., the number of depth layers and the width of the disparity interval are zero in these cases).

The unique solution is for both parameters. This implies that neither R nor ω can optimize numbers of spatial samples alone. On the other hand, assuming that, for example, R is fixed for a given panoramic camera, but ω is variable, this allows a "viewfinder" to be defined for a stereoscopic panoramic camera: different values of ω correspond to different RoIs.

7.3.2 Examples

We present the results of control values for R and ω for four commonly occurring situations: (1) a close-range indoor scene covering an area of about $36\,\text{m}^2$; (2) a far-range indoor scene of an area of about $400\,\text{m}^2$; (3) an outdoor scene with an RoI with radii of 6 and 50 meters; and (4) an outdoor scene with radii of 20 and 200 meters. A floor plan and aerial images are provided in Figures 7.9 and 7.10.

Results for R and ω are shown in Table 7.1. The WAAC camera used is as specified in Section 7.2.4, with an effective focal length of $f = 21.7\,\text{mm}$, pixel size $\tau = 0.007\,\text{mm}$, image height $H = 5{,}184$ pixels (i.e., values of W follow accordingly for the four examples), an assumed 17″ display screen (with $1{,}024 \times 768$ pixels) for stereo viewing at 0.4 m distance (frontal and centered position of viewer). Input parameters D_1, D_2, and H_1 are measured (estimated) in meters, as is the resulting value of R; θ_w and ω are given in degrees, and W in pixels.

To allow for comfortable stereo viewing under the specified conditions (let $H_S = 768$), equation (7.2) becomes

$$\theta_w = \frac{2\pi d_w H}{W H_S},$$

where $d_w = 70$ pixels is used in all four cases for the assumed 17″ display. The term H/H_S allows the vertical perspective acquired to be fully rendered on the screen. For interactive

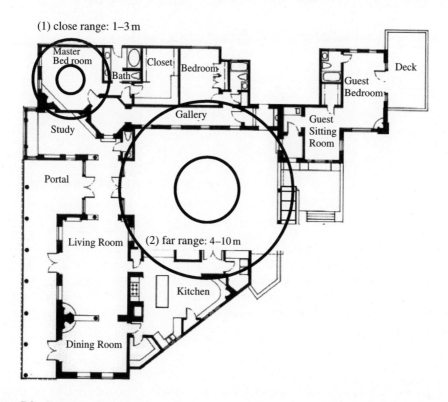

Figure 7.9 Example of a floor plan: case (1) is a close-range indoor scene (top left-hand corner), and case (2) a far-range indoor scene.

Figure 7.10 Two aerial views (captured by the digital WAAC) of central Auckland: case (3) is a close-range outdoor scene (in Albert Park), and case (4) a far-range outdoor scene (America's Cup Village in 2001).

	D_1	D_2	H_1	W	θ_w	R	ω
Example (1)	1	3	1.2	16232	10.48	0.2499	146.88
Example (2)	4	10	4.2	18550	9.17	0.5809	113.92
Example (3)	6	50	5.5	21249	8.00	0.6768	44.66
Example (4a)	20	200	20	19478	8.74	1.6942	92.43
Example (4b)	20	200	20	19478	5.00	0.9695	91.39

$$H_P = 5,184 \text{ (pixels)} \qquad \tau = 0.007 \text{ (mm)}$$
$$H_S = 768 \text{ (pixels)} \qquad f = 21.7 \text{ (mm)}$$

Table 7.1 Calculated values for R and ω using Theorem 7.1. We assume stereo viewing on a screen (resulting in a disparity limit of 70 pixels) and the four examples in Figures 7.9 and 7.10. Case (4b) is defined by the additional constraint that R is restricted to 1.0 m at most. (Note that all the distance measurements are in meters, angle measurements are in degrees, and image width is in pixels.)

zooming within stereoscopic visualization, "stereo shifts" between left and right panoramic image must be dynamically adapted such that the desired quality of stereo perception is ensured.

Results for R and ω in Table 7.1 show the actual values which correspond to scene composition and optimize stereoacuity.

Values of the principal value ω for the first two examples illustrate typical results for adapting a stereoscopic panoramic camera to indoor conditions. Values $\omega > 90°$ (i.e., the inward case) illustrate the benefit of increasing H_1 compared to D_1.

Values of R computed for examples (1)–(3) were actually realizable for the setup in Huang et al. (2001a) which allowed $R \leq 1.0$ m. Case (4a) leads to a value of R greater than 1.0 m. In this case we demonstrate a tradeoff with stereoacuity: we reduce the value of θ_w from 8.74° to 5.00° (i.e., fewer spatial samples) and obtain a result with $R < 1.0$ m.

7.4 Error Analysis

Measurements (or estimates) of the three parameters D_1, D_2, and H_1 (the latter one instead of the more difficult measurement of V_1) are likely to be inaccurate. This section analyzes how errors in these measurements impact the image quality of resultant stereo panoramas. First we define error measures, then we explain parameter dependency for the analysis of error propagation, and finally we provide an error analysis for each of the three input parameters.

7.4.1 Definitions and Notation

We assume three independent error variables because measurements of D_1, D_2, and H_1 are independent. They are denoted by ε_{D_1}, ε_{D_2}, and ε_{H_1}, respectively, and are real numbers. Measured values of D_1, D_2, and H_1 are $\hat{D}_1 = D_1 + \varepsilon_{D_1}$, $\hat{D}_2 = D_2 + \varepsilon_{D_2}$, and $\hat{H}_1 = H_1 + \varepsilon_{H_1}$. The symbol '^' indicates that a parameter may contain error, and symbols without '^' stand for true values.

Consequently, calculated parameters based on (potentially inaccurate) measurements are $\hat{R} = R + \varepsilon_R$ and $\hat{\omega} = \omega + \varepsilon_\omega$. Finally, these affect the disparity interval width $\hat{d}_w = d_w + \varepsilon_{d_w}$ (which characterizes stereoacuity in the resultant panorama) and the height $\hat{V}_1 = V_1 + \varepsilon_{V_1}$ (which is a resultant parameter of the RoI). We analyze how measurement errors propagate to d_w and V_1.

7.4.2 Parameter Dependencies

In order to study how errors propagate from a measurement to a 'final parameter', we clarify dependencies among the parameters. This subsection lists algebraic dependencies following individual steps during an image acquisition procedure. The function notation $Y = f_Y(X_1, X_2, \ldots, X_n)$ means that the parameter Y potentially depends upon parameters $X_1, X_2, \ldots, X_n$, and (within the scope of this study) upon no others.

The number W of image columns is calculated based on f, τ, D_1, and H_1; see equation (7.4). We write this dependency as

$$W = f_W(f, \tau, D_1, H_1).$$

The width θ_w of the angular disparity interval is defined by the width of the image disparity interval d_w and the number of image columns W; see equation (7.2). Accordingly we have

$$\theta_w = f_{\theta_w}(d_w, W).$$

Theorem 7.1 defines dependencies between R or ω on the one hand, and D_1, D_2, H_1, and θ_w on the other. Thus, we also have

$$R = f_R(D_1, D_2, H_1, \theta_w)$$

and

$$\omega = f_\omega(D_1, D_2, H_1, \theta_w).$$

Now consider a situation where parameters R and ω are used for the acquisition of a symmetric panorama (e.g., calculated by optimization using the formulas in Theorem 7.1), and the RoI is defined by radii D_1 and D_2. These four parameter values allow the values of H_1 and θ_w to be recovered as follows: for the value of H_1 we have that

$$H_1 = \sqrt{D_1^2 + R^2 \cos(2\alpha) + 2R \cos(\alpha)\sqrt{D_1^2 - R^2 \sin^2(\alpha)}} \qquad (7.12)$$

where $\alpha = 180° - \omega$. In function notation we express this as

$$H_1 = f_{H_1}(D_1, R, \omega).$$

Furthermore, the value of θ_w can be calculated by the formula

$$\theta_w = 2 \arcsin\left(\frac{R}{D_1 D_2}\sin(\omega)\left[\sqrt{D_2^2 - R^2\sin^2(\omega)} - \sqrt{D_1^2 - R^2\sin^2(\omega)}\right]\right). \qquad (7.13)$$

Thus, we have

$$\theta_w = f_{\theta_w}(D_1, D_2, R, \omega).$$

Assuming an error-free process, the calculated values of H_1 and θ_w should be identical to those values defined by the image acquisition requirements.

For a conversion from the (backtracked) width of the angular disparity interval to the width of the image disparity interval, see equation (7.2). In more abstract form, we also have

$$d_w = f_{d_w}(\theta_w, W).$$

Similarly, a conversion from (the backtracked) H_1 to the height of the RoI can be obtained through equation (7.1); we have

$$V_1 = f_{V_1}(H_1, \upsilon).$$

7.4.3 Error in the Distance to Inner Border of RoI

When the measurement of distance D_1 contains an error part, namely ε_{D_1}, we denote this estimated distance by $\hat{D}_1$. Because the number of image columns depends on D_1 it will also be erroneous, and is denoted by $\hat{W}$. Given $\hat{W}$, we obtain $\hat{\theta}_w$ due to the dependency $\hat{\theta}_w = f_{\theta_w}(d_w, \hat{W})$. Both parameters R and ω in equations (7.9) and (7.11) depend on D_1, D_2, H_1 and θ_w. Thus, the erroneous values $\hat{W}$ and $\hat{\theta}_w$ cause erroneous values $\hat{R}$ and $\hat{\omega}$, and the (additive) errors are respectively defined as

$$\varepsilon_R = \hat{R} - R$$
$$= f_R(\hat{D}_1, D_2, H_1, \hat{\theta}_w) - f_R(D_1, D_2, H_1, \theta_w)$$

and

$$\varepsilon_\omega = \hat{\omega} - \omega$$
$$= f_\omega(\hat{D}_1, D_2, H_1, \hat{\theta}_w) - f_\omega(D_1, D_2, H_1, \theta_w).$$

The actual width of the angular disparity interval for an acquired symmetric panorama, using control values $\hat{R}$ and $\hat{\omega}$, can be computed by equation (7.13); the erroneous value is defined by

$$\hat{\hat{\theta}}_w = f_{\theta_w}(D_1, D_2, \hat{R}, \hat{\omega})$$

where $\hat{\hat{\theta}}_w$ is used in distinction to $\hat{\theta}_w$ because $\hat{\hat{\theta}}_w$ contains an error propagated from $\hat{R}$ and $\hat{\omega}$ whose errors were originally introduced by $\hat{D}_1$. In contrast, $\hat{\theta}_w$ is erroneous directly and exclusively due to the error in $\hat{D}_1$.

Similarly, for computing the distance $\hat{H}_1$ to the target range, equation (7.12) is used, and the parameter dependency is denoted by $\hat{H}_1 = f_{H_1}(D_1, \hat{R}, \hat{\omega})$.

Finally, we use the width $\hat{\hat{\theta}}_w$ of the angular disparity interval to calculate the width $\hat{d}_w$ of the image disparity interval, and $\hat{H}_1$ for calculating $\hat{V}_1$. We have $\hat{d}_w = f_{d_w}(\hat{\hat{\theta}}_w, \hat{W})$ using equation (7.2), and $\hat{V}_1 = f_{V_1}(\hat{H}_1, \upsilon)$ using equation (7.1).

Numerical results for the propagation of errors in D_1, first to R and ω, and then to d_w and V_1, are presented in Table 7.2 using scene specification values as in Table 7.1. Errors are measured as percentages, and definitions are shown at the top of each column. $\hat{D}_1$, $\hat{R}$ and $\hat{V}_1$ are measured in meters $\hat{\omega}$ is given in degrees, and $\hat{d}_w$ in pixels.

We consider an error interval of $[-10\%, +10\%]$ for D_1. A positive (negative) sign means that the erroneous value is greater (smaller) than the true value. The shaded middle row in each of the four tables in Table 7.2 shows error-free values for reference. With respect to absolute values, the greater is ε_{D_1}, the greater are ε_R and ε_ω in general. However, exceptions occur in two cases. First, in example (2), the smallest (absolute) value of ε_R appears for the considered maximum error ε_{D_1} (see the bottom row for the 10% error of ε_{D_1}). Secondly, in example (3), a -10% error in ε_{D_1} leads (only) to a -28.43% error in ε_R which is less than the -30.48% error in ε_R caused by a -8% error in ε_{D_1} (see the top two rows). To understand this behavior, see the investigations reported in the next chapter.

$\frac{\varepsilon_{D_1}}{D_1}$ (%)	$\hat{D}_1$	$\hat{R}$	$\frac{\varepsilon_R}{R}$ (%)	$\hat{\omega}$	$\frac{\varepsilon_\omega}{\omega}$ (%)	$\hat{d}_w$	$\frac{\varepsilon_{d_w}}{d_w}$ (%)	$\hat{V}_1$	$\frac{\varepsilon_{V_1}}{V_1}$ (%)
−10	0.90	0.3356	34.28	157.21	7.03	60.0	−14.35	2.1755	8.41
−8	0.92	0.3178	27.14	155.60	5.94	61.9	−11.59	2.1417	6.73
−6	0.94	0.3002	20.11	153.80	4.71	63.9	−8.78	2.1079	5.05
−4	0.96	0.2830	13.23	151.77	3.33	65.9	−5.91	2.0742	3.36
−2	0.98	0.2662	6.52	149.48	1.77	67.9	−2.98	2.0404	1.68
0	1.00	0.2499	0.00	146.88	0.00	70.0	0.00	2.0067	0.00
2	1.02	0.2343	−6.27	143.92	−2.01	72.1	3.05	1.9729	−1.68
4	1.04	0.2193	−12.25	140.55	−4.31	74.3	6.15	1.9391	−3.36
6	1.06	0.2053	−17.87	136.69	−6.94	76.5	9.33	1.9054	−5.05
8	1.08	0.1923	−23.06	132.28	−9.94	78.8	12.56	1.8716	−6.73
10	1.10	0.1807	−27.72	127.27	−13.35	81.1	15.87	1.8378	−8.41

Example (1)

$\frac{\varepsilon_{D_1}}{D_1}$ (%)	$\hat{D}_1$	$\hat{R}$	$\frac{\varepsilon_R}{R}$ (%)	$\hat{\omega}$	$\frac{\varepsilon_\omega}{\omega}$ (%)	$\hat{d}_w$	$\frac{\varepsilon_{d_w}}{d_w}$ (%)	$\hat{V}_1$	$\frac{\varepsilon_{V_1}}{V_1}$ (%)
−10	3.60	0.8066	38.85	141.90	24.56	59.0	−15.69	7.6984	9.61
−8	3.68	0.7496	29.04	137.74	20.91	61.1	−12.71	7.5635	7.69
−6	3.76	0.6973	20.04	132.93	16.69	63.2	−9.65	7.4286	5.76
−4	3.84	0.6509	12.04	127.38	11.81	65.4	−6.52	7.2936	3.84
−2	3.92	0.6116	5.28	121.04	6.25	67.7	−3.30	7.1587	1.92
0	4.00	0.5809	0.00	113.92	0.00	70.0	0.00	7.0237	0.00
2	4.08	0.5603	−3.55	106.14	−6.83	72.4	3.39	6.8887	−1.92
4	4.16	0.5509	−5.17	97.94	−14.03	74.8	6.88	6.7537	−3.84
6	4.24	0.5532	−4.77	89.62	−21.33	77.3	10.46	6.6187	−5.77
8	4.32	0.5671	−2.38	81.55	−28.41	79.9	14.15	6.4836	−7.69
10	4.40	0.5918	1.88	74.02	−35.02	82.6	17.94	6.3486	−9.61

Example (2)

$\frac{\varepsilon_{D_1}}{D_1}$ (%)	$\hat{D}_1$	$\hat{R}$	$\frac{\varepsilon_R}{R}$ (%)	$\hat{\omega}$	$\frac{\varepsilon_\omega}{\omega}$ (%)	$\hat{d}_w$	$\frac{\varepsilon_{d_w}}{d_w}$ (%)	$\hat{V}_1$	$\frac{\varepsilon_{V_1}}{V_1}$ (%)
−10	5.40	0.4844	−28.43	104.37	133.68	62.1	−11.23	6.1018	10.95
−8	5.52	0.4705	−30.48	89.98	101.47	63.7	−9.01	5.9814	8.76
−6	5.64	0.4869	−28.05	75.68	69.44	65.3	−6.78	5.8610	6.57
−4	5.76	0.5308	−21.56	63.03	41.12	66.8	−4.53	5.7405	4.38
−2	5.88	0.5962	−11.91	52.73	18.05	68.4	−2.27	5.6202	2.19
0	6.00	0.6768	0.00	44.66	0.00	70.0	0.00	5.5000	0.00
2	6.12	0.7678	13.45	38.41	−14.00	71.6	2.28	5.3794	−2.19
4	6.24	0.8660	27.96	33.53	−24.82	73.2	4.58	5.2590	−4.38
6	6.36	0.9692	43.21	29.67	−33.58	74.8	6.89	5.1387	−6.57
8	6.48	1.0759	58.98	26.56	−40.54	76.4	9.21	5.0183	−8.75
10	6.60	1.1852	75.13	24.02	−46.23	78.1	11.54	4.8980	−10.94

Example (3)

$\frac{\varepsilon_{D_1}}{D_1}$ (%)	$\hat{D}_1$	$\hat{R}$	$\frac{\varepsilon_R}{R}$ (%)	$\hat{\omega}$	$\frac{\varepsilon_\omega}{\omega}$ (%)	$\hat{d}_w$	$\frac{\varepsilon_{d_w}}{d_w}$ (%)	$\hat{V}_1$	$\frac{\varepsilon_{V_1}}{V_1}$ (%)
−10	18.0	2.6674	57.57	141.17	52.73	62.3	−11.01	22.008	10.04
−8	18.4	2.3708	40.05	135.01	46.06	63.8	−8.83	21.606	8.03
−6	18.8	2.1092	24.59	127.21	37.62	65.4	−6.64	21.205	6.02
−4	19.2	1.8971	12.07	117.44	27.05	66.9	−4.43	20.803	4.01
−2	19.6	1.7525	3.53	105.66	14.31	68.4	−2.22	20.402	2.01
0	20.0	1.6928	0.00	92.43	0.00	70.0	0.00	20.000	0.00
2	20.4	1.7268	2.01	79.01	−14.52	71.6	2.23	19.600	−2.01
4	20.8	1.8493	9.24	66.74	−27.80	73.1	4.47	19.197	−4.01
6	21.2	2.0444	20.77	56.40	−38.98	74.7	6.73	18.796	−6.02
8	21.6	2.2937	35.49	48.08	−47.98	76.3	8.99	18.395	−8.03
10	22.0	2.5814	52.49	41.51	−55.10	77.9	11.26	17.994	−10.03

Example (4)

Table 7.2 Propagation of errors from D_1 to $\hat{d}_w$ or $\hat{V}_1$. A positive (negative) sign means that the erroneous value is greater (smaller) than the true value.

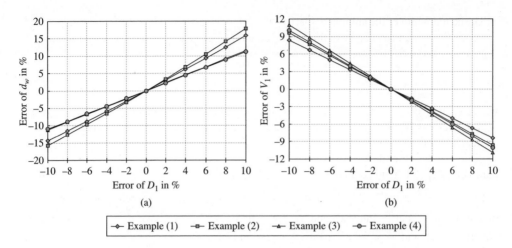

Figure 7.11   Error of D_1 versus errors of (a) $\hat{d}_w$ and (b) $\hat{V}_1$.

The interpretation of these experimental error propagations is as follows. In general, errors in ε_R and ε_ω increase (in absolute value) exponentially with respect to ε_{D_1}; however, exception may occur in some special situations. Furthermore, the rate of increase varies for positive or negative errors, or from example to example. (See the next chapter for a more general analysis.)

Besides exponential changes in ε_R or ε_ω, errors ε_{d_w} and ε_{V_1} change with respect to ε_{D_1} (surprisingly!) approximately linearly. Figure 7.11 plots propagation results for examples (1)–(4). Figure 7.11(a) shows ε_{D_1} versus ε_{d_w}. Lines for examples (1) and (2) have steeper slopes than those for examples (3) and (4), and this indicates that indoor acquisition might be more sensitive (with respect to ε_{d_w}) to errors in ε_{D_1} than outdoor acquisition. Figure 7.11(b) (ε_{D_1} versus ε_{V_1}) shows steeper slopes for (3) and (4) compared to (1) and (2).

In terms of magnitude, the error interval $[-10\%, +10\%]$ of D_1 introduces maximum errors of 75.13% for ε_R, and 133.68% for ε_ω, but (only) about 10% for both ε_{d_w} and ε_{V_1}. This indicates that parameters d_w and V_1 are less strongly influenced by error ε_{D_1}.

7.4.4 Error in the Distance to Outer Border of RoI

Analogously to the previous subsection, we now investigate the propagation of the independent error ε_{D_2}, first to calculated parameters R and ω, and then to d_w and V_1.

Unlike parameter $\hat{D}_1$, parameter $\hat{D}_2$ affects neither the number W of image columns nor the width θ_w of the angular disparity interval. But both parameters R and ω are affected, and we denote resulting erroneous values by $\hat{R}$ and $\hat{\omega}$. The corresponding errors are defined as

$$\varepsilon_R = \hat{R} - R$$
$$= f_R(D_1, \hat{D}_2, H_1, \theta_w) - f_R(D_1, D_2, H_1, \theta_w)$$

and

$$\varepsilon_\omega = \hat{\omega} - \omega$$
$$= f_\omega(D_1, \hat{D}_2, H_1, \theta_w) - f_\omega(D_1, D_2, H_1, \theta_w).$$

The width $\hat{\theta}_w$ of the angular disparity interval, assuming the use of previously determined parameters $\hat{R}$ and $\hat{\omega}$, can be computed by equation (7.13). This is abbreviated as

$$\hat{\theta}_w = f_{\theta_w}(D_1, D_2, \hat{R}, \hat{\omega}).$$

Equation (7.12) is used to compute the distance $\hat{H}_1$ to the target range, and the parameter dependency is given by $\hat{H}_1 = f_{H_1}(D_1, \hat{R}, \hat{\omega})$.

To analyze the results of error propagation, we convert the width $\hat{\theta}_w$ of the angular disparity interval into the width $\hat{d}_w$ of the image disparity interval, and $\hat{H}_1$ into $\hat{V}_1$. We have $\hat{d}_w = f_{d_w}(\hat{\theta}_w, W)$ using equation (7.2), and $\hat{V}_1 = f_{V_1}(\hat{H}_1, \upsilon)$ using equation (7.1).

For numerical data on the error propagation (of the D_2 error into errors of R and ω, and then into errors of d_w and V_1), see Table 7.3. Again we use the four scene examples as in Table 7.1. All the errors are measured as percentages, and definitions are shown at the top of each column. $\hat{D}_2$, $\hat{R}$ and $\hat{V}_1$ are measured in meters, $\hat{\omega}$ is given in degrees, and $\hat{d}_w$ in pixels. We provide data for the error interval $[-10\%, +10\%]$ for D_2. A positive (negative) sign means the value is greater (smaller) than the true value. The shaded middle rows of the four tables show error-free values for reference.

The greater the magnitude of ε_{D_2} the greater the magnitudes of ε_R, ε_ω and ε_{d_w}, in all four cases. Although the errors change exponentially, the quantities are in fact very small for the interval $[-10\%, +10\%]$ in comparison to those for ε_{D_1}. The maximum errors of ε_R, ε_ω and ε_{d_w} are 7.02%, 1.05% and 7.97%, compared to 75.13%, 133.68% and 17.94% for ε_{D_1}.

Figure 7.12 shows steeper slopes for examples (1) and (2) compared to those for examples (3) and (4); this indicates that ε_{d_w} might be more sensitive to ε_{D_2} for indoor acquisition than for outdoor cases. Moreover, in general, rates of change of ε_R, ε_ω and ε_{d_w} with respect to the change in ε_{D_2} are larger for negative errors than for positive errors. Note that for the indoor case, the worst case is about 7.97% ε_{d_w} for the interval $[-10\%, +10\%]$ of ε_{D_2}; but for the outdoor cases, the worst case is about 1.54% ε_{d_w} (i.e., about one pixel difference to the true width of the disparity interval, which only causes a minor loss of stereoacuity).

$\frac{\varepsilon_{D_2}(\%)}{D_2}$	$\hat{D}_2$	$\hat{R}$	$\frac{\varepsilon_R(\%)}{R}$	$\hat{\omega}$	$\frac{\varepsilon_\omega(\%)}{\omega}$	$\hat{d}_w$	$\frac{\varepsilon_{d_w}(\%)}{d_w}$	$\hat{V}_1$	$\frac{\varepsilon_{V_1}(\%)}{V_1}$
-10	2.70	0.2553	2.14	145.53	-0.92	74.1	5.86	2.0067	0.00
-8	2.76	0.2540	1.65	145.83	-0.72	73.2	4.53	2.0067	0.00
-6	2.82	0.2529	1.19	146.12	-0.52	72.3	3.28	2.0067	0.00
-4	2.88	0.2518	0.76	146.39	-0.34	71.5	2.12	2.0067	0.00
-2	2.94	0.2509	0.37	146.64	-0.16	70.7	1.03	2.0067	0.00
0	3.00	0.2499	0.00	146.88	0.00	70.0	0.00	2.0067	0.00
2	3.06	0.2491	-0.35	147.11	0.16	69.3	-0.97	2.0067	0.00
4	3.12	0.2483	-0.67	147.33	0.31	68.7	-1.88	2.0067	0.00
6	3.18	0.2475	-0.97	147.54	0.45	68.1	-2.74	2.0067	0.00
8	3.24	0.2468	-1.26	147.74	0.58	67.5	-3.56	2.0067	0.00
10	3.30	0.2461	-1.53	147.93	0.71	67.0	-4.33	2.0067	0.00

Example (1)

$\frac{\varepsilon_{D_2}(\%)}{D_2}$	$\hat{D}_2$	$\hat{R}$	$\frac{\varepsilon_R(\%)}{R}$	$\hat{\omega}$	$\frac{\varepsilon_\omega(\%)}{\omega}$	$\hat{d}_w$	$\frac{\varepsilon_{d_w}(\%)}{d_w}$	$\hat{V}_1$	$\frac{\varepsilon_{V_1}(\%)}{V_1}$
-10	9.0	0.6217	7.02	112.85	-0.94	75.6	7.97	7.0237	0.00
-8	9.2	0.6123	5.39	113.08	-0.74	74.3	6.13	7.0237	0.00
-6	9.4	0.6035	3.89	113.30	-0.54	73.1	4.43	7.0237	0.00
-4	9.6	0.5955	2.50	113.51	-0.36	72.0	2.85	7.0237	0.00
-2	9.8	0.5879	1.21	113.72	-0.17	71.0	1.37	7.0237	0.00
0	10.0	0.5809	0.00	113.92	0.00	70.0	0.00	7.0237	0.00
2	10.2	0.5744	-1.13	114.11	0.17	69.1	-1.29	7.0237	0.00
4	10.4	0.5682	-2.18	114.30	0.33	68.3	-2.49	7.0237	0.00
6	10.6	0.5625	-3.17	114.47	0.49	67.5	-3.63	7.0237	0.00
8	10.8	0.5571	-4.11	114.65	0.64	66.7	-4.69	7.0237	0.00
10	11.0	0.5520	-4.98	114.81	0.78	66.0	-5.70	7.0237	0.00

Example (2)

$\frac{\varepsilon_{D_2}(\%)}{D_2}$	$\hat{D}_2$	$\hat{R}$	$\frac{\varepsilon_R(\%)}{R}$	$\hat{\omega}$	$\frac{\varepsilon_\omega(\%)}{\omega}$	$\hat{d}_w$	$\frac{\varepsilon_{d_w}(\%)}{d_w}$	$\hat{V}_1$	$\frac{\varepsilon_{V_1}(\%)}{V_1}$
-10	45.0	0.6815	0.70	45.13	1.05	71.1	1.54	5.5000	0.00
-8	46.0	0.6804	0.55	45.03	0.82	70.8	1.20	5.5000	0.00
-6	47.0	0.6795	0.40	44.93	0.60	70.6	0.88	5.5000	0.00
-4	48.0	0.6785	0.26	44.84	0.39	70.4	0.57	5.5000	0.00
-2	49.0	0.6776	0.13	44.75	0.19	70.2	0.28	5.5000	0.00
0	50.0	0.6768	0.00	44.66	0.00	70.0	0.00	5.5000	0.00
2	51.0	0.6759	-0.12	44.58	-0.18	69.8	-0.27	5.5000	0.00
4	52.0	0.6752	-0.24	44.50	-0.36	69.6	-0.52	5.5000	0.00
6	53.0	0.6744	-0.35	44.43	-0.53	69.5	-0.77	5.5000	0.00
8	54.0	0.6737	-0.45	44.35	-0.69	69.3	-1.00	5.5000	0.00
10	55.0	0.6730	-0.55	44.28	-0.85	69.1	-1.22	5.5000	0.00

Example (3)

$\frac{\varepsilon_{D_2}(\%)}{D_2}$	$\hat{D}_2$	$\hat{R}$	$\frac{\varepsilon_R(\%)}{R}$	$\hat{\omega}$	$\frac{\varepsilon_\omega(\%)}{\omega}$	$\hat{d}_w$	$\frac{\varepsilon_{d_w}(\%)}{d_w}$	$\hat{V}_1$	$\frac{\varepsilon_{V_1}(\%)}{V_1}$
-10	180.0	2.6674	1.25	92.46	0.03	70.9	1.25	20.000	0.00
-8	184.0	2.3708	0.97	92.46	0.03	70.7	0.97	20.000	0.00
-6	188.0	2.1092	0.71	92.45	0.02	70.5	0.71	20.000	0.00
-4	192.0	1.8971	0.46	92.44	0.01	70.3	0.46	20.000	0.00
-2	196.0	1.7525	0.23	92.44	0.01	70.2	0.23	20.000	0.00
0	200.0	1.6928	0.00	92.43	0.00	70.0	0.00	20.000	0.00
2	204.0	1.7268	-0.22	92.43	-0.01	69.8	-0.22	20.000	0.00
4	208.0	1.8493	-0.42	92.42	-0.01	69.7	-0.42	20.000	0.00
6	212.0	2.0444	-0.62	92.42	-0.02	69.6	-0.62	20.000	0.00
8	216.0	2.2937	-0.81	92.41	-0.02	69.4	-0.82	20.000	0.00
10	220.0	2.5814	-1.00	92.41	-0.03	69.3	-1.00	20.000	0.00

Example (4)

Table 7.3 Propagation of errors from D_2 to $\hat{d}_w$ and $\hat{V}_1$. A positive (negative) sign means that the erroneous value is greater (smaller) than the true value.

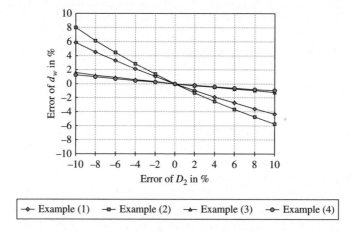

Figure 7.12   Error of D_2 versus error of d_w.

In Table 7.3 we have $\varepsilon_{V_1} = 0$, which suggests that ε_{D_2} introduces no error to V_1 at all. This result can be interpreted as follows. Since $\hat{V}_1 = f_{V_1}(\hat{H}_1, \upsilon)$ and $V_1 = f_{V_1}(H_1, \upsilon)$, we know that $\hat{V}_1 = V_1$ if $\hat{H}_1 = H_1$. Furthermore, $\hat{H}_1 = H_1$ if $f_{H_1}(D_1, \hat{R}, \hat{\omega}) = f_{H_1}(D_1, R, \omega)$, where f_{H_1} is defined by equation (7.12).

It is very complicated to show $\hat{H}_1 = H_1$ algebraically. We can use the fact that if two circles $\hat{\mathcal{H}}$ and $\mathcal{H}$ (see Figure 7.7) with radii $\hat{H}_1$ and H_1, respectively, coincide then $\hat{H}_1 = H_1$. Since $R = f_R(D_1, D_2, H_1, \theta_w)$ and $\omega = f_\omega(D_1, D_2, H_1, \theta_w)$, $f_{H_1}(D_1, R, \omega)$ defines a circle $\mathcal{H}$ whose center is D_1 away from the center **O** of the base circle, and the radius is H_1. Furthermore, since $\hat{R} = f_R(D_1, \hat{D}_2, H_1, \theta_w)$ and $\hat{\omega} = f_\omega(D_1, \hat{D}_2, H_1, \theta_w)$, $f_{H_1}(D_1, \hat{R}, \hat{\omega})$ defines a circle $\hat{\mathcal{H}}$ whose center is D_1 away from the center **O** of the base circle, and the radius is H_1.

Both circles $\hat{\mathcal{H}}$ and $\mathcal{H}$ may coincide if both centers are in the inner circle of radius D_1, and the radii are equal to H_1. In this case we have $\hat{H}_1 = H_1$, and thus $\hat{V}_1 = V_1$.

To conclude, ε_{D_2} has an impact on image quality results only with respect to stereoacuity (assuming the parameter specification as in Theorem 7.1).

7.4.5 Error in the Distance to Target Range

Error ε_{H_1} causes error in $\hat{W}$, and this leads to an erroneous $\hat{\theta}_w = f_{\theta_w}(d_w, \hat{W})$. Both parameters R and ω in equations (7.9) and (7.11) depend on the four variables D_1, D_2, H_1 and θ_w. Thus, erroneous values $\hat{W}$ and $\hat{\theta}_w$ cause erroneous values $\hat{R}$ and $\hat{\omega}$, and the errors are respectively defined as

$$\varepsilon_R = \hat{R} - R$$
$$= f_R(D_1, D_2, \hat{H}_1, \hat{\theta}_w) - f_R(D_1, D_2, H_1, \theta_w)$$

and

$$\varepsilon_\omega = \hat{\omega} - \omega$$
$$= f_\omega(D_1, D_2, \hat{H}_1, \hat{\theta}_w) - f_\omega(D_1, D_2, H_1, \theta_w).$$

The width $\hat{\theta}_w$ of the angular disparity interval is now assumed to be calculated from equation (7.13), using previously calculated control values $\hat{R}$ and $\hat{\omega}$; we use

$$\hat{\theta}_w = f_{\theta_w}(D_1, D_2, \hat{R}, \hat{\omega})$$

in the error propagation experiments. Because

$$\hat{R} = f_R(D_1, D_2, \hat{H}_1, \hat{\theta}_w), \quad \hat{\omega} = f_\omega(D_1, D_2, \hat{H}_1, \hat{\theta}_w)$$

and

$$\hat{\theta}_w = f_{\theta_w}(D_1, D_2, \hat{R}, \hat{\omega})$$

share the same parameters D_1, D_2, $\hat{H}_1$, $\hat{R}$, and $\hat{\omega}$, it can be shown that

$$\hat{\theta}_w = \hat{\theta}_w.$$

Equation (7.12) is used to compute the distance $\hat{H}_1$ between optical centers and the inner cylinder of the RoI. Parameter dependency is expressed as $\hat{H}_1 = f_{H_1}(D_1, \hat{R}, \hat{\omega})$. We also need to convert the width $\hat{\theta}_w$ of the angular disparity interval into the width $\hat{d}_w$ of the image disparity interval, and $\hat{H}_1$ into $\hat{V}_1$. We have

$$\hat{d}_w = f_{d_w}(\hat{\theta}_w, \hat{W})$$

using equation (7.2), and $\hat{V}_1 = f_{V_1}(\hat{H}_1, \upsilon)$ using equation (7.1).

For numerical data of error propagation (first error ε_{H_1} to errors in R and ω, and then to errors in d_w and V_1), see Table 7.4. We use the four examples of Table 7.1. As before, an error

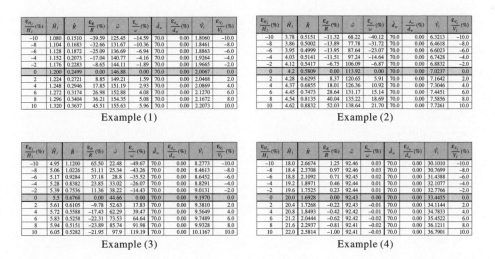

Example (1)

$\frac{\varepsilon_{H_1}}{H_1}$ (%)	$\hat{H}_1$	$\hat{R}$	$\frac{\varepsilon_R}{R}$ (%)	$\hat{\omega}$	$\frac{\varepsilon_\omega}{\omega}$ (%)	$\hat{d}_w$	$\frac{\varepsilon_{d_w}}{d_w}$ (%)	$\hat{V}_1$	$\frac{\varepsilon_{V_1}}{V_1}$ (%)
-10	1.080	0.1510	-39.59	125.45	-14.59	70.0	0.00	1.8060	-10.0
-8	1.104	0.1683	-32.66	131.67	-10.36	70.0	0.00	1.8461	-8.0
-6	1.128	0.1872	-25.09	136.69	-6.94	70.0	0.00	1.8863	-6.0
-4	1.152	0.2073	-17.04	140.77	-4.16	70.0	0.00	1.9264	-4.0
-2	1.176	0.2283	-8.65	144.11	-1.89	70.0	0.00	1.9665	-2.0
0	1.200	0.2499	0.00	146.88	0.00	70.0	0.00	2.0067	0.0
2	1.224	0.2721	8.85	149.21	1.59	70.0	0.00	2.0468	2.0
4	1.248	0.2946	17.85	151.19	2.93	70.0	0.00	2.0869	4.0
6	1.272	0.3174	26.98	152.88	4.08	70.0	0.00	2.1270	6.0
8	1.296	0.3404	36.21	154.35	5.08	70.0	0.00	2.1672	8.0
10	1.320	0.3637	45.51	155.63	5.96	70.0	0.00	2.2073	10.0

Example (2)

$\frac{\varepsilon_{H_1}}{H_1}$ (%)	$\hat{H}_1$	$\hat{R}$	$\frac{\varepsilon_R}{R}$ (%)	$\hat{\omega}$	$\frac{\varepsilon_\omega}{\omega}$ (%)	$\hat{d}_w$	$\frac{\varepsilon_{d_w}}{d_w}$ (%)	$\hat{V}_1$	$\frac{\varepsilon_{V_1}}{V_1}$ (%)
-10	3.78	0.5151	-11.32	68.22	-40.12	70.0	0.00	6.3213	-10.0
-8	3.86	0.5002	-13.89	77.78	-31.72	70.0	0.00	6.4618	-8.0
-6	3.95	0.4999	-13.95	87.64	-23.07	70.0	0.00	6.6023	-6.0
-4	4.03	0.5141	-11.51	97.24	-14.64	70.0	0.00	6.7428	-4.0
-2	4.12	0.5417	-6.75	106.09	-6.87	70.0	0.00	6.8832	-2.0
0	4.2	0.5809	0.00	113.92	0.00	70.0	0.00	7.0237	0.0
2	4.28	0.6295	8.37	120.65	5.91	70.0	0.00	7.1642	2.0
4	4.37	0.6855	18.01	126.36	10.92	70.0	0.00	7.3046	4.0
6	4.45	0.7473	28.64	131.17	15.14	70.0	0.00	7.4451	6.0
8	4.54	0.8135	40.04	135.22	18.69	70.0	0.00	7.5856	8.0
10	4.62	0.8832	52.03	138.64	21.70	70.0	0.00	7.7261	10.0

Example (3)

$\frac{\varepsilon_{H_1}}{H_1}$ (%)	$\hat{H}_1$	$\hat{R}$	$\frac{\varepsilon_R}{R}$ (%)	$\hat{\omega}$	$\frac{\varepsilon_\omega}{\omega}$ (%)	$\hat{d}_w$	$\frac{\varepsilon_{d_w}}{d_w}$ (%)	$\hat{V}_1$	$\frac{\varepsilon_{V_1}}{V_1}$ (%)
-10	4.95	1.1200	65.50	22.48	-49.67	70.0	0.00	8.2773	-10.0
-8	5.06	1.0226	51.11	25.34	-43.26	70.0	0.00	8.4613	-8.0
-6	5.17	0.9284	37.18	28.8	-35.52	70.0	0.00	8.6452	-6.0
-4	5.28	0.8382	23.85	33.02	-26.07	70.0	0.00	8.8291	-4.0
-2	5.39	0.7536	11.36	38.22	-14.43	70.0	0.00	9.0131	-2.0
0	5.5	0.6768	0.00	44.66	0.00	70.0	0.00	9.1970	0.0
2	5.61	0.6105	-9.78	52.63	17.83	70.0	0.00	9.3810	2.0
4	5.72	0.5588	-17.43	62.29	39.47	70.0	0.00	9.5649	4.0
6	5.83	0.5258	-22.31	73.53	64.64	70.0	0.00	9.7489	6.0
8	5.94	0.5151	-23.89	85.74	91.98	70.0	0.00	9.9328	8.0
10	6.05	0.5282	-21.95	97.9	119.19	70.0	0.00	10.1167	10.0

Example (4)

$\frac{\varepsilon_{H_1}}{H_1}$ (%)	$\hat{H}_1$	$\hat{R}$	$\frac{\varepsilon_R}{R}$ (%)	$\hat{\omega}$	$\frac{\varepsilon_\omega}{\omega}$ (%)	$\hat{d}_w$	$\frac{\varepsilon_{d_w}}{d_w}$ (%)	$\hat{V}_1$	$\frac{\varepsilon_{V_1}}{V_1}$ (%)
-10	18.0	2.6674	1.25	92.46	0.03	70.0	0.00	30.1010	-10.0
-8	18.4	2.3708	0.97	92.46	0.03	70.0	0.00	30.7699	-8.0
-6	18.8	2.1092	0.71	92.45	0.02	70.0	0.00	31.4388	-6.0
-4	19.2	1.8971	0.46	92.44	0.01	70.0	0.00	32.1077	-4.0
-2	19.6	1.7525	0.23	92.44	0.01	70.0	0.00	32.7766	-2.0
0	20.0	1.6928	0.00	92.43	0.00	70.0	0.00	33.4455	0.0
2	20.4	1.7268	-0.22	92.43	-0.01	70.0	0.00	34.1144	2.0
4	20.8	1.8493	-0.42	92.42	-0.01	70.0	0.00	34.7833	4.0
6	21.2	2.0444	-0.62	92.42	-0.02	70.0	0.00	35.4522	6.0
8	21.6	2.2937	-0.81	92.41	-0.02	70.0	0.00	36.1211	8.0
10	22.0	2.5814	-1.00	92.41	-0.03	70.0	0.00	36.7901	10.0

Table 7.4 Propagation of errors from H_1 to $\hat{d}_w$ and $\hat{V}_1$. The positive (negative) sign means that the erroneous value is greater (smaller) than the true value.

interval of $[-10\%, +10\%]$ is considered for ε_{H_1}. All the errors are measured as percentages, and definitions are shown at the top of each column. $\hat{D}_1$, $\hat{R}$ and $\hat{V}_1$ are measured in meters, $\hat{\omega}$ is in degrees, and $\hat{d}_w$ in pixels.

The larger the magnitude of ε_{H_1}, the larger the magnitude of ε_R and ε_ω (in general). Exceptions do occur: see (for example) ε_R for positive or negative errors of ε_{H_1} in examples (2) and (3). The rates of change of ε_R and ε_ω with respect to ε_{H_1} are exponential, and vary (with different speed) for positive or negative errors in the four examples. (The general behavior of such parameter interactions is discussed in the next chapter.)

Although the rates of change of ε_R and ε_ω are exponential, those of ε_{d_w} and ε_{V_1} with respect to ε_{H_1} are surprisingly "systematic". First, there is no impact of ε_{H_1} on d_w, that is, $\hat{d}_w = d_w$. This explains why

$$\hat{\hat{\theta}}_w = \hat{\theta}_w$$

and $\hat{W}$, as used in $\theta_w = f_{\theta_w}(d_w, W)$, is the same as used in $d_w = f_{d_w}(\theta_w, W)$. Second, the impact of ε_{H_1} on V_1 shows a constant error percentage. This can be understood based on equation (7.1), which shows that the relation between V_1 and H_1 is only defined by the scale factor $2\tan(\upsilon/2)$.

To conclude, based on our image quality control method, ε_{H_1} has no effect on the resultant stereoscopic panoramas with respect to stereoacuity (i.e., numbers of depth layers), but it has an impact, characterized by identical source-error percentages, on other parameters of scene composition.

7.5 Exercises

7.1. Assume that we wish to take a photo of a tree, 5 meters tall, using a camera whose vertical field of view is equal to $40°$. How far from the tree should the camera be positioned to ensure that both of the following specifications are satisfied: first, the tree is fully shown in the photo (assuming no occlusion); and second, the projection of the tree on the image covers 80% of the image's height? (Hint: use equation (7.1).)

7.2. Suppose that we have a 17-inch screen of resolution $1,024 \times 768$, and a stereo viewing distance that is equal to 0.4 m. Show that the upper disparity limit is about 70 pixels.

7.3. Prove equations (7.12) and (7.13).

7.4. [Possible lab project] This project is a continuation of Exercise 4.2. Having calculated corresponding points for the symmetric stereo panorama, apply the formula for calculating depth D based on disparity or angular disparity. This allows a set of 3D points to be generated in the 3D space, basically one 3D point for every pair of corresponding points. Try to obtain a fairly dense population of 3D points, and visualize them by mapping color values of original pixels onto those 3D points. Allow a fly-through visualization of the calculated set of colored 3D points.

7.5. [Possible lab project] This project is a continuation of Exercise 6.4. For your symmetric pairs of panoramic images, calculate corresponding points, the depth of projected 3D points, and evaluate the accuracy of recovered surface points in comparison to your synthetic 3D object.

7.6 Further Reading

On the health and comfort issues of stereo viewing, such as dipodia, see Viire (1997), Siegel et al. (1999) and Mayer et al. (2000). Insufficient stereoacuity produces a cardboard effect; see Yamanoue et al. (2000). Cardboard effects and dipodia are mentioned, for example, in Shum and He (1999) and Peleg and Ben-Ezra (1999).

The maximum disparity limit for human stereo fusion is discussed in Valyrus (1966), resulting in the suggestion of the maximum value of

$$0.03 \times \text{viewing distance.}$$

This value corresponds to discussions of "fusibility" of corresponding points in other references; see Howard and Rogers (1995) for particular viewing conditions, or Roberts and Slattery (2000) and Ware et al. (1998) for viewing distances in relation to the size of a computer screen, with special calculations in Siegel and Nagata (2000) leading to the 70 pixel value mentioned. Calculations based on the upper disparity limit, as suggested in Valyrus (1966), also lead to this value of $d \approx 70$ pixels. Stelmach et al. (2000) discuss stereo image quality (for non-panoramic images).

For discussions of binocular "fusibility", see Howard and Rogers (1995), Roberts and Slattery (2000) and Ware et al. (1998).

The following publications take image quality issues into account in their studies of panorama acquisition: Ishiguro et al. (1992), Shum and He (1999), Shum et al. (1999), Shum and Szeliski (1999), Peleg and Ben-Ezra (1999), Peleg et al. (2000) and Huang and Pajdla (2000). The geometric relation between camera and RoI was studied for the first time in Wei et al. (2002b). The possible range of R is $0 < R < D_1$ in general; in practice the value of R is subject to equipment constraints (e.g., at most 1.0 m in (Huang et al., 2001a).

Approaches which restrict camera parameters to constant values, as, for example, in Shum et al. (1999), cannot adapt to image quality. In the case of single-center panoramic acquisition (see, for example, Chen, 1995), we can choose the number W of image columns independently of the 3D scene complexity.

For the panoramic image in Figure 7.1, see Huang et al. (2001a). For the WAAC camera, see Reulke and Scheele (1998).

8

Sensor Analysis and Design

The control method, as provided in the previous chapter, does not allow solutions for all possible quadruples of four individual input values for D_1, D_2, H_1, and θ_w. There are geometric constraints (e.g., $D_1 < D_2$) or relations between these parameters which restrict the set of possible input values. Furthermore, such constraints or relations also restrict the set of all 6-tuples $(D_1, D_2, H_1, \theta_w, R, \omega)$. If R is only between 0 and 0.2 m, then this (eventually) results in constraints for the other five parameters.

This chapter investigates dependencies and interactions between these six parameters. To be more precise, we analyze validity ranges, interrelationships, and characteristics of value changes for these six sensor parameters. The aim is to study what can be achieved with respect to camera (or sensor) analysis and design.

8.1 Introduction

Cameras or sensors are limited with respect to possible parameter settings. A value of $R = 1.0$ m, as available to us in some experiments, can certainly not be guaranteed in general. There might be a customer expectation that R should be (say) 0.4 m at most. The question occurs, for which scenes such a sensor still would allow optimum (or "near-optimum") parameter settings, and how such a limitation of R interacts with other sensor or scene parameters.[1] In order to answer these questions, sensor (or camera) parameter analysis is essential; this is the main subject of this chapter.

Consider another example; assume that there are various ranges of interest within some 3D scene, such that the distances D_1 (between sensor and closest scene objects) and the number of depth levels (calculated based on the given value of θ_w) may be specified by two intervals of appropriate values, respectively, rather than by two specific constants. The question arises: what values of the off-axis distance R and principal angle ω are sufficient to satisfy

---

[1] Applications can create extreme parameter settings. For example, when a line camera was placed at a window of a revolving restaurant (in Auckland, spinning at maximum possible speed), R was actually equal to about 12 m.

Panoramic Imaging: Sensor-Line Cameras and Laser Range-Finders F. Huang, R. Klette, and K. Scheibe
© 2008 John Wiley & Sons, Ltd

the demands for this particular application? This type of question leads to a particular sensor design issue, and will also be discussed in this chapter.

This chapter draws attention to fundamental issues of sensor analysis and design with respect to the capability and controllability for scene composition and stereoacuity (over a dynamic range of 3D scenes). Such a study provides fundamentals for the understanding of acquisition geometry of stereoscopic panoramic imaging (e.g., decisions for multi-view sensor positions).

The chapter is divided into three main sections. Sections 8.2 and 8.3 provide a comprehensive sensor parameter analysis with respect to scene composition and stereoacuity, respectively. The discussion of both issues (i.e., with respect to analysis methods and steps) follows a similar pattern, and the presentations of results are actually very similar in these two sections. Section 8.4 focuses on the issue of sensor (or camera) parameter design.

8.2 Scene Composition Analysis

Our interest is focused on six parameters in total. If any four of them are fixed, then the other two are either uniquely defined or, in some cases, have exactly two possible values. For example, in the previous chapter we showed that when the values of parameters D_1, D_2, H_1 and σ_w are specified, then the values of R and ω are always uniquely determined.

If any three of these six parameters are fixed, then the values of the other three parameters are constrained to lie in certain intervals, in order to satisfy $0 \leq R < D_1 < D_2$. More specifically, Figure 7.9 leads to the conclusion that if any three of the six parameters are bounded (i.e., by lower and upper bounds), then valid intervals follow for the remaining three parameters. For example, if the application-specific parameters D_1, D_2 and θ_w are bounded, then (only) this fact already implies that the values of parameters H_1, R, and ω have to lie in particular intervals. We call such intervals *conditionally valid*.

By specifying bounds (i.e., *endpoints* of intervals) for any three application-specific parameters, we are able to compute the conditionally valid intervals of the remaining (fourth) application-specific parameter and of both sensor parameters. This allows more flexibility in image quality control.

8.2.1 Simplifications, Main Focus, and Layout

Throughout this chapter, let (for the purpose of simplifying formal expressions)

$$\sigma_w = \frac{\theta_w}{2}$$

be half the width of a given angular disparity interval, satisfying $0° < \sigma_w < 90°$. Figure 8.1 demonstrates an invalid case with $\sigma_w > 90°$. Because $\sigma_w > 90°$ implies $\angle 1 < 90°$, which then implies $\angle 2 > 90°$, the value of R would always be greater than D_1 in this case. Thus we conclude that the value of σ_w has to be less than 90°.

Because geometric relations between the six parameters are symmetric with respect to the normal vector of the base circle, we can always assume that $0° \leq \omega \leq 180°$.

In this section, we study how the sensor parameters R and ω interact with the application-specific parameter H_1 while the values of the parameters D_1, D_2, and θ_w (and hence σ_w) are fixed. In practice, this means that if the location of panorama acquisition is determined (i.e., the

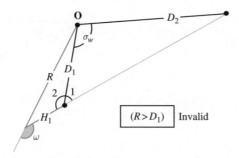

Figure 8.1   Example of an invalid case, with $\sigma_w > 90°$.

scene of interest is given) and the stereo display method is known (thus, an estimated θ_w value is also given), then with the availability of an extension slider or bar (i.e., the off-axis distance R) and various values of ω, what possible scene composition can be offered (i.e., possible range of H_1)?

We specify the following:

- conditionally valid intervals for H_1, R, and ω;
- *mappings* characterizing convergence relations between endpoints of these three intervals; and
- the influence of value changes for one of the parameters H_1, R, or ω on values of the remaining two parameters.

For these purposes, a *geometric analysis* (i.e., a scheme for qualitative analysis) is always followed by an *algebraic analysis* (i.e., a scheme for quantitative analysis).

A geometric analysis will always proceed as follows: first we specify preconditions of the three constants (i.e., to ensure that valid intervals exist for the three variables), then we study *interactions* (i.e., how a change of one variable influences the others) between the three variables, calculate the conditionally valid intervals of the variables, and finally discuss mappings between endpoints of valid intervals of different variables.

In our algebraic analysis, we express relations between the six parameters based on both functions (for R and ω) provided earlier for image quality control. Using these functions, we derive formulas for computing endpoints as well as extrema identified before in the geometric analysis.

8.2.2 Geometric Analysis

In the following, parameters H_1, R, and ω are variables, and D_1, D_2, and σ_w are constants, such that $D_1 < D_2$ and $0° < \sigma_w < 90°$.

Interaction between Variables

When the value of one of the variables changes, then the values of the other two will change accordingly. We discuss this first from a qualitative point of view; see Figure 8.2.

In Figure 8.2(a), the value of D_1 is less than the value of D_2; the value of σ_w is less than $90°$; both values of R and H_1 are less than D_1; and the value of ω is greater than $90°$. In the

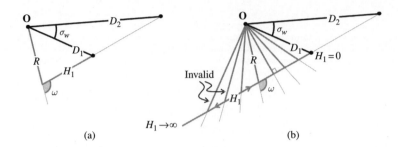

Figure 8.2   Qualitative (geometric) interaction among changes in H_1, R, and ω, while values of D_1, D_2, and σ_w are kept constant. (a) A simple example showing a valid combination of the variables. (b) Seven different combinations of the variables, including two invalid cases.

figure we use thicker black lines for constant parameters and thicker gray lines for variable parameters.

In Figure 8.2(b), a few combinations of values of H_1, R, and ω are shown to illustrate valid and invalid cases. Based on studying such combinations of values of H_1, R, and ω, we are able to conclude the following general rules:

- As H_1 increases from zero to infinity, the corresponding value of R starts at D_1, first decreases, and then increases towards infinity.
- The value of ω increases as H_1 increases.
- As ω increases towards $180°$, the value of R starts at D_1, decreases until $\omega = 90°$, and then increases towards infinity.

These results also explain a phenomenon discovered within the error analysis for H_1 in the previous chapter. We recall example (2) in Table 7.1. We noticed in this case that, while the error of H_1 increases, the value of $\hat{R}$ decreases, regardless of the sign of the error (see the upper three rows of example (2) in Table 7.4). For a similar phenomenon, see the bottom two rows of example (3) in Table 7.4.

In order to explain these phenomena, it is necessary to introduce a new notation, $H_{1\min R}$, in which the subscript min R is added to indicate the value of H_1 when R is minimum. The formula for calculating $H_{1\min R}$ will be derived and given later in equation (8.4) of Section 8.2.3.

Both aforementioned phenomena occur because the true value of H_1 is very close to $H_{1\min R}$ such that the error interval $[-10\%, +10\%]$ of ε_{H_1} contains the value $H_{1\min R}$. For example, the true value of H_1 is 4.2 m, the $[-10\%, +10\%]$ error interval as shown in example (2) of Table 7.4 is [3.78 m, 4.62 m]. This interval contains $H_{1\min R}$, which is equal to 3.96 m. Thus, as the value of $\hat{H}_1$ decreases, the error of H_1, ε_{H_1}, increases, we can see that $\hat{R}$ decreases at first, namely until $\hat{H}_1 = H_{1\min R}$, and increases afterwards. Therefore, correspondingly, the value of ε_R first increases and then starts to decrease, after the value of $\hat{H}_1$ reaches $H_{1\min R}$.

Conditionally Valid Intervals

There are infinitely many solutions of H_1, R, and ω for constants D_1, D_2, and σ_w under the constraints $D_1 < D_2$ and $\sigma_w < 90°$. However, these valid values of H_1, R, and ω are limited to smaller intervals, due to the constraint that $R < D_1$.

This means that the values of H_1, R, and ω cannot be specified arbitrarily for image acquisition. The range of the valid values for each variable is specified by a *conditionally valid interval*. If any value of one of the three variables is given and lies within its conditionally valid interval, then there exist corresponding values of the other two variables that lie within their own conditionally valid intervals.

The following subsections address the endpoints of the conditionally valid intervals of all three variables and the mapping between endpoints among those variables.

Endpoints of the conditionally valid interval of H_1

For H_1 we can initially assume any non-zero positive real. For inferring endpoints of a conditionally valid interval of H_1, we may refer to Figure 8.2(b).

First, we have that $\lim_{H_1 \to 0+} R = D_1$ and $\lim_{H_1 \to \infty} R = \infty$: as the value of H_1 increases from zero to infinity, the corresponding value of R starts at D_1, first decreases, and then increases towards infinity. We also have the constraint $R < D_1$. Therefore, there exists a conditional maximum value of H_1 denoted by H_{1+}, and the conditionally valid interval of H_1 becomes $(0, H_{1+})$.[2] This implies that $\lim_{H_1 \to H_{1+}} R = D_1$.

Endpoints of the conditionally valid interval of R

The upper endpoint of the conditionally valid interval of R is equal to D_1. Figure 8.2(b) also shows that there exists a minimum value $R_{\min} > 0$ of R, when $\omega = 90°$. Thus the conditionally valid interval of R is $[R_{\min}, D_1)$.

Endpoints of the conditionally valid interval of ω

We already have that $0° < \omega < 180°$. From Figure 8.2(b) we can also see that $0° < \lim_{H_1 \to 0+} \omega < 180°$ and $\lim_{H_1 \to \infty} \omega = 180°$. From previous conclusions we know that ω increases if H_1 increases. Thus, ω is always in the permitted range for any finite value of H_1. Because there exists a maximum value of H_1, there also exists an upper endpoint ω_+ of the conditionally valid interval of ω. Moreover, we know that when the value of H_1 goes to zero, the limit of ω equals to a minimum $\omega_{\min}$. Thus, the conditionally valid interval of ω is $(\omega_{\min}, \omega_+)$.

Mapping of Endpoints

We discuss all three pairs of two (out of the three) variables. Mappings specify convergence relations.

Off-axis distance R and distance H_1 to target range

When H_1 lies in its conditionally valid interval, the corresponding value of R is always less than the constant D_1. When the value of H_1 increases from zero to H_{1+}, the corresponding value of R starts at constant D_1, decreases until it reaches a minimum at $R_{\min}$, and then increases towards constant D_1. Therefore we have the following mappings of endpoints:

$$\lim_{H_1 \to 0+} R = \lim_{H_1 \to H_{1+}} R = D_1.$$

---

[2] The symbol plus "+" used in the subscript implies the "upper bound" of the conditionally valid interval.

$$
\begin{array}{ccccc}
R: & D_1 & > & R_{\min} & < & D_1 \\[2pt]
   & \updownarrow & & \updownarrow & & \updownarrow \\[2pt]
H_1: & 0 & < & H_{1\min R} & < & H_{1+} \\[2pt]
   & \updownarrow & & \updownarrow & & \updownarrow \\[2pt]
\omega: & \omega_{\min} & < & 90^\circ & < & \omega_+
\end{array}
$$

Table 8.1 Sketch of endpoint-value mappings and relations among H_1, R, and ω.

Note that these mappings are also reversible but not unique (i.e., $\lim_{R \to D_1^-} H_1 = 0$ or $= H_{1+}$). For the minimum $R_{\min}$, there exists a corresponding value of H_1, denoted by $H_{1\min R}$, such that

$$
\lim_{H_1 \to H_{1\min R}} R = R_{\min}.
$$

The inverse relation is also true (i.e., $\lim_{R \to R_{\min}} H_1 = H_{1\min R}$).

These mappings are summarized in the two top rows of Table 8.1, based on the following interpretations. The variable is shown in the leftmost column, followed (to the right of the colon) by its endpoints and/or extrema. Symbols "<" or ">" between endpoints and/or extrema have obvious meanings. Dotted bidirectional arrows between rows (three in each column) symbolize direct dependencies. For example, the top leftmost arrow in this diagram stands for "when the value of R goes to D_1 then the value of H_1 goes to zero, and vice versa".

In practice, conditionally valid intervals of the three variables normally have to be reduced due to existing conditions (imposed by camera, 3D scene geometry, and so on). Studies of interactions between the three variables (within their conditionally valid intervals) provide information on how a change in a conditionally valid interval for one variable affects the intervals of the other two. For example, if the upper bound for R needs to be reduced (e.g., for system-stabilization reasons), then it follows that both boundaries of the conditionally valid intervals of H_1 and ω need to be altered in a well-defined way (note that the formulas for calculating those conditionally valid intervals' boundary values are given in Section 8.2.3).

Principal angle ω and distance to target range H_1

The conditionally valid interval of ω is $(\omega_{\min}, \omega_+)$. The conditionally valid interval of H_1 is $(0, H_{1+})$. We already know that the value of ω increases as the value of H_1 increases, which implies that

$$
\lim_{H_1 \to 0^+} \omega = \omega_{\min} \quad \text{and} \quad \lim_{H_1 \to H_{1+}} \omega = \omega_+.
$$

Both mappings are uniquely invertible.

If $H_1 = H_{1\min R}$, then $\omega = 90^\circ$. This is not obvious just by considering the relation between H_1 and ω; it will be clear when we look at the endpoint mapping between R and ω.

Note that $\omega_{\min}$ is always greater than zero and less than 90°: the value of ω would be equal to zero if and only if the value of σ_w is also equal to zero, thus $\theta_w = 0^\circ$ and there would be no stereo effect at all; on the other hand, because of $D_1 < D_2$ it follows that $\omega_{\min} < 90^\circ$.

Figure 8.2(b) indicates that $\omega_+ > 90^\circ$ and $\omega_{\min} + \omega_+ = 180^\circ$. A proof follows in Section 8.2.3. Endpoint mappings for H_1 and ω are illustrated by the second and third row of Table 8.1.

Off-axis distance R and principal angle ω

The conditionally valid interval of R is $[R_{\min}, D_1)$, and that of ω is $(\omega_{\min}, \omega_+)$. If ω increases from $\omega_{\min}$ to ω_+, then R starts at D_1, decreases until $\omega = 90°$, and then increases again towards D_1. This is in accordance with our previous conclusion that $0° < \omega_{\min} < 90° < \omega_+$. This implies that

$$\lim_{\omega \to \omega_{\min}^+} R = \lim_{\omega \to \omega_+^-} R = D_1.$$

These relations are also reversible but not unique.

Figure 8.2(b) illustrates that R reaches its minimum when $\omega = 90°$. Table 8.1 summarizes the results.

We return to the previously mentioned statement: when $H_1 = H_{1\min R}$, why is it that $\omega = 90°$? The answer can be obtained as follows: because $H_1 = H_{1\min R}$ when $R = R_{\min}$, and $R = R_{\min}$ when $\omega = 90°$, we can conclude that $\lim_{H_1 \to H_{1\min R}} \omega = 90°$.

8.2.3 Algebraic Analysis

This subsection specifies algebraic representations of previously introduced geometric relations. Two basic functions are given first, followed by derivations of formulas that allow endpoints and extrema to be computed.

Basic Function

We use both functions for R and ω as derived for image quality control, but consider only H_1 as a variable. We denote these unary functions as f_R and f_ω, and have (as a direct consequence of the general control functions)

$$f_R(H_1) = \sqrt{D_1^2 + H_1^2 + 2D_1 H_1 \frac{D_1 - D_2 \cos \sigma_w}{\sqrt{A}}}, \tag{8.1}$$

$$f_\omega(H_1) = \arccos\left(\frac{D_1 D_2 \cos \sigma_w - D_1^2 - H_1 \sqrt{A}}{\sqrt{A(D_1^2 + H_1^2) + 2D_1 H_1(D_1 - D_2 \cos \sigma_w)\sqrt{A}}}\right), \tag{8.2}$$

where $A = D_1^2 + D_2^2 - 2D_1 D_2 \cos \sigma_w$.

Endpoints of Intervals and Extreme Variables

Formulas for those endpoints or extrema, namely, $R_{\min}$, $H_{1\min R}$, H_{1+}, $\omega_{\min}$, and ω_+, will be derived in terms of the constant parameters D_1, D_2, and σ_w.

$R_{\min}$ and $H_{1\min R}$

The minimum value $R_{\min}$ of R and the corresponding value $H_{1\min R}$ of H_1 can be found by setting the first derivative $f'_R(H_1)$ equal to zero, and solving it with respect to H_1.

Since the function $f_R(H_1)$ is rather complicated to differentiate we choose to work instead with the function of R^2, denoted by $f_{R^2}(H_1)$. We have

$$f_{R^2}(H_1) = D_1^2 + H_1^2 + 2D_1 H_1 \frac{D_1 - D_2 \cos \sigma_w}{\sqrt{D_1^2 + D_2^2 - 2D_1 D_2 \cos \sigma_w}}.$$

Thus, it follows that

$$\frac{df_{R^2}}{dH_1} = 2H_1 + 2D_1 \frac{D_1 - D_2 \cos \sigma_w}{\sqrt{D_1^2 + D_2^2 - 2D_1 D_2 \cos \sigma_w}}$$

$$= \frac{2\left(H_1 \sqrt{D_1^2 + D_2^2 - 2D_1 D_2 \cos \sigma_w} + D_1(D_1 - D_2 \cos \sigma_w)\right)}{\sqrt{D_1^2 + D_2^2 - 2D_1 D_2 \cos \sigma_w}}. \tag{8.3}$$

We set equation (8.3) equal to zero and solve it with respect to H_1. We obtain

$$H_{1 \min R} = \frac{D_1(D_2 \cos \sigma_w - D_1)}{\sqrt{D_1^2 + D_2^2 - 2D_1 D_2 \cos \sigma_w}}. \tag{8.4}$$

When $H_1 = H_{1 \min R}$, the value of $f_R(H_1)$ is a minimum of the function f_R, that is, $R = R_{\min}$. The value of $R_{\min}$ can be obtained from equation (8.1) by replacing H_1 by the value of $H_{1 \min R}$:

$$R_{\min} = f_R(H_{1 \min R})$$

$$= \sqrt{D_1^2 + \frac{D_1^2(D_1 - D_2 \cos \sigma_w)^2}{D_1^2 + D_2^2 - 2D_1 D_2 \cos \sigma_w} - \frac{2D_1^2(D_1 - D_2 \cos \sigma_w)^2}{D_1^2 + D_2^2 - 2D_1 D_2 \cos \sigma_w}}$$

$$= D_1 \sqrt{\frac{D_1^2 + D_2^2 - 2D_1 D_2 \cos \sigma_w - (D_1^2 + D_2^2 \cos^2 \sigma_w - 2D_1 D_2 \cos \sigma_w)}{D_1^2 + D_2^2 - 2D_1 D_2 \cos \sigma_w}}$$

$$= \frac{D_1 D_2 \sin \sigma_w}{\sqrt{D_1^2 + D_2^2 - 2D_1 D_2 \cos \sigma_w}}.$$

$\omega_{\min}$

The minimum value $\omega_{\min}$ of ω is equal to the limit of the values of function $f_\omega(H_1)$ as H_1 approaches to zero. It can be calculated from equation (8.2) by setting $H_1 = 0$. We have

$$\omega_{\min} = \lim_{H_1 \to 0^+} f_\omega(H_1)$$

$$= \arccos\left(\frac{D_1 D_2 \cos \sigma_w - D_1^2}{\sqrt{(D_1^2 + D_2^2 - 2D_1 D_2 \cos \sigma_w)D_1^2}}\right)$$

$$= \arccos\left(\frac{D_2 \cos \sigma_w - D_1}{\sqrt{D_1^2 + D_2^2 - 2D_1 D_2 \cos \sigma_w}}\right). \tag{8.5}$$

H_{1+}

A previous conclusion based on Table 8.1 says that the upper bound H_{1+} of H_1 can be found by setting the value of R equal to the value of D_1, which is equivalent to considering $f_R(H_1) = D_1$ and solving it for H_1. We have

$$D_1^2 + H_1^2 + 2D_1H_1 \frac{D_1 - D_2 \cos\sigma_w}{\sqrt{D_1^2 + D_2^2 - 2D_1D_2\cos\sigma_w}} = D_1^2$$

$$H_1\left(H_1 + \frac{2D_1(D_1 - D_2\cos\sigma_w)}{\sqrt{D_1^2 + D_2^2 - 2D_1D_2\cos\sigma_w}}\right) = 0$$

$$H_1 = 0 \quad \text{or} \quad H_1 = \frac{2D_1(D_2\cos\sigma_w - D_1)}{\sqrt{D_1^2 + D_2^2 - 2D_1D_2\cos\sigma_w}}.$$

Obviously, the only permissible solution is

$$H_{1+} = \frac{2D_1(D_2\cos\sigma_w - D_1)}{\sqrt{D_1^2 + D_2^2 - 2D_1D_2\cos\sigma_w}}. \tag{8.6}$$

From equations (8.4) and (8.6) we can conclude that $H_{1+} = 2H_{1\min R}$. This fact can also be observed from Figure 8.2(b).

ω_+

The upper bound ω_+ of ω can be found by replacing H_1 by H_{1+} in equation (8.2). We obtain

$$\omega_+ = f_\omega(H_{1+})$$

$$= \arccos\left(\frac{D_1D_2\cos\sigma_w - D_1^2 - 2D_1D_2\cos\sigma_w + 2D_1^2}{\sqrt{AD_1^2 + 4D_1^2(D_2\cos\sigma_w - D_1)^2 - 4(D_1^2 - D_1D_2\cos\sigma_w)^2}}\right)$$

where $A = D_1^2 + D_2^2 - 2D_1D_2\cos\sigma_w$. It follows that

$$\omega_+ = \arccos\left(\frac{D_1 - D_2\cos\sigma_w}{\sqrt{D_1^2 + D_2^2 - 2D_1D_2\cos\sigma_w}}\right). \tag{8.7}$$

From equations (8.5) and (8.7) we may conclude that $\omega_+ = 180° - \omega_{\min}$. This proves our previous statement that $\omega_{\min}$ is complementary to ω_+ with respect to $180°$ (see the previous analysis of the mapping of endpoints of ω and H_1).

Validity Checking

The basic idea of a validity check is to determine for any given quadruple $(D_1, D_2, H_1, \sigma_w)$ (which might be called *the user input*) whether it is acceptable for the image quality control process (i.e., leading to well-defined values of R and ω), or not.

From the foregoing we know that we are able to obtain conditionally valid intervals of R, ω, and H_1 when values of D_1, D_2, and σ_w are specified. Similarly, in the next section, we will study conditionally valid intervals of R, ω, and σ_w when values of D_1, D_2, and H_1 are specified. These conditionally valid intervals allow a validity check on values of parameters σ_w or H_1, when the other three application-specific parameters are already known to be in specified (finitely bounded) intervals.

Recall the four examples used in Chapter 7 where the values of D_1, D_2, H_1, and σ_w have to be provided in order to determine optimum values of sensor parameters R and ω. Now, based on the analysis and calculations carried out in this section, we address the question

	D_1	D_2	θ_w	H_{1-}	H_{1+}
Example (1)	1	3	10.48	0.0	1.9813
Example (2)	4	10	9.17	0.0	7.9292
Example (3)	6	50	8.00	0.0	11.9622
Example (4)	20	200	8.74	0.0	39.8567

$H_P = 5, 184$ (pixels) $\tau = 0.007$ (mm)
$H_S = 768$ (pixels) $f = 21.7$ (mm)

Table 8.2   Table of conditionally valid intervals of H_1 for the examples of Table 7.1.

whether the quantities provided are geometrically or mathematically valid for sensor parameter determination. Here, we answer this question by checking whether the value of H_1 lies within its conditionally valid interval when the specifications of D_1, D_2, and σ_w have to satisfy some given constraints. Table 8.2 shows all conditionally valid intervals of H_1 for a validity check (the same four examples as in Table 7.1).

Summary

Figure 8.3 shows the graphs of the functions $f_R(H_1)$ and $f_\omega(H_1)$. Note that they confirm the initial geometric analysis. We summarize the curve behaviors as follows. For the graph of $f_R(H_1)$, the value of f_R starts at D_1, first decreases and then increases towards ∞ as the value of H_1 increases from 0 to ∞. For the graph of $f_\omega(H_1)$, the value of ω increases as the value of H_1 increases. In other words, the function $f_\omega(H_1)$ increases monotonically on H_1 (i.e., on the interval $(0, \infty)$). Note that $\omega = 180°$ is the horizontal asymptote of the curve $\omega = f_\omega(H_1)$.

Although the graphs are provided for one particular combination of specified constants, their asymptotic or extreme value behavior remains basically the same for all valid values of the constant parameters (basically it only differs by scaling factors). Note that the graphs in Figure 8.3 allow us to see how a change in the values of R or ω will impact the value of H_1, because basic relations or general changing behavior do not alter. Note that there is no unique inverse function for $f_R(H_1)$.

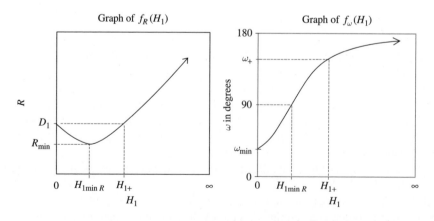

Figure 8.3   Graphs of functions $f_R(H_1)$ and $f_\omega(H_1)$.

The endpoints of the conditionally valid intervals of the three variables are also illustrated by both graphs in Figure 8.3. The conditionally valid interval $[R_{min}, D_1)$ of R and the conditionally valid interval $(0, H_{1+})$ of H_1 are indicated along the axes of the graph of f_R. For the graph of f_ω, the conditionally valid interval of ω is (ω_{min}, ω_+)

Mappings between endpoint values of these intervals are represented by dashed lines, namely for the mapping between H_{1+} and D_1, and between $H_{1\min R}$ and R_{min} (in the graph of $f_R(H_1)$), and for the mapping between H_{1+} and ω_+ (in the graph of $f_\omega(H_1)$). These mappings are (summarized) as follows:

$$\lim_{H_1 \to 0^+} R = \lim_{H_1 \to H_{1+}} R = D_1,$$

$$\lim_{H_1 \to H_{1\min R}} R = R_{min},$$

$$\lim_{H_1 \to 0^+} \omega = \omega_{min},$$

$$\lim_{H_1 \to H_{1+}} \omega = \omega_+.$$

As concluded earlier, inverse relations of the second, third, and forth mapping are also true; the inverse of the first mapping is not uniquely defined.

From these mappings and an earlier conclusion from equations (8.4) and (8.6), we know that $H_{1+} = 2H_{1\min R}$. Together with Figure 8.3 we can conclude that the curve of the function $f_R(H_1)$ on the conditionally valid interval $(0, H_{1+})$ of H_1 is symmetric with respect to the straight line $H_1 = H_{1\min R}$. Moreover, from the graph of the function $f_\omega(H_1)$ we see that $\omega_{min} = 180° - \omega_+$, as previously obtained when computing ω_+.

8.3 Stereoacuity Analysis

This section studies how the sensor parameters R and ω interact with the stereoacuity parameter σ_w while D_1, D_2, and H_1 are constant. By analogy with the previous case, we focus on:

- conditionally valid intervals for σ_w, R, and ω while the other three parameters are kept constant;
- mappings between endpoints of these intervals; and
- the influence of changes in σ_w, R, or ω on the values of the other two parameters.

The overall structure of the section is similar to that of the previous section.

8.3.1 Geometric Analysis

As in the previous section, we start with a qualitative geometric analysis, followed in the next subsection by an algebraic analysis.

Geometric model

Qualitatively there are five different possibilities for specifying H_1: we may take $H_1 < D_1$, $H_1 = D_1$, $D_1 < H_1 < D_2$, $H_1 = D_2$, or $H_1 > D_2$. It turns out that the cases $H_1 \leq D_1$ and $H_1 > D_1$

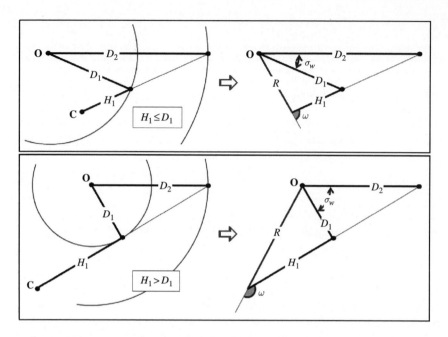

Figure 8.4 Formation of graphs for the analysis of σ_w, R, and ω.

are of particular interest, and we discuss both in detail. Geometries of both cases are shown in Figure 8.4.

Figure 8.4 (left) represents the three constant parameters by three line segments of lengths D_1, D_2 and H_1. The two circular gray arcs (with center at **O**) have radii D_1 and D_2 respectively, and illustrate the top views of inner and outer cylinders for sense range of interest. They are used as a constraint for possible positions of focal point **C**. The focal point should be located at distance H_1 from the object coinciding with the inner cylinder (associated with radius D_1). Figure 8.4 (right) illustrates the geometric relationships among those variables by representing a valid combination of σ_w, R, and ω for $H_1 \leq D_1$ (top) and $H_1 > D_1$ (bottom). Note that the angle σ_w between two thicker line segments (related to D_1 and D_2, respectively) has to be less than 90°.

Based on the chosen values of H_1, D_1, and D_2 in Figure 8.4, Figure 8.5 illustrates a few more supporting combinations of values of σ_w, R, and ω, used in the following studies.

Interaction between Variables

The parameter σ_w is measured in degrees. It is constrained to be less than 90°, and we do not consider $\sigma_w = 0°$ or $\sigma_w = 90°$.

Qualitative results on interactions between variables can be obtained by geometric reasoning based on Figure 8.5. We summarize the conclusions as follows. For $H_1 \leq D_1$ (Figure 8.5(a)):

- R increases as σ_w increases, and vice versa.
- ω increases as σ_w increases, and vice versa.
- ω increases as R increases, and vice versa.

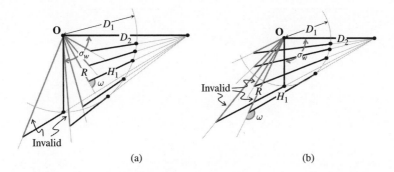

Figure 8.5 Qualitative (geometric) interaction among changes in σ_w, R, and ω, while values of D_1, D_2, and H_1 are kept constant. (a) $H_1 \leq D_1$, showing six different combinations of the variables, including two invalid cases. (b) $H_1 > D_1$, showing five different combinations of the variables, including three invalid cases.

For $H_1 > D_1$ (Figure 8.5(b)):

- R increases as σ_w increases, and vice versa.
- σ_w increases from $0°$ to $90°$; the value of ω starts from $180°$, first decreases, and then increases again as σ_w goes to $90°$.
- As R increases, ω starts from $180°$, first decreases, and then increases towards $180°$.

Conditionally Valid Intervals

The values of σ_w, R, and ω cannot be specified arbitrarily in image acquisition due to the given limitations (e.g., type of camera), and we also have the constraint that $R < D_1$. This subsection analyzes the endpoints of the conditionally valid intervals for the three variables σ_w, R, and ω.

Endpoints of the conditionally valid interval of σ_w
The value of σ_w is a real in the open interval $(0°, 90°)$. For a geometric analysis of the case where $H_1 \leq D_1$, see Figure 8.5(a). We know that R increases as σ_w increases. We can also obtain that $\lim_{\sigma_w \to 0+} R = D_1 - H_1$, and also that $R > D_1$ as σ_w approaches $90°$. Thus, there is an upper bound σ_{w+} for σ_w, where $\sigma_{w+} < 90°$, such that for any value of σ_w within the interval $(0°, \sigma_{w+})$, the corresponding value of R is less than D_1. As a result, $(0°, \sigma_{w+})$ is the conditionally valid interval for σ_w.

For a geometric analysis of the case where $H_1 > D_1$, see Figure 8.5(b). The value of R increases as σ_w increases. Similarly to case $H_1 \leq D_1$, due to the constraint that $R < D_1$, there exists an upper bound σ_{w+} for σ_w with $\sigma_{w+} < 90°$, such that $\lim_{\sigma_w \to \sigma_{w+}^-} R = D_1$. Thus, for $H_1 > D_1$, $(0°, \sigma_{w+})$ is also the conditionally valid interval for σ_w.

A summary of these qualitative results (for both cases) is given in the top two rows of Table 8.3.

Endpoints of the conditionally valid interval of R
The off-axis distance R is a non-negative real number having D_1 as its upper bound.

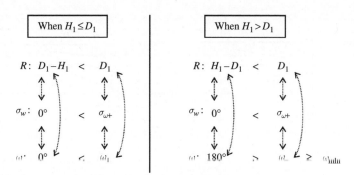

Table 8.3 Sketch of endpoint-value mappings and relations among σ_w, R, and ω.

For $H_1 \leq D_1$ (Figure 8.5(a)), it follows that $R > D_1 - H_1$, which defines a lower bound. The conditionally valid interval of R is $(D_1 - H_1, D_1)$ in this case.

For $H_1 > D_1$ (Figure 8.5(b)), we observe that $R > H_1 - D_1$, which defines a lower bound. The conditionally valid interval of R is $(H_1 - D_1, D_1)$ in this case.

Endpoints of the conditionally valid interval of ω

The principal angle ω is a positive real less than 360°; due to the symmetry of left and right camera, we only need to consider values of ω less than 180°.

The value of ω increases as σ_w increases. Thus, if $H_1 \leq D_1$, the maximum value of σ_w defines an upper bound ω_+ of the conditionally valid interval of ω. Figure 8.5(a) shows that the limit of ω equal to zero as the value of σ_w goes to zero. Therefore, the conditionally valid interval of ω is $(0°, \omega_+)$ in this case.

If $H_1 > D_1$, as the value of σ_w increases, the value of ω starts at 180°, first decreases, and then increases back towards 180°. It follows that there exists a minimum $\omega_{\min}$ of ω. Thus, in this case, $(\omega_{\min}, 180°)$ is a conditionally valid interval of ω. This interval may be denoted differently if the value of σ_w reaches its upper bound of σ_{w+} before the value of ω reaches its minimum $\omega_{\min}$. In this special situation, the conditionally valid interval of ω becomes $(\omega_-, 180°)$,[3] where $\omega_- \geq \omega_{\min}$.

Mapping of Endpoints

We now examine mappings between endpoints of conditionally valid intervals (three different pairings of parameters).

Off-axis distance R and angular disparity width σw

For $H_1 \leq D_1$, we identified the conditionally valid interval of R as $(D_1 - H_1, D_1)$, and the conditionally valid interval of σ_w as $(0°, \sigma_{w+})$. The value of R increases if σ_w increases. We have the following results:

$$\lim_{\sigma_w \to 0^+} R = D_1 - H_1 \quad \text{and} \quad \lim_{\sigma_w \to \sigma_{w+}^-} R = D_1.$$

Both mappings are uniquely invertible.

---

[3] The symbol minus "−" used in the subscript implies "lower bound" of the conditionally valid interval.

For $H_1 > D_1$, the conditionally valid interval of R is $(H_1 - D_1, D_1)$, and the conditionally valid interval of σ_w is $(0°, \sigma_{w+})$. The "second" resulting mapping in this case is the same as in $H_1 \leq D_1$ case, and the "first" is

$$\lim_{\sigma_w \to 0^+} R = H_1 - D_1.$$

This mapping is also uniquely invertible.

The above three mappings are visualized in the top and the middle rows of Table 8.3.

Principal angle ω and angular disparity width σ_w

For $H_1 \leq D_1$, the conditionally valid interval of ω is $(0°, \omega_+)$. The conditionally valid interval of σ_w is $(0°, \sigma_{w+})$. The value of ω increases as σ_w increases; we have

$$\lim_{\sigma_w \to 0^+} \omega = 0° \quad \text{and} \quad \lim_{\sigma_w \to \sigma_{w+}^-} \omega = \omega_+.$$

Both mappings are uniquely invertible.

For $H_1 > D_1$, the conditionally valid interval of ω is either $(\omega_{\min}, 180°)$ or $(\omega_-, 180°)$, where $\omega_- \geq \omega_{\min}$. The conditionally valid interval of σ_w is $(0°, \sigma_{w+})$. As the value of σ_w increases from $0°$ to σ_{w+}, there are two possibilities for the change in ω. It may start from $180°$, decrease to $\omega_{\min}$, and then increase to ω_-, or it may decrease from $180°$ until reaching ω_- without going down to $\omega_{\min}$ and also not going up afterwards. In both situations, we can have the following unified summaries:

$$\lim_{\sigma_w \to 0^+} \omega = 180° \quad \text{and} \quad \lim_{\sigma_w \to \sigma_{w+}^-} \omega = \omega_-.$$

The first mapping is uniquely invertible, but the inverse mapping is not always uniquely defined for the second.

The middle and the bottom rows of Table 8.3 illustrate these endpoint mappings.

Off-axis distance R and principal angle ω

For $H_1 \leq D_1$, the conditionally valid interval of R is equal to $(D_1 - H_1, D_1)$, and the conditionally valid interval of ω is equal to $(0°, \omega_+)$. Because ω increases as R increases, we have

$$\lim_{\omega \to 0^+} R = (D_1 - H_1) \quad \text{and} \quad \lim_{\omega \to \omega_+^-} R = D_1.$$

These two mappings are uniquely invertible.

For $H_1 > D_1$, the conditionally valid interval of R is $(H_1 - D_1, D_1)$, and the conditionally valid interval of ω is either $(\omega_{\min}, 180°)$ or $(\omega_-, 180°)$. In both situations, we can have the following unified summaries:

$$\lim_{\omega \to 180^-} R = (H_1 - D_1) \quad \text{and} \quad \lim_{R \to D_1^-} \omega = \omega_-.$$

The mapping on the left is uniquely invertible, but the inverse is not always uniquely defined for the one on the right.

All these results are illustrated in the top and the bottom rows of Table 8.3.

8.3.2 Algebraic Analysis

The qualitative geometric analysis for σ_w is now followed by computations of endpoints or extrema. We also discuss links between qualitative and quantitative results.

Basic Functions

Let R be defined by a unary function f_R, dependent only on the variable σ_w. We have

$$f_R(\sigma_w) = \sqrt{D_1^2 + H_1^2 + 2D_1 H_1 \frac{D_1 - D_2 \cos \sigma_w}{\sqrt{D_1^2 + D_2^2 - 2D_1 D_2 \cos \sigma_w}}}. \tag{8.8}$$

Let ω be defined by a unary function g_ω with a single independent variable R. We have

$$g_\omega(R) = \arccos\left(\frac{D_1^2 - H_1^2 - R^2}{2H_1 R}\right). \tag{8.9}$$

We also consider ω to be defined by the composite function $f_\omega = g_\omega \circ f_R$, which is a function with a single independent variable σ_w. We obtain

$$f_\omega(\sigma_w) = \arccos\left(\frac{D_1 D_2 \cos \sigma_w - D_1^2 - H_1 \sqrt{A}}{\sqrt{A(D_1^2 + H_1^2) + 2D_1 H_1 (D_1 - D_2 \cos \sigma_w)\sqrt{A}}}\right) \tag{8.10}$$

where $A = D_1^2 + D_2^2 - 2D_1 D_2 \cos \sigma_w$.

Endpoints and Extrema

In our geometric analysis we identified endpoint and extreme values of variables. They are σ_{w+}, ω_+, ω_-, and $\omega_{\min}$. We now show how to compute these values in terms of given constant parameters D_1, D_2, and H_1.

σ_{w+}

The value of σ_{w+} can be found by setting $f_R(\sigma_w) = D_1$ and then solving with respect to σ_w. We obtain

$$D_1^2 + H_1^2 + 2D_1 H_1 \frac{D_1 - D_2 \cos \sigma_w}{\sqrt{D_1^2 + D_2^2 - 2D_1 D_2 \cos \sigma_w}} = D_1^2,$$

$$H_1 \sqrt{D_1^2 + D_2^2 - 2D_1 D_2 \cos \sigma_w} = 2D_1 D_2 \cos \sigma_w - 2D_1^2.$$

Squaring both sides and we reach the following three conclusions:

$$H_1^2 D_1^2 + H_1^2 D_2^2 - 2D_1 D_2 H_1^2 \cos \sigma_w - 4D_1^2 D_2^2 \cos^2 \sigma_w - 4D_1^4 + 8D_1^3 D_2 \cos \sigma_w = 0,$$

$$4D_1^2 D_2^2 \cos^2 \sigma_w + 2\left(D_1 D_2 H_1^2 - 4D_1^3 D_2\right) \cos \sigma_w + \left(4D_1^4 - H_1^2 D_1^2 - H_1^2 D_2^2\right) = 0,$$

$$\cos^2 \sigma_w + 2\left(\frac{D_1 D_2 H_1^2 - 4D_1^3 D_2}{4D_1^2 D_2^2}\right) \cos \sigma_w + \left(\frac{4D_1^4 - H_1^2 D_1^2 - H_1^2 D_2^2}{4D_1^2 D_2^2}\right) = 0.$$

Thus, the value of $\cos \sigma_w$ is given by

$$
\cos \sigma_w = \frac{4D_1^2 - H_1^2}{4D_1 D_2} \pm \sqrt{\left(\frac{H_1^2 - 4D_1^2}{4D_1 D_2}\right)^2 - \left(\frac{4D_1^4 - H_1^2 D_1^2 - H_1^2 D_2^2}{4D_1^2 D_2^2}\right)}
$$

$$
= \frac{4D_1^2 - H_1^2}{4D_1 D_2} \pm \sqrt{\frac{H_1^4 - 4H_1^2 D_1^2 - 4H_1^2 D_2^2}{16 D_1^2 D_2^2}}
$$

$$
= \frac{4D_1^2 - H_1^2 \pm H_1 \sqrt{H_1^2 - 4D_1^2 + 4D_2^2}}{4D_1 D_2}.
$$

Both solutions of σ_w can be calculated by applying the inverse cosine, and the positive solution

$$
\sigma_{w+} = \arccos\left(\frac{4D_1^2 - H_1^2 + H_1 \sqrt{H_1^2 - 4D_1^2 + 4D_2^2}}{4D_1 D_2}\right)
$$

is valid.

ω_+ *and* ω_-

The values of both ω_+ and ω_- can be obtained by substituting D_1 for R in equation (8.9). This leads to

$$
\omega_+ = g_\omega(D_1)
$$

$$
= \arccos\left(\frac{D_1^2 - H_1^2 - D_1^2}{2H_1 D_1}\right)
$$

$$
= \arccos\left(\frac{-H_1}{2D_1}\right)
$$

and

$$
\omega_- = \arccos\left(\frac{-H_1}{2D_1}\right).
$$

Formulas for these two endpoints are the same, but they correspond to the different cases (i.e., different values of H_1 and D_1); thus in general $\omega_+ \neq \omega_-$.

$\omega_{\min}$

The value of $\omega_{\min}$ can be found by setting the first derivative of the function $f_\omega(\sigma_w)$ to zero (i.e., $f_\omega' = 0$), and solving with respect to σ_w. Because differentiation of the function f_ω (specified in equation (8.10)) is rather complicated, we decided (for the experiments to be reported in Figure 8.6) to use mathematical software in this case to find the desired values. Note that this is the only extremum for which we did not provide an explicit formula to calculate its value.

Summary

The graphs of the functions $f_R(\sigma_w)$ and $f_\omega(\sigma_w)$ are given in Figure 8.6. The graph of $f_R(\sigma_w)$ (Figure 8.6(a)) shows that R increases as σ_w increases, for both $H_1 \leq D_1$ and $H_1 > D_1$. The graph of $f_\omega(\sigma_w)$ for $H_1 \leq D_1$ (Figure 8.6(b)) shows that ω increases as σ_w increases. The graph of

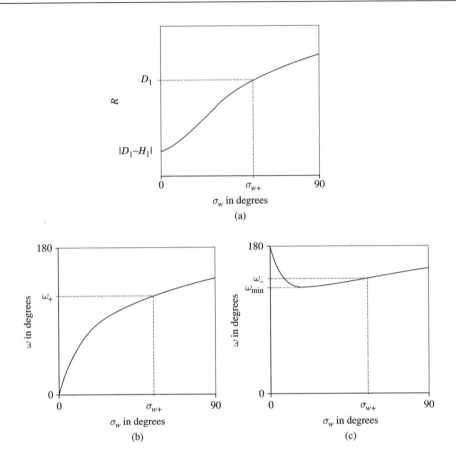

Figure 8.6 Graphs of (a) $f_R(\sigma_w)$, (b) $f_\omega(\sigma_w)$ for $H_1 \leq D_1$, (c) $f_\omega(\sigma_w)$ for $H_1 > D_1$.

$f_\omega(\sigma_w)$ for $H_1 > D_1$ (Figure 8.6(c)) shows that ω decreases from 180°, reaches its minimum, and then increases afterwards as σ_w keeps increasing. These results confirm our earlier conclusions drawn just from geometric interpretations.

Although the graphs in Figure 8.6 are for a particular combination of the three constants, the given characterization is invariant to value changes (except for different scales). Those graphs also illustrate changes in σ_w depending on changes of R or ω, except that $f_\omega(\sigma_w)$ is not uniquely invertible in the case $H_1 > D_1$.

Endpoints of the conditionally valid intervals of the three variables are also illustrated in Figure 8.6. In the graph of f_R (Figure 8.6(a)), the conditionally valid intervals of R and σ_w are indicated along the axes. For both cases, we have the same conditionally valid interval $(|D_1 - H_1|, D_1)$ for R, and the same conditionally valid interval $(0°, \sigma_{w+})$ for σ_w. Hence, only one graph is shown for f_R instead of two separate graphs for different cases.

Figures 8.6(b) and 8.6(c) show function f_ω. The conditionally valid intervals $(0°, \omega_+)$ and $(\omega_-, 180°)$ of ω, and $(0°, \sigma_{w+})$ of σ_w are labeled along the axes. Additionally, the extreme ω_{min} is indicated for the case $H_1 > D_1$.

The following mappings between those endpoints or extrema are shown in this figure by dashed lines: a mapping between σ_{w+} and D_1 in the graph of f_R, a mapping between σ_{w+} and ω_+ in the graph of f_ω for case $H_1 \leq D_1$, and a mapping between σ_{w+} and ω_- in the graph of f_ω for case $H_1 > D_1$. The mapping results may be summarized as follows:

$$\lim_{\sigma_w \to 0^+} R = |D_1 - H_1|, \quad \lim_{\sigma_w \to \sigma_{w+}^-} R = D_1$$

for both cases,

$$\lim_{\sigma_w \to 0^+} \omega = 0°, \quad \lim_{\sigma_w \to \sigma_{w+}^-} \omega = \omega_+$$

for $H_1 \leq D_1$, and

$$\lim_{\sigma_w \to 0^+} \omega = 180°, \quad \lim_{\sigma_w \to \sigma_{w+}^-} \omega = \omega_-$$

for $H_1 > D_1$.

8.4 Specification of Camera Parameters

The design of stereo panorama sensors is typically concerned with epipolar geometry or optics optimization. This section discusses two additional criteria: controllabilities of scene composition and stereoacuity (depth levels) over dynamic ranges of 3D scenes. Neglecting these two criteria in camera design may lead to increased cost, poor stereo quality such as cardboard effects, dipodia at the time of stereo viewing, and so forth.

Parameter optimization is feasible after the search space for solutions is reduced based on our studies. We reduce the original (unconstrained) search space, which is a four-dimensional bounded set, to a set defined by eight value comparisons at most.

The specifications of application requirements for stereo panorama image acquisition can be described by defining intervals for all four application-specific parameters. Formally, we define these intervals as $[D_{1\min}, D_{1\max}]$, $[D_{2\min}, D_{2\max}]$, $[H_{1\min}, H_{1\max}]$, and $[\theta_{\min}, \theta_{\max}]$, where $D_{1\max} < D_{2\min}$. Values within these intervals are assumed to be valid according to our previous definition of conditionally valid intervals. In other words, for any given combination of these values within the selected intervals, there have to exist values of R and ω which satisfy the specification. Using our analysis results, the validity of the intervals can be checked beforehand, and the intervals can be adjusted if required.

In theory, the value of R can be any positive real and the value of ω can be any positive real less than $360°$. In practice, bearing in mind system stabilization or cost issues, the intervals of both sensor parameters should be as small as possible for any given application.

Without loss of generality, let both $R_{\min}$ and $\omega_{\min}$ be zero. The problem is to find minimum values of $R_{\max}$ and $\omega_{\max}$ that fully satisfy specifications of the given application requirements.

8.4.1 Analysis of Scene Ranges

Geometrically the problem involves finding maximum values in a bounded 4D space. Obviously it is difficult to imagine a hypersurface in a 5D space. The method of analysis follows the method we used for parameter analysis for scene composition and stereoacuity. We will restrict our discussions of geometry and provide just summaries of analysis results.

Analysis of the Distance to Inner Border of RoI

We analyze interactions between $D_1 > 0$ and both sensor parameters R and ω. It is assumed that $D_1 < D_2$.

Figure 8.7 illustrates the geometry of changes in D_1 versus changes in R or ω, for constant D_2, H_1, and θ_w. We illustrate seven different situations of values of D_1, R, and ω. Basically we infer continuous transitions from such a finite set of states. Conclusions use just one additional precondition, namely that $\sigma_w < 90°$, where $\sigma_w = \frac{1}{2}\theta_w$.

From the geometry of endpoint conditions, for parameters D_1, R and ω, we conclude the following:

(1) $\lim_{D_1 \to 0^+} R = H_1$ and $\lim_{D_1 \to 0^+} \omega = 180°$;

(2)

$$\lim_{D_1 \to D_2^-} R = \sqrt{D_2^2 + H_1^2 + 2 D_2 H_1 \sin\left(\frac{\sigma_w}{2}\right)}$$

and

$$\lim_{D_1 \to D_2^-} \omega = \arccos\left(\frac{-H_1 - D_2 \sin\left(\frac{\sigma_w}{2}\right)}{\sqrt{D_2^2 + H_1^2 + 2 D_2 H_1 \sin\left(\frac{\sigma_w}{2}\right)}}\right);$$

(3) by (1) and (2), we also have that $\lim_{D_1 \to 0^+} R < \lim_{D_1 \to D_2^-} R$ and $\lim_{D_1 \to 0^+} \omega > \lim_{D_1 \to D_2^-} \omega$.

Furthermore, the geometric situation also leads to the conclusion that, when D_1 increases from zero to D_2, both values of R and ω first decrease and then increase until D_1 approaches D_2. Thus, for $0 < D_1 < D_2$, we have the following:

- there exists a uniquely defined minimum of R; and
- there exists a uniquely defined minimum of ω.

Note that the minimum value of R does not correspond to the minimum value of ω.

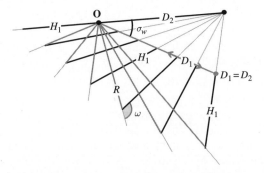

Figure 8.7   Geometry of changes in D_1, R and ω when D_2, H_1, and σ_w are assumed to be constant.

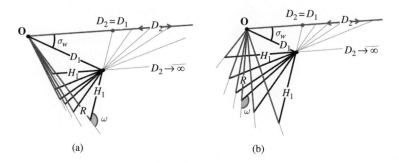

Figure 8.8   Geometry of changes in D_2, R and ω when D_1, H_1, and σ_w are kept constant: (a) $D_1 \geq H_1$; (b) $D_1 < H_1$.

Analysis of the Distance to Outer Border of RoI

We analyze interactions between D_2 and both sensor parameters R and ω. D_2 is potentially any non-zero positive real greater than D_1. Figure 8.8 illustrates the geometry of changes in D_2 versus changes in R and ω, while the values of D_1, H_1, and σ_w are kept constant. Figure 8.8(a) is for $D_1 \geq H_1$, and Figure 8.8(b) is for $D_1 < H_1$.

For any given values of D_1, H_1, and σ_w, where $\sigma_w < 90°$, we have that:

- for both $D_1 \geq H_1$ and $D_1 < H_1$, the value of R increases (decreases) while D_2 decreases (increases);
- for $D_1 \geq H_1$, the value of ω increases (decreases) while the value of D_2 decreases (increases);
- for $D_1 < H_1$, as D_2 increases from D_1 to ∞, the value of ω first decreases and then increases again. Thus, there exists a minimum of ω.

This completes the information needed to specify the sensor parameters in the following subsections.

8.4.2 Specification of Off-Axis Distance R

Figure 8.9 presents graphs of R versus each of the parameters D_1, H_1, D_2, and σ_w, and illustrates how changing the value of just one of the parameters D_1, H_1, D_2, and σ_w affects the value of R.

Let f_R be the function for calculating R with respect to the single variable D_1. Conclusions drawn in the previous subsection imply that the function f_R is continuous, has a single minimum, and its graph is concave (upward) on $(0, D_2)$.

Now, let f_R be the function with respect to the single variable H_1; see equation (8.1). According to our results on scene composition, this function f_R is also continuous, has a single minimum, and its graph is concave (upward) on $(0, \infty)$.

Let f_R be defined with respect to the single variable D_2. It follows that this function is continuous and decreasing on (D_1, ∞); its graph has a horizontal asymptote.

Finally, the function f_R defined in equation (8.8) with respect to the single variable σ_w is continuous and increasing on $(0°, 90°)$.

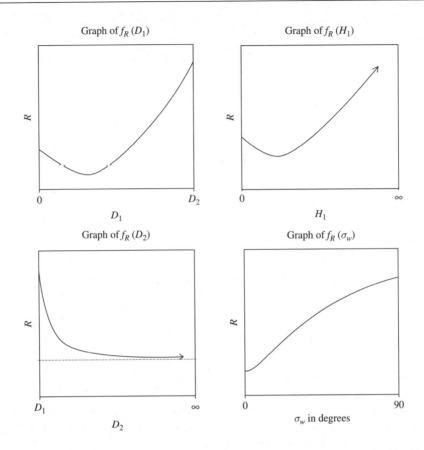

Figure 8.9 Graphs of R, in each case with respect to exactly one of the parameters D_1, D_2, H_1, or σ_w.

Because the four graphs of f_R in Figure 8.9 are either decreasing, increasing or concave (upward), we have that the maximum value of R lies within the 4D hypercube defined by the cross product of intervals $[D_{1\min}, D_{1\max}]$, $[D_{2\min}, D_{2\max}]$, $[H_{1\min}, H_{1\max}]$, and $[\sigma_{w\min}, \sigma_{w\max}]$. We may also conclude that the maximum value of R lies at one of the corners of this region. This reduces the search space for the maximum of R from a 4D bounded region to only 16 (search) points (i.e., each corner of this 4D region can be calculated by a combination of those endpoints, thus four pairs of endpoints give 16 different combinations).

Furthermore, because the function $f_R(D_2)$ is decreasing, the maximum of R is taken at $D_2 = D_{2\min}$. Because $f_R(\sigma_w)$ is increasing, the maximum of R is taken at $\sigma_w = \sigma_{w\max}$. The search space is thus further reduced to four points when searching for the maximum value of R.

8.4.3 Specification of Principal Angle ω

Figure 8.10 shows function graphs for changes in ω with respect to each of the parameters D_1, H_1, D_2, and σ_w.

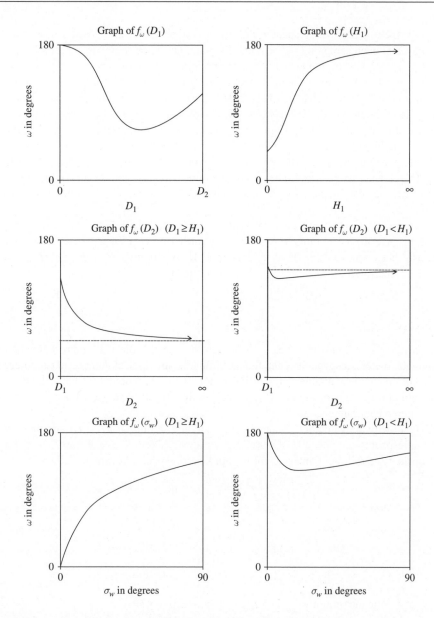

Figure 8.10 Graphs of functions defining ω, each with respect to exactly one of the parameters D_1, D_2, H_1, and σ_w.

Let f_ω be the function for specifying ω with respect to the single variable D_1. Conclusions drawn about the distance D_1 versus principal angle show that f_ω is continuous and has a single minimum on $(0, D_2)$.

Now consider the function f_ω with respect to the single variable H_1; see equation (8.2). This function is continuous and increasing on $(0, \infty)$.

Let f_ω be the function with respect to the single variable D_2. If $D_1 \geq H_1$, it follows that f_ω is continuous and decreasing on (D_1, ∞), and its graph has a horizontal asymptote. If $D_1 < H_1$, the function $f_\omega(D_2)$ is continuous and has a single minimum on (D_1, ∞).

Finally, consider function f_ω with respect to σ_w; see equation (8.10). If $D_1 \geq H_1$ it follows that f_ω is continuous and increasing on $(0°, 90°)$. If $D_1 < H_1$, the function f_ω is continuous and has a single minimum on $(0°, 90°)$.

Because all the graphs of f_ω in Figure 8.10 are either decreasing, increasing, or continuous and have a single minimum, we can again conclude that the maximum value of ω lies within the bounded region defined by the cross product of the four intervals $[D_{1\min}, D_{1\max}]$, $[D_{2\min}, D_{2\max}]$, $[H_{1\min}, H_{1\max}]$, and $[\sigma_{w\min}, \sigma_{w\max}]$. The maximum value of ω lies at one corner of the hypercube. This reduces the search space for the maximum of ω to 16 possible combinations of these endpoints.

Because $f_\omega(H_1)$ is increasing, the maximum of ω is taken at $H_1 = H_{1\max}$. This further reduces the search space from 16 points to eight points. Furthermore, if $D_{1\min} \geq H_{1\max}$, then just one comparison between two values is sufficient to find the maximum of ω. If $D_{1\min} \leq H_{1\max} \leq D_{1\max}$ then the cardinality of the search space is equal to just 5.

8.5 Exercises

8.1. Consider a situation where the closest and furthest scene objects of interest are at about 5 and 100 meters, respectively, and the preferable width of the angular disparity interval θ_w (according to the available stereo visualization method) is about $8°$. What is the minimum length of the extension slider (which specifies the off-axis distance R) needed to provide for stereo panorama acquisition?

8.2. Explain why the value of σ_{w+} can be calculated by setting $f_R(\sigma_w) = D_1$ and then solving with respect to σ_w.

8.3. Consider the following intervals of the application-specific parameters:

$$[D_{1\min}, D_{1\max}] = [5, 7] \text{ meters},$$

$$[D_{2\min}, D_{2\max}] = [60, 150] \text{ meters},$$

$$[H_{1\min}, H_{1\max}] = [5, 6] \text{ meters, and}$$

$$[\theta_{\min}, \theta_{\max}] = [6, 10] \text{ degrees}.$$

What are the maximum values of R and ω, respectively?

8.4. [Possible lab project] This project requires a tripod with a slider; the slider is mounted on top of the tripod and should allow the camera to slide by about 40 cm.

Generate a symmetric panorama by rotating a matrix-sensor camera at about $R = 40$ cm (or the maximum allowed by your slider). Optimize the angle ω by taking the recorded scene geometry into account (hint: use the formula for ω as provided in this chapter). The final task is to generate a panoramic anaglyph by combining the red channel of the $+\omega$ image with the green and blue channels of the $-\omega$ image, for stereo visualization on a computer screen.

8.6 Further Reading

In the 1990s, some studies paid great attention to how a proposed imaging approach could support a chosen area of application such as stereoscopic visualization (Huang and Hung, 1998; Peleg and Ben-Ezra, 1999), stereo reconstruction (Ishiguro et al., 1992; Kang and Szeliski, 1997), or image-based rendering (Chen, 1995; Kang and Desikan, 1997). Regarding the design of stereo panorama sensors, these studies were mainly focused on epipolar geometry, optics optimization, or other realization-related or practical issues (Peleg et al., 2000). However, issues of camera analysis and design in relation to capability and controllability for scene composition and stereoacuity were not dealt with in those studies, and also not in more current studies.

The studies in this chapter are based on a more general and flexible sensor model compared to the one proposed in Peleg and Ben-Ezra (1999).

For cylindrical panoramas for stereoscopic visualization, see Huang and Hung (1998) and Peleg and Ben-Ezra (1999); for stereo reconstruction, see Ishiguro et al. (1992), Murray (1995), Kang and Szeliski (1997), Huang et al. (2001c) and Shum and Szeliski (1999); and for image-based rendering, see Chen (1995), McMillan and Bishop (1995), Kang and Desikan (1997), Rademacher and Bishop (1998) and Shum and He (1999).

Regarding the design of stereo panorama cameras, see Ishiguro et al. (1992), Shum et al. (1999), Peleg et al. (2000), Huang et al. (2001b) and Seitz (2001).

The reported parameter optimization results can be incorporated into approaches published in Ishiguro et al. (1992) and Shum et al. (1999) because our sensor model covers these more specific studies.

9

3D Meshing and Visualization

So far this book has discussed data acquisition and the photogrammetric interpretation of captured data. This chapter deals with fundamentals in 3D modeling and visualization. 3D computer graphics is more diverse, as briefly reported in this chapter. This book is not about 3D computer graphics, and this chapter is not intended as an introduction into this topic. This chapter assumes that the reader is already familiar with (basics of) 3D computer graphics, and discusses those techniques which are in particular used for LRF and panorama data.

9.1 3D Graphics

To visualize 3D objects (by identifying object coordinates) it is necessary to project each 3D point (i.e., if in the regular orthogonal 3D grid, then a *voxel*) from Euclidean $\mathbb{R}^3$ space into a pixel on a 2D surface, which means, in a more general sense, into a point in $\mathbb{R}^2$. This is done by projections. Projections can be comfortably implemented with OpenGL (the Open Graphics Library). OpenGL is an application programming interface which allows the user to *render* simple 3D primitives (e.g., points, lines, polygons) into a selected *buffer*; rendering in this context means the projection of each primitive under given geometric and radiometric conditions. The targeted projection manifold is not necessarily the final "screen"; a buffer, for example a front, back, stencil or *z*-buffer, may also be selected. This allows the user the freedom to use such a buffer to create visual effects (e.g., shadow calculations, blending, and so forth).

9.1.1 The Graphics Pipeline

We briefly recall that OpenGL is designed as a state machine. All primitives of a given list go through a graphics pipeline where their states are initialized by the user. In this way the user can render "thousands" of primitives using the same conditions, change the states of the machine or the states of modules within this pipeline, and call the next list of primitives. As an example, a state can be the color or material property with which the primitives should be

Panoramic Imaging: Sensor-Line Cameras and Laser Range-Finders   F. Huang, R. Klette, and K. Scheibe
© 2008 John Wiley & Sons, Ltd

rendered. The properties of the *virtual camera* (the camera through which we see the rendered scene) are defined by matrices, and are another example of a state of the pipeline.

An important module of this pipeline is lighting. This module can be manipulated by various states to obtain the desired effect. The result depends on light properties (e.g., the ambient or diffuse part of light), the location of the light source, the material properties, the attitude of a primitive (i.e., position and the normal vector), the attitude of the virtual camera, and so forth. This part of the rendering process is called *shading*, and includes the calculation of an intensity or color which is returned by an illuminated surface, based on a *shading model* (e.g., flat, Gouraud,[1] or Phong[2]). These shading models are based on Lambert's cosine (emission) law and extended to also take specularities into account. We implemented the Gouraud ("smooth") shading model because it is fast and has adequate visual results, but it does not provide varying lighting effects within a single polygon.

Note that shading procedures do not include complex volume-based algorithms (e.g., modeling reflections or shadows). For such "effects" the scene has to be raytraced, which has two meanings in this context: energy tracing from a light source through the scene, which actually also includes the second meaning, namely tracing for occluded objects from a ray's origin. The reason is the simplification of both approaches for real-time applications. Note that movie productions often use more realistic physics-based models (with necessary extensions), but these rendering engines do not (in general) support real-time visualizations.

We use, for example, Crow's approach for shadow calculations. Briefly, this approach renders all object silhouettes into a stencil buffer in such a way as to mask the visible scene. This mask is then used to cast (blend) shadows into the scene. Figure 9.1 shows a rendered image using the Gouraud shading model and Crow's shadow algorithm.

Important information can be saved in arrays (e.g., vertex coordinates, vertex index lists defining how vertices are connected, normal vectors, or texture coordinates). Another possibility is to use precompiled display lists. In such a list, all 3D objects are defined (or built using OpenGL base functions such as glRotate()). The list is then compiled by starting the program, and these precompiled objects will only be called by changing a state of the pipeline (e.g., changing the camera view). We use vertex arrays because these allow the possibility of (geometry-based) dynamic scenes (e.g., opening or closing a window or door).

Transformations and projection parameters are stored in different types of matrices. The rendering pipeline multiplies all matrices into a common transformation matrix, then it transforms each object coordinate by multiplying the current transformation matrix with the vector of the object coordinates. Different kinds of matrices can be stored in stacks to manipulate different objects. The main transformation matrix $\mathbf{M}_T$ is given by

$$\mathbf{M}_T = \mathbf{M}_V \cdot \mathbf{M}_N \cdot \mathbf{M}_P \cdot \mathbf{M}_M. \tag{9.1}$$

$\mathbf{M}_V$ is the view port matrix, which is the transformation to the final window coordinates (window scaling and proportions), $\mathbf{M}_N$ is the normalization matrix (perspective division) of the device coordinates, $\mathbf{M}_P$ is the projection matrix, and $\mathbf{M}_M$ is the model matrix to transform the coordinates (e.g., a rotation, scaling, or translation).

---

[1] Invented by H. Gouraud in 1971.
[2] Invented by B. Phong in 1975.

Figure 9.1   Rendered LRF data with shadows cast behind the pillars.

We use only basic functions of OpenGL (and no utility libraries such as GLUT).[3] This requires a good knowledge of the graphics pipeline but also allows us to render our scenes (characterized by a huge data size, both of LRF and texture data) in a reasonable time, and also provides the freedom to implement novel algorithms and data management routines.

In the rest of this section we describe possible projections as applied to our data, and characterize or illustrate the results achieved.

9.1.2 Central Projection

A central perspective projection (as discussed in Chapter 3) considers the projection of an object or scene on a perspective plane. The actual homogeneous perspective matrix $\mathbf{M}_P$ (including clipping coordinates) for this projection is given in OpenGL by

$$\mathbf{M}_P = \begin{pmatrix} \cot\frac{\theta}{2} \cdot \frac{H}{W} & 0 & 0 & 0 \\ 0 & \cot\frac{\theta}{2} & 0 & 0 \\ 0 & 0 & \frac{z_F + z_N}{z_F - z_N} & -\frac{2 \cdot z_F \cdot z_N}{z_F - z_N} \\ 0 & 0 & -1 & 0 \end{pmatrix}. \tag{9.2}$$

---

[3] The use of a utility library would restrict our solution to the functionality of such a library.

This differs from the projection matrix in Chapter 3 in the inclusion of clipping coordinates; note that the projection formulas in Chapter 3 assume unbounded perspective projection. The clipping planes are shown as *far* (we use the symbol Z_F) and *near* (we use the symbol Z_N). The clipping planes can be seen as a bounding box which specifies the depth of the scene. Furthermore the "sensor size", or here the rendering window, is limited by a maximum count of pixels $H \times W$. The focal length can be expressed by the field of view of the current window.

Figure 9.2 depicts a 3D model rendered by central projection based on image data as shown in Figure 5.24. It shows a measured cloud of 3D points with gray levels (as measured by an LRF at a single attitude).

9.1.3 Orthogonal Projection

An orthogonal projection considers the projection of each 3D point orthogonally to a specified plane. For each point $\mathbf{p} = (p_x, p_y, p_z)^{\mathrm{T}}$, the transformed point would be:

$$\mathbf{p}' = \begin{pmatrix} 1 & 0 & 0 \\ 0 & 1 & 0 \\ 0 & 0 & 0 \end{pmatrix} \begin{pmatrix} p_x \\ p_y \\ p_z \end{pmatrix} = \begin{pmatrix} p_x \\ p_y \\ 0 \end{pmatrix}. \tag{9.3}$$

In computer graphics it is common practice to define matrices used for orthographic projection by a 6-tuple (left, right, bottom, top, near, far), which defines the clipping planes. These planes form a cube with one corner at (left, bottom, near) and another corner at (right, top, far). The

Figure 9.2   Central projection as measured by an LRF, for the throne room as shown in Figure 5.24.

cube is translated so that its center is at the origin, then it is scaled to the unit cube which is defined by having one corner at $(-1, -1, -1)$ and another corner at $(1, 1, 1)$. The orthographic transformation can then be given by the matrix

$$
\mathbf{M}_P = \begin{pmatrix} \frac{2}{R-L} & 0 & 0 & -\frac{R+L}{R-L} \\ 0 & \frac{2}{T-B} & 0 & -\frac{T+B}{T-B} \\ 0 & 0 & \frac{2}{F-N} & -\frac{F+N}{F-N} \\ 0 & 0 & 0 & 1 \end{pmatrix} \tag{9.4}
$$

which is defined by the chosen values for F (far) and N (near), L (left) and R (right), and T (top) and B (bottom). Figure 9.3 and equation (9.4) illustrate and represent the dependencies.

9.1.4 Stereo Projection

Model viewing can be modified by changing the matrix $\mathbf{M}_V$; in this way the 3D object can rotate or translate in any direction. The virtual camera attitude can also be modified (i.e., by applying the inverse of $\mathbf{M}_V$). It is possible to fly into the 3D scene and look around from any attitude within it.

Furthermore, it is possible to render more than one image from a virtual camera position in the same rendering context, and to create (e.g., anaglyphic) stereo pairs in this way. There are different methods for setting up a virtual camera, and for rendering stereo pairs. Actually, many methods are basically incorrect since they introduce an "artificial" vertical parallax. One example is the *toe-in method*, see Figure 9.4(a). In the toe-in projection, both cameras have a fixed and symmetric aperture, and are directed into a single point. Images created using

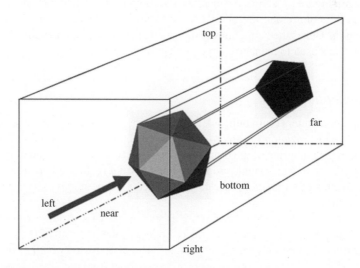

Figure 9.3 Orthogonal parallel projection: the screen (window) can be assumed at any intersection coplanar to the front (or back) side of the visualized cuboidal scene.

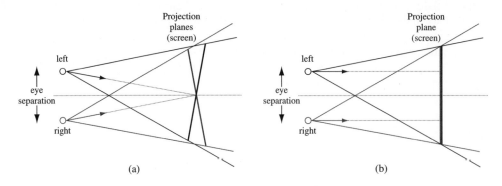

Figure 9.4 (a) Incorrect toe-in stereo projection (b) Correct stereo projection based on asymmetric camera frustums.

the toe-in method still allow a stereoscopic effect, but it introduces a vertical parallax which causes visual discomfort. The vertical parallax increases with the distance to the center of the projection plane, and becomes more disturbing when the camera's field of view increases.

The correct way to create stereo pairs is the *asymmetric frustum* method (Figure 9.4(b)). It introduces no vertical parallax. A stereo rendered anaglyphic image is shown in Plate 10.

9.2 Surface Modeling

In Section 9.1.2 we processed voxel data. After the projection of each voxel into the perspective plane, including all interpolation routines, the final result is produced. In computer graphics this is called *rasterization* because the projected voxels are sorted in a non-floating point raster (the final image). Figure 9.2 shows the measured 3D points with (LRF) gray levels. High point density makes the point cloud look like a surface, but individual 3D points become visible when zooming in. This is the common way to visualize data captured in medical applications. Special hardware is used to manage a large number of voxels.

9.2.1 Triangulation

Standard graphics adapters (with built-in 3D acceleration) support fast rendering of triangles. This includes the rasterization (as a part of the graphics pipeline discussed in Section 9.1.1) of each triangle, the rendering of the triangulated surface (interpolation) and of triangle edges (anti-aliasing), depending on triangle properties (e.g., normal vectors, color, or texture) of each *vertex*[4] (i.e., a vertex of a triangle). Therefore, each given 3D point has to be triangulated into the triangle mesh.

Of course, it is possible to use other primitives such as polygons. But, because of the need for unified primitives for hardware accelerated rendering, the first step of the pipeline would be the *tessellation* of these polygons into triangles. The fastest way, which also reduces data

---

[4] In computer graphics, a vertex is a corner point of a primitive.

size, is to render a scene by using "triangle strips". They are defined by one triangle (the first), and each next 3D point defines a new triangle with the last two vertices of the triangle before.

The main challenge within 3D model generation is data reduction, namely to triangulate a given cloud of points by a reduced number of triangles or triangle stripes without loosing geometric resolution. This process is called *meshing*, and is discussed in the next sections of this chapter.

9.2.2 Creating a Digital Surface Model

A *digital surface model* (DSM) is a simplified 3D model. We call it 2.5D instead of 3D: each voxel in $\mathbb{R}^3$ is projected orthogonally into an orthoplane; this orthoplane has two dimensions, like a normal image, but in addition each pixel has a value which represents the orthogonal distance of a 3D point to the orthoplane. This distance is called *depth*, and the DSM is also called *depth map*.[5] Each pixel has just one depth value. Objects which are occluding one another cannot be represented in such a depth map.

To create the DSM an orthoplane needs to be defined. This can be done by selecting at least three 3D points in an LRF scan, or by estimation of an adjusting plane in a point cloud (characterized by minimization of each point distance to this adjusting plane, as described in detail in Section 9.3.4). Figure 9.5 (left) shows a defined orthoplane "behind" a surface of interest.

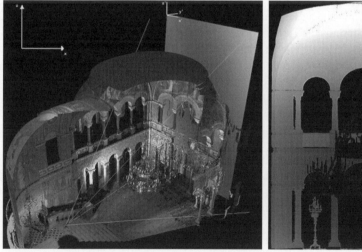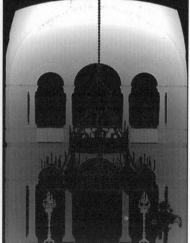

Figure 9.5 Creating a digital surface model (depth map). A defined orthoplane "behind" the generated 3D data (left). Gray-value encoded and orthogonally projected range data for those surface points which are at a distance of 2 metres from the defined plane (right) (see Plate 31).

---

[5] Klette et al. (1998) (and other books in computer vision) make a distinction between *depth* (distance between camera and a 3D point) and *height* (orthogonal distance between a 3D point and an assumed background plane). In the latter case, we refer to an orthoplane (with possible parallel translation in 3D space), and use 'depth' rather than 'height'.

The cross product of the vectors between the corner points of the orthoplane gives the normal vector **n** which is written as a quaternion as follows:

$$\bar{q}_n = (0, n_x, n_y, n_z)^T. \tag{9.5}$$

To align the normal $\bar{q}_n$ of the orthoplane to the y-axis of the reference system – as quaternion form, $\bar{q}_y = (0, 0, -1, 0)^T$ – the transformation $\bar{q}_t$ needs to be determined by the quaternions product as follows:

$$\bar{q}_t = \bar{q}_y \circ \bar{q}_n^*. \tag{9.6}$$

Each LRF point **p** of all LRF scans is transformed into the reference system by means of equation (9.6) – in quaternion form, $\bar{r} = (0, r_x, r_y, r_z)^T = (0, \mathbf{r}_v^T)^T$ – and it is projected into the orthoplane by using the equations

$$\bar{r}' = \bar{q}_t \circ \bar{r} \circ \bar{q}_t^* \tag{9.7}$$

and

$$\mathrm{DSM} = \begin{pmatrix} s_x & 0 & 0 & -c_x \\ 0 & s_y & 0 & c_y \\ 0 & 0 & s_z & -c_z \\ 0 & 0 & 0 & 1 \end{pmatrix} \cdot \begin{pmatrix} r'_{vx} \\ -r'_{vy} \\ r'_{vz} \\ 1 \end{pmatrix}, \tag{9.8}$$

with the transformed upper left corner point $\mathbf{c} = (c_x, c_y, c_z)^T$ of the aligned orthoplane for clipping the resulting image, and s_x, s_y and s_z as scaling factors to specify the resolution of the DSM. The resulting coordinates DSM_x, DSM_z correspond to the image coordinates i, j in this DSM raster, and the DSM_y-value is the "depth" (the orthogonal distance of a 3D point from the orthoplane defined).

In a DSM, the data are rectified and perspective-free. This DSM is the basis for further calculations (e.g., creating orthophotos). Figure 9.5 shows an example of a depth map. The image is gray-value encoded, and darker values are closer to the surface. The big chandelier (in the middle of the room) is not projected into the depth map (i.e., the black values, due to the use of a single scan), and this is an example of an occluded object. A maximum distance limit between orthoplane and surface is necessary for providing a clear depth map, without unwanted certainties (far objects are ignored this way).

Furthermore, an object at the shortest distance from the reference plane is always chosen by the algorithm (e.g., if using multiple scans to create the DSM). This approach has the advantage that, for example, a painting on a wall is not overwritten by a second scan within an object which is near to this painting, but further away from the reference plane.

9.2.3 Bump Mapping

The meshing algorithm allows a level of detail to be set. A planar surface with small structure (e.g., woodchip wallpaper, wrinkled surfaces) does not have to be included in the modeled geometry. In fact, if the distance of a 3D point to an approximated surface is smaller than a chosen threshold ε, the point will not be meshed. But, for example, an absolutely planar wall will not look realistic.

In 1973, Ed Catmull introduced the idea of using the parameter values of parametrically defined surfaces to index a texture definition function which scales the intensity of the reflected light. In 1978, J. Blinn introduced the method of *bump mapping*, which simulates wrinkled surfaces. In computer graphics, bump mapping is a technique where at each pixel, a perturbation to the surface normal of the object being rendered is looked up in a bump map and applied before the illumination calculation is done. The result is a richer, more detailed surface representation that more closely resembles the detail inherent in the natural world, without increasing the geometric complexity. Instead of calculating the perturbation of each pixel to its surface normal, we use the rectified DSM as a bump map. Figure 9.6 depicts an instance where bump mapping is useful, and Figure 9.7 shows its result.

Figure 9.6 A digital surface model (left) used for bump mapping for the timber panel (right) around the fireplace (see Plate 32).

Figure 9.7 Shaded scene with bump mapping (left of fireplace) and without (right of fireplace).

9.2.4 Improving Triangulated Surfaces

As an initial step towards generating a final composite mesh, it is first necessary to generate a dense triangular mesh. The triangulation of points in a *regular point cloud* (i.e., sorted points) poses little difficulty because of the known neighborhood relationship. The neighborhood relationship is defined by the movement of the LRF.

A calibrated LRF scan is rasterized in a 2.5D image, similar to a DSM, with dimensions $W \times H$, indexed by i and j. Unlike a DSM, the triangulated LRF scan is defined in terms of polar coordinates. As is typical in computer graphics, triangulation is determined counterclockwise, whereby the relation is given by the image indices (i, j), $(i, j + 1)$, $(i + 1, j)$, and so on (Figure 9.8).

After triangulating all points it is necessary to separate objects from the foreground or background (or both) via two distinct filter procedures. In the first, the lengths of the triangle's sides are compared to the mesh resolution as given by the angular resolution $\Delta\varphi$ and $\Delta\vartheta$ of the LRF, and by the object's distance. A second filter for removing incorrectly triangulated points is to check the side ratio; an optimal mesh should have equilateral sides. Figure 9.9 depicts these simple steps: the left-hand image shows all triangulated points,

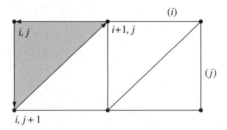

Figure 9.8   Simple triangulation of a regular point cloud.

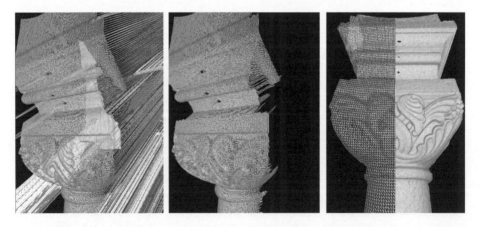

Figure 9.9   Triangulated regular point cloud, unfiltered (left). Filtered mesh after applying the first filter (middle). Final result, depicted here as half shaded and half meshed (right).

unfiltered, the middle image the same sculpture after applying the first filter, and the right-hand image the final mesh of the sculpture after applying both filters. Both filters are necessary for the final mesh. As depicted in the figure, the removal of points or triangles from the mesh creates holes. The detection and darning of such holes will be addressed in Section 9.3.3.

The discussion until now has addressed regular point clouds. When we combine different LRF scans to form one point cloud, however, we build an irregular point cloud. It is possible to build an irregular point cloud via a composite of multiple regular clouds, first meshing each LRF scan separately and then merging all meshes into one. This method, however, has disadvantages, first of excessive data overflow, and second of a complicated challenge of merging edges and overlapping parts of distinct meshes.

A preferable method is to build an initially dense mesh, and then insert points from different scans, point by point, into the mesh. Originally, this approach (by Bodenmüller) was developed for online processing of unorganized data from hand-guided scanner systems (tactile sensors). Nevertheless, the method is also highly suitable for the processing of our LRF data, as it uses a sparse, dynamic data structure that can hold larger data sets. It is also able to generate a single mesh from multiple scans. (See also the filters discussed in Section 9.3.)

9.2.5 Delaunay Triangulation

The following procedure briefly lists the steps of triangulation:

- thinning of 3D points (point density);
- normal approximation (local approximation of surface);
- 3D point insertion (dependent on normals and density);
- estimation of Euclidean neighborhood relations;
- neighborhood projection to tangent planes (i.e., from 3D to 2D points); and
- calculation of local Delaunay triangulations for the points inserted.

The final step of Delaunay triangulation actually calculates the triangles.

In mathematics and computational geometry, the Delaunay triangulation for a set S of points in the Euclidean plane is the unique triangulation $D_T(S)$ of S such that no point in S is inside the circumcircle of any triangle in $D_T(S)$. $D_T(S)$ is the dual of the Voronoi diagram of S. Delaunay triangulations maximize the minimum angle of all angles in the triangulation's triangles.[6] A typical Delaunay triangulation (characterized by regular triangles), applied to our LRF data, is illustrated in Figure 9.10. By virtue of estimating the surfaces, with resultant tangential projections, Delaunay triangulations are not immune to error. Such an error is depicted in the figure; the colored triangles shown here stand in 3D space perpendicular to the mesh.

---

[6] The triangulation was invented in 1934 by Boris Delaunay, and Voronoi diagrams are named for Georgii Voronoi.

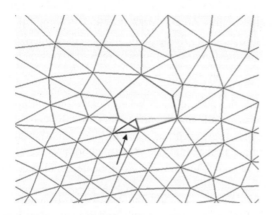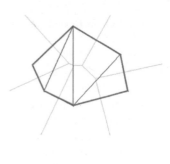

Figure 9.10   Left: triangulated point cloud (Delaunay) with a hole in the mesh at the center; the arrow shows an edge which remains to be filtered (see text). Right: Delaunay triangulated hole; the green lines are the Voronoi diagram (see Plate 33).

9.3 More Techniques for Dealing with Digital Surfaces

The subject of improving, correcting or filling holes in digital surfaces arose in Section 9.2.4. This section goes into more detail, and also touches on the time efficiency of algorithms.

9.3.1 Basic Filters

In essence we differentiate two types of filters for optimizing the LRF data. Firstly, the noise of every 3D point in the point cloud is reduced by various filters (e.g., median filter, histogram filters, spike filters, building Gaussian pyramids, and so on), before we turn to the meshing algorithm. Errors not caused by the system noise (e.g., errors caused by an unfavorable incident angle) and known object geometry are expected. They are fixed subsequent to generating the first initial dense mesh in a second filtering process. This is in fact the step of repairing the mesh.

The error ε discussed in Section 5.5 is determined on a standard dark gray surface. In practice we have specular highlights (e.g., on objects of gold metal). The error on such objects is much higher ($0 < \varepsilon < 2$ m, say, in a large room of up to 50 m diameter). The candlestick in Figure 9.11 is an example of such disturbances. To solve this problem we use a histogram filter, which automatically detects the first and last peak. All depth values before and after these peaks are made invalid and are then interpolated.

9.3.2 Fast Connectivity

Connectivity is defined as the transitive closure of edge adjacency between polygons. If an edge E of a polygon is the same E of another polygon, then both polygons are connected to each other. An edge E is defined as two connected 3D points p_n and p_m, where n and m are the indices of these points. Each 3D point is unique in the point cloud and the meshed 3D model. Note that the point indices have no systematic order in the mesh. Each triangle has three edges; we count them counterclockwise and name them side 1, side 2, and side 3.

In some computer graphics applications it is not necessary to improve the available algorithms for calculating connected components, because models only have a feasible number

Figure 9.11   Illustration of the use of a histogram filter to optimize the LRF data. A candlestick with golden surface, caused noisy LRF data (left), and the candlestick after applying the histogram filter (right).

of polygons. To check every edge of a polygon versus each edge of all the other polygons (by proper subdivision of search spaces) is straightforward but time-consuming. In our case we have many millions of polygons in just one LRF scan. The implementation of the common connectivity algorithm is based on Lee (2000), and leads to connected component detections, a process that takes more than an hour.

One means of improving the processing speed is to hash the point indices to one edge index. Figure 9.12 illustrates such hashing of edges. It is important that (by the sorting of indices) the first column represents the smaller point index p_n. Let p_m be the larger index. Every pair p_n, p_m then has a unique address E by pushing the p_n value into the higher part of a register and p_m into the lower part. Now we can sort our structure (left-hand table in Figure 9.12) by the first column E. The result is shown in the right-hand table. One loop is sufficient to identify all dependencies. If row i and row $i + 1$ have the same E-value, then the dependencies are directly given by the second and third column of our structure. For example, rows 1 and 2 in Figure 9.12 must have the same E-value, and the connectivity can be identified in columns 2 and 3: triangle I, side 3 is connected to triangle II, side 1.

Secondly, the *point–triangle* connectivity is stored in a map within this loop. This means that this map contains, for each point, a list of triangles which are connected to this point. Using this algorithm, we needed only a few seconds, compared to the one hour or more before.

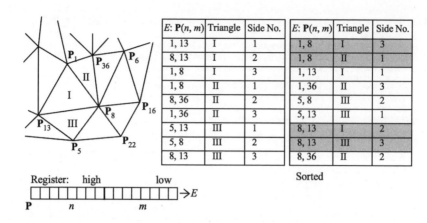

E: $P(n, m)$	Triangle	Side No.	E: $P(n, m)$	Triangle	Side No.
1, 13	I	1	1, 8	I	3
8, 13	I	2	1, 8	II	1
1, 8	I	3	1, 13	I	1
1, 8	II	1	1, 36	II	3
8, 36	II	2	5, 8	III	2
1, 36	II	3	5, 13	III	1
5, 13	III	1	8, 13	I	2
5, 8	III	2	8, 13	III	3
8, 13	III	3	8, 36	II	2

Sorted

Figure 9.12   Fast connectivity calculation of triangles.

Knowing these connections is an important element in most mesh-based algorithms. If, for example, a polygon has only (completely) connected sides, then it is a part inside of a mesh. If more than one side of a polygon is unconnected, then it is a border polygon, which may signal a hole in the mesh. If no side of a polygon is connected, then this polygon is a single polygon, or it is just connected about one point of the polygon to the mesh. Such polygons are mostly caused by errors in the LRF scan, or by deleted points (basic filters), and they need to be filtered. We use the connectivity information to also accelerate the shadow calculation (determination of border polygons).

Figure 9.10 illustrates an example in which connectivity is used as a mesh-based filter (mesh repairing). The triangles marked blue or green in this figure are invalid. The connectivity algorithm finds the marked edges in this figure by virtue of the fact that these are edges at which more than two triangles are connected. The blue triangle, however, connects to a triangle which also has one open side. The decision here, which triangle is invalid remains undefined at this point.

After using a filter that deletes erroneous triangles, we then always use a second filter which deletes all triangles that are connected on one or fewer sides. The first filter deletes the blue triangle and that triangle which is also connected to the same edge and that also has open sides, the second the green triangle. These filters are used iteratively and mostly more than once; this drawback defines an open problem for future work. The polygon marked red is a hole in the mesh and must be darned, and this is dealt with in the next section.

9.3.3 Detection and Darning of Holes

In order to detect holes in the mesh we use the connectivity information. The algorithm begins with an arbitrary polygon in the mesh and checks whether the polygon is connected on all sides. If a polygon is found which does not fulfill this condition, the algorithm grows from this starting polygon to the next unconnected side from the polygon itself, or the adjacent polygon (also determined by connectivity).

A hole is successfully detected when the growing algorithm reaches the start polygon. The hole size is defined by the number of sides of the triangles that describe this hole. The red polygon in Figure 9.10 therefore has a hole size of 6. A threshold parameter aborts the growth of the algorithm if the hole size becomes too large, or if the polygons describe an exterior border of the mesh.

The edges found together describe a new polygon which is projected onto a tangential plane. Therefore, the average normal vector is estimated by the normal vectors of all polygons surrounding the discovered hole. The projected polygon is then triangulated by the same Delaunay algorithm as described before.

The general cavity problem (i.e., connection of those border points whose surface would not be inside of the hole), arising through the triangulation of points, is then solved by the following simple trick. First, all points are connected, whereby all resulting triangles are then checked for complete connectivity. Only those triangles with complete connectivity are inside the polygon, which defines the hole. The connectivity calculation is here optimized, and considers only the new triangles from the triangulation and the triangles passed over during the hole detection procedure. Figure 9.13 shows a triangulated LRF scan with holes (left), and the filled mesh (right) using the algorithm described. Some holes are not detected, however, because they are too big. In these cases, additional LRF scans should first be used to fill the larger holes. Of course, if we were to choose a bigger threshold parameter, the algorithm would also fill openings (also called "holes") in surfaces, such as existing open windows or door, or incorrect gaps in surfaces.

9.3.4 Detection of Planar Surface Patches

The floor of the Neuschwanstein castle throne room (see Figure 9.15), especially in the area of the stairs, is a good example of where normal filters are not sufficient to fit planar surfaces. The reason here is that the floor is made of marble. With such a material, the LRF causes many errors as a result of the high shininess of the marble. For a case such as this, a filter can be used that detects the mean planar surface and projects each LRF point onto this approximated surface. Figure 9.14 depicts such a case. It illustrates the profile of a floor, where $\mathbf{P}_i$ are the 3D points of the triangles, and $\mathbf{S}_i$ are the main points of the triangles with their normal vectors $\mathbf{v}_i^\circ$. The mean planar surface is given by the normal vector $\mathbf{n}^\circ$ and the distance a to the origin of the

Figure 9.13   LRF scan with holes in the triangulated mesh (left), and the mesh after hole-filling (right).

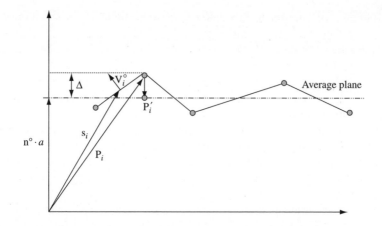

Figure 9.14   Geometric dependencies when approximating a planar surface.

reference system. To determine $\mathbf{n}°$, the mean of all normal vectors $\mathbf{v}_i°$ (assuming k triangles) has to be calculated:

$$\mathbf{n} = \sum_{i=0}^{k-1} \mathbf{v}_i° \quad \text{and} \quad \mathbf{n}° = \frac{\mathbf{n}}{|\mathbf{n}|}.$$

a is the mean of the projections of each $\mathbf{s}_i$ onto the surface orientation $\mathbf{n}°$ and is calculated by the scalar products:

$$a = \frac{1}{k} \sum_{i=0}^{k-1} (\mathbf{n}° \cdot \mathbf{s}_i).$$

Finally each Δ_i, which is the distance of each 3D point $\mathbf{p}$ to the approximated mean surface, can be determined as follows:

$$\Delta_i = (\mathbf{p}_i \cdot \mathbf{n}°) - a.$$

Therefore, each projected 3D point $\mathbf{p}_i'$ is given by

$$\mathbf{p}_i' = \mathbf{p}_i - \Delta_i \cdot \mathbf{n}°.$$

Note that this correction only adjusts the "z-values" along the normal vector of the approximated plane. Figure 9.15 (right) illustrates the stairs after approximating mean planar surfaces for each stair. The stairs are now flat, and they do not have disturbances. The pattern on the upper floor is caused by the marble structure as well. In reality the floor is absolutely flat, as seen after the correction shown in Figure 9.15 (right). In order to ensure that real objects do not become flat, a threshold parameter ϵ for a maximum distance Δ and a maximum angle $< (\mathbf{v}_i°, \mathbf{n}°)$ can be chosen.

For a correction of the edges, an orthogonal view to this plane is needed, which can be rendered from the panoramic image (see Section 9.3.5).

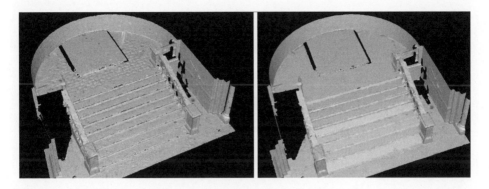

Figure 9.15 Shaded floor of the throne room at Neuschwanstein with errors caused by the reflection of the marble (left). Corrected geometry of the floor by approximating mean planar surface patches (right).

If we do not have triangulated points, a useful equalization procedure involves applying a covariance matrix. By finding the eigenvalues and eigenvectors of the covariance matrix,

$$s_{xy} = \frac{1}{n-1} \left(\sum_{i=0}^{n-1} (x_i \cdot y_i) - n \cdot \overline{x} \cdot \overline{y} \right),$$

$$\mathbf{S} = \begin{pmatrix} s_{xx} & s_{xy} & s_{xz} \\ s_{yx} & s_{yy} & s_{yz} \\ s_{zx} & s_{zy} & s_{zz} \end{pmatrix},$$

we see that the eigenvectors with the smallest eigenvalues correspond to the dimensions that have the strongest correlation (smallest variance) in the point set. This implies that the eigenvector with the smallest eigenvalue is the normal vector of the approximated plane. Therefore the plane coefficients are given as

$$0 = A \cdot x + B \cdot y + C \cdot z + D$$

with A, B, and C as the three components of this vector. The coefficient D is given by

$$D = -(A \cdot x + B \cdot y + C \cdot z).$$

9.3.5 Edge Estimation and Edge Correction

For an ideal setup where rotating line camera and LRF have exactly the same attitude, the correction of edges is straightforward. The image of the panoramic camera can be used to find edges and correct the LRF data, because both images have the same perspective.

In reality, when the panoramic sensor and the LRF have different attitudes, the perspectives are different. For a correct co-registration of the data, however, a correct DSM was in fact already needed. Errors that may be caused by the meshing of the LRF data are therefore already projected into the mapped panoramic image. A correct straight edge in the panoramic image will not be preserved after mapping the data if an incorrect DSM is used.

Figure 9.16 Panoramic input image: perspective visualization (left); orthogonal visualization (right) (see Plate 34).

A perspective-free panoramic image must therefore be used for image-edge-based corrections. One possibility for overcoming this problem is to use approximated planar surfaces (see Section 9.3.4). After the estimation of such a surface, the panoramic image can be orthogonally projected onto this plane. This projection is the orthogonal view of this plane, and Figure 9.16 illustrates this on the marble stairs of the Neuschwanstein throne room. The left-hand image is the original panoramic image, whereas the right-hand image is the rectified, orthogonal image. Because all the stairs are rendered onto the same plane (the floor), the upper stairs look larger than the lower ones. If have more than one depth layer, we must project the panoramic image onto each layer (the approximated mean plane) separately. Figure 22 depicts the panoramic image (in blue, rendered from different layers for each stair) overlaid on the DSM. Then, each 3D point in the 3D model can be shifted along an edge.

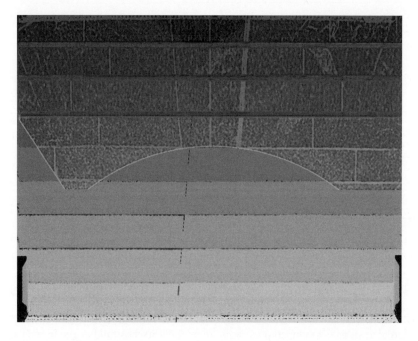

Figure 9.17 DSM (gray) with an overlayed rectified panoramic image (blue) (see Plate 35).

Figure 9.18 A low-resolution overview of individualy triangulated surfaces generated for Neuschwanstein castle within a project of Illustrated Architecture, Berlin (see Plate 36).

The 3D points will only be moved in a direction within the xy plane of the DSM. All these processing steps (i.e., calculation of the DSM, the rectified panorama related to the DSM, edge estimation) are done step by step, possibly with manual interaction. A full automatic pipeline, also using further image-based algorithms, remains an open problem for future work. However, the available approach was sufficient for finishing major reconstruction projects, such as the one illustrated in Figure 9.18.

Plate 11 shows the corrected stairs after the following two stages: first, we have adjusted the z-values (by approximation of mean planar surface patches); and secondly, we have corrected the edges (by using edges in the panoramic image) in the base plane.

9.4 Exercises

9.1. [possible lab project] Implement a (simple) program which opens an OpenGL window and draws some simple geometric primitives, for example, a set of triangles which all together define a rectangle.

    (1) Draw these primitives "manually".
    (2) Save the vertices in a display list and call this list by drawing.
    (3) Save the vertices in a vertex array and call this array by drawing. What is the advantage of having such an array?

9.2. [possible lab project] Load an image as active texture into the memory and bind it. Assign texture coordinates to the vertices of Exercise *9.1*. (See the code example and readme on the books's website.)

9.3. [possible lab project] Take a sample LRF file (available with this exercise on the book's website) and implement a program which visualizes these data as a 3D point cloud. Secondly, mesh the points by (simply) using a neighborhood relation. Delete incorrect triangles which result from using only the neighborhood relations, and visualize your result. This method is possible for one 2.5D LRF scan. What is the procedure for combining several LRF scans together?

9.4. [possible lab project] Load a more complex 3D model and visualize it (see the book's website). See also the sample code for this exercise and use the VRML loader, which supports loading simple VRML 2.0 files.

1. Create your own simple VRML 2.0 file including the vertices and texture coordinates of Exercise *9.1.*
2. Load the test room (provided for this and the following exercises). The test room is available in two versions, untextured as an optimized triangulated LRF scan, and textured as the synthetic "original" scene scanned by the LRF or a specified camera. Panoramic images from this test room are also available within this data package.
3. Fly through the room and make screen dumps, which will be used later for mapping such camera views back to the untextured 3D model. When making a screen dump notice the camera attitude available in the inverse $\mathbf{M}_M$ model matrix. Use the perspective matrix to record relevant information about the camera properties.

9.5. [possible lab project] Modify the program in Exercise *9.4* to render two virtual camera attitudes into the same rendering context. In so doing, mask the rendering buffer for the first camera location for red = true, blue = false and green = false. Then render the second camera attitude with inverse masking into the same buffer. Use red-cyan anaglyph eyeglasses for viewing the resulting anaglyph image. Experiment with different base distances between both cameras.

9.5 Further Reading

A comprehensive treatment of 3D computer graphics can be found, for example, in Watt (2002), Baker (2004) and Hill and Kelley (2007).

Fundamental information for starting programming of 3D graphics, from setting up an OpenGL development environment, to creating realistic textures and shadows, is given, for example, in Silicon Graphics (2007) and Wright and Lipchak (2005). For a step-by-step tutorial, see *http://nehe.gamedev.net/*.

The common Gouraud shading model, which is supported by OpenGL, is described in Gouraud (1971). For Phong's shading model, see Phong (1975).

Further details of fast shadow implementations can be found in Crow (1977). State-of-the-art algorithms, especially based on current hardware support, are also described in standard textbooks on OpenGL programming (references as above).

J. Blinn's bump mapping algorithm was described in Blinn (1978).

For stereo rendering methods, see Bourke (2004). Figure 9.4 is from Bourke (2004).

Bodenmueller and Hirzinger (2004) introduced a pointwise insertion meshing algorithm which supports triangulation of huge data sets (basically: build an initial dense mesh, then insert points from different scans, point by point). Rui Lui et al. (2007) also address topics of mesh generations, mesh simplification, mesh cutting, mesh merging, and texture mapping.

10

Data Fusion

The fusion of data sets (LRF data, panoramic images) starts with the transformation of coordinate systems (i.e., those of LRF and camera attitudes) into a uniform reference coordinate system (i.e., a world coordinate system). For this step, the attitudes of both systems need to be known (see Chapter 5). A transformation of LRF data into the reference coordinate system applies an affine transform between 3D coordinate systems (see Chapter 3). The known 3D object points $\mathbf{P}_w$ (vertices of the triangulation) are given by the LRF system, and are textured with color information provided by the recorded panoramic images. Therefore, panoramic image coordinates are calculated in relation to object points $\mathbf{P}_w$, and this is the main subject in this chapter.

10.1 Determination of Camera Image Coordinates

This section is organized into four subsections, each addressing one possible option for an "acquisition set-up". As discussed before, we differentiate between single-projection-center panoramas, multi-projection-center panoramas with $\omega = 0$, and multi-projection-center panoramas with $\omega \neq 0$ (typically used as a stereo setup). A sensor-line camera used should provide "sufficient" rotational accuracy (i.e., on a circular path), even for (minor) mechanical variations. However, in the fourth subsection we also consider the general case, also taking inaccuracies into account, thus moving further away from idealizing assumptions.

We start with some definitions using the calculated (calibrated) exterior attitude parameters $\mathbf{R}$ and $\mathbf{t}_0$. This allows us to define (in the reference system) a difference vector between object coordinate $\mathbf{P}_w$ and camera location $\mathbf{t}_0$, written as

$$(\mathbf{P}_w - \mathbf{t}_0)' = \mathbf{R}^{-1}(\mathbf{P}_w - \mathbf{t}_0).$$

We also use the abbreviation

$$\tilde{\mathbf{P}}_w = \mathbf{R}^{-1}(\mathbf{P}_w - \mathbf{t}_0)'.$$

Panoramic Imaging: Sensor-Line Cameras and Laser Range-Finders   F. Huang, R. Klette, and K. Scheibe
© 2008 John Wiley & Sons, Ltd

By applying both rotations (i.e., $\mathbf{R}$ describes a rotation of the exterior attitude of the camera, and $\mathbf{R}_{\varphi(i)}$ for the CCD line rotation) to the difference vector, we have

$$(\mathbf{P}_w - \mathbf{t}_0)'' = \mathbf{R}_{\varphi(i)}^{-1}\mathbf{R}^{-1}(\mathbf{P}_w - \mathbf{t}_0).$$

By applying the tilt parameters (i.e., $\mathbf{R}_\xi$ and $\mathbf{R}_\omega$ of the camera head, and the interior attitude $\mathbf{R}_i$ and Δ of the CCD line), we transform the image vector $\mathbf{v}_j = \mathbf{v}_{j,\Delta} + \mathbf{v}_f$ into

$$\mathbf{v}'_{j,\Delta} = (a, b, c)^{\mathrm{T}} \cdot j = \mathbf{R}_i \cdot \mathbf{v}_{j,\Delta}$$

and

$$\mathbf{v}'_f = (u, v, w)^{\mathrm{T}} = \mathbf{R}_\xi \mathbf{R}_\omega \mathbf{v}_f.$$

This results in the substituted image vector

$$\mathbf{v}'_j = \begin{pmatrix} a \\ b \\ c \end{pmatrix} j + \begin{pmatrix} u \\ v \\ w \end{pmatrix}$$

for the general case.

10.1.1 Single-Projection-Center Panoramas

The panoramic images are preprocessed as described in Section 3.7. Therefore, all image coordinates are in rectified ideal cylindrical coordinates, and we can use the basic equation (3.11). The simple truth is that accuracy depends on calibration. (Single-projection-center panoramas are rectified as accurate as supported by the calculated calibration parameters.) The use of equation (3.11) is not just a simplification.

The required viewing direction $\mathbf{R}_{\varphi(i)}\mathbf{v}_j$ of the panoramic sensor (for a "normal" panorama with $R = 0$ and $\omega = 0$) is described by the following equation (assuming that the sensor is calibrated):

$$\mathbf{R}^{-1}(\mathbf{P}_w - \mathbf{t}_0) = \lambda \mathbf{R}_{\varphi(i)}\mathbf{v}_j. \tag{10.1}$$

The application of the calculated exterior attitude to the difference vector provides $(\mathbf{P}_w - \mathbf{t}_0)'$. This allows us to specify the pixel column i in the panoramic image. In more detail, we have that

$$\left(\mathbf{P}_{wx} - \mathbf{t}_{x_0}\right)' = \lambda \sin(i \cdot \Delta\varphi) f,$$

$$\left(\mathbf{P}_{wy} - \mathbf{t}_{y_0}\right)' = \lambda j\tau,$$

$$\left(\mathbf{P}_{wz} - \mathbf{t}_{z_0}\right)' = -\lambda \cos(i \cdot \Delta\varphi) f,$$

with

$$i = -\arctan\left(\frac{\left(\mathbf{P}_{wx} - \mathbf{t}_{x_0}\right)'}{\left(\mathbf{P}_{wz} - \mathbf{t}_{z_0}\right)'}\right)\frac{1}{\Delta\varphi}. \tag{10.2}$$

By substituting $\mathbf{R}_{\varphi(i)}$ into equation (10.1), and due to the fact that the rotation $\mathbf{R}_{\varphi(i)}$ of the CCD line corresponds to index i (i.e., $\mathbf{R}_\varphi(i)$) given by equation (10.2), the vertical pixel row j can now be calculated as follows:

$$\left(\mathbf{P}_{wy} - \mathbf{t}_{y_0}\right)'' = \lambda j \tau,$$

$$\left(\mathbf{P}_{wz} - \mathbf{t}_{z_0}\right)'' = \lambda f,$$

with

$$j = \frac{\left(\mathbf{P}_{wy} - \mathbf{t}_{y_0}\right)''}{\left(\mathbf{P}_{wz} - \mathbf{t}_{z_0}\right)''} \frac{f}{\tau}. \tag{10.3}$$

10.1.2 Multi-Projection-Center Panoramas with $\omega = 0$

We now consider the more general case where the camera is mounted on a lever and we have an off-axis parameter $R > 0$. (In fact $R > 0$ may also be caused by the eccentricity of a setup, as described in Section 10.1.1.) The following equations are based on an ideal cylinder, which is preprocessed only for the $R = 0$ case (see Section 3.7). In fact, the following determination of the image coordinates assumes a rotation of an ideal sensor line (i.e., the sensor line is not tilted). This restriction is sufficient for many applications, and a solution for the more general case is given below. Equation (3.12), solved for the required viewing direction, gives the following (here we still assume that $\omega = 0$):

$$\mathbf{R}^{-1}\left(\mathbf{P}_w - \mathbf{t}_0\right) = \mathbf{R}_{\varphi(i)}(\lambda \mathbf{v}_j + R\mathbf{z}^\circ). \tag{10.4}$$

By analogy with the case in Section 10.1.1, the pixel column i is determined by applying the exterior attitude $\mathbf{R}^{-1}$ to the difference vector. The known off-axis parameter, depending on the rotation $\mathbf{R}_{\varphi(i)}$, and the three components of equation (10.4) are given as follows:

$$\left(\mathbf{P}_x - \mathbf{t}_{x_0}\right)' = \sin(i \cdot \Delta\varphi)(\lambda f + R),$$

$$\left(\mathbf{P}_y - \mathbf{t}_{y_0}\right)' = \lambda j \tau,$$

$$\left(\mathbf{P}_z - \mathbf{t}_{z_0}\right)' = -\cos(i \cdot \Delta\varphi)(\lambda f + R).$$

It follows that the off-axis parameter $R > 0$ has no influence on image coordinate i, and equation (10.2) is still valid.

For the calculation of image coordinate j (for an ideal hollow cylinder) we utilize the known image coordinate i. We have that

$$\left(\mathbf{P}_y - \mathbf{t}_{y_0}\right)'' = \lambda j \tau,$$

$$\left(\mathbf{P}_z - \mathbf{t}_{z_0}\right)'' = \lambda f + R.$$

It follows that the panoramic image coordinate j is then given as follows:

$$j = \frac{\left(\mathbf{P}_{wy} - \mathbf{t}_{y_0}\right)''}{\left(\mathbf{P}_{wz} - \mathbf{t}_{z_0}\right)'' - R} \cdot \frac{f}{\tau}. \tag{10.5}$$

10.1.3 Multi-Projection-Center Panoramas with $\omega \neq 0$

Equations (10.4) and (10.5) describe an ideal off-axis case. The camera is rotating on an ideal circle around a rotation center. For panoramic stereo data acquisition it is necessary to change

ω for a second acquisition to $-\omega$; therefore, we consider $\omega \neq 0$ in this section. The image vector $\mathbf{v}_j$ has to be transformed by this angle ω. Equation (10.4) becomes

$$\mathbf{R}^{-1}(\mathbf{P}_w - \mathbf{t}_0) = \mathbf{R}_\varphi(\mathbf{R}_\omega \lambda \mathbf{v}_j + R\mathbf{z}^\circ).$$

By also applying the exterior attitude $\mathbf{R}^{-1}$ to the difference vector, and by some algebraic manipulation of the above equation (we take $\mathbf{R}_{\varphi(i)}$ and R to the left), we obtain the following detailed equations:

$$\cos(i \cdot \Delta\varphi)(\mathbf{P}_{wx} - \mathbf{t}_{x_0})' + \sin(i \cdot \Delta\varphi)(\mathbf{P}_{wz} - \mathbf{t}_{z_0})' = \sin(\omega)\lambda f, \tag{10.6}$$

$$(\mathbf{P}_{wy} - \mathbf{t}_{y_0})' = \lambda j \tau, \tag{10.7}$$

$$-\sin(i \cdot \Delta\varphi)(\mathbf{P}_{wx} - \mathbf{t}_{x_0})' + \cos(i \cdot \Delta\varphi)(\mathbf{P}_{wz} - \mathbf{t}_{z_0})' - R = -\cos(\omega)\lambda f. \tag{10.8}$$

Therefore, ω may be calculated based on known values R and i as follows:[1]

$$\omega = -\arctan\left(\frac{\cos(i \cdot \Delta\varphi)(\mathbf{P}_{wx} - \mathbf{t}_{x_0})' + \sin(i \cdot \Delta\varphi)(\mathbf{P}_{wz} - \mathbf{t}_{z_0})'}{-\sin(i \cdot \Delta\varphi)(\mathbf{P}_{wx} - \mathbf{t}_{x_0})' + \cos(i \cdot \Delta\varphi)(\mathbf{P}_{wz} - \mathbf{t}_{z_0})' - R}\right).$$

To obtain image coordinates i and j, λ needs to be calculated first. By taking the sum of the squares of equations (10.6) and (10.8) (with R taken to the right for the latter equation), the unknown $i \cdot \varphi$ is eliminated. Thus we have

$$(\mathbf{P}_{wx} - \mathbf{t}_{x_0})'^2 + (\mathbf{P}_{wz} - \mathbf{t}_{z_0})'^2 = \sin^2(\omega)\lambda^2 f^2 + \cos^2(\omega)\lambda^2 f^2 + 2\cos(\omega)\lambda f R + R^2.$$

Consequently, a quadratic equation in λ is given as follows:

$$\lambda^2 + \frac{2\cos(\omega)\lambda R}{f} + \frac{R^2 - (\mathbf{P}_{wx} - \mathbf{t}_{x_0})'^2 - (\mathbf{P}_{wz} - \mathbf{t}_{z_0})'^2}{f^2} = 0.$$

The solutions of this equation are as follows:

$$\lambda_{1,2} = \frac{-R\cos(\omega)}{f} \pm \sqrt{\frac{R^2\cos^2(\omega)}{f^2} - \frac{R^2 - (\mathbf{P}_{wx} - \mathbf{t}_{x_0})'^2 - (\mathbf{P}_{wz} - \mathbf{t}_{z_0})'^2}{f^2}}$$

$$= \frac{1}{f}\left(-R\cos(\omega) \pm \sqrt{(\mathbf{P}_{wx} - \mathbf{t}_{x_0})'^2 + (\mathbf{P}_{wz} - \mathbf{t}_{z_0})'^2 - R^2\sin^2(\omega)}\right).$$

A real solution requires a non-negative radicand. Therefore we assume that

$$(\mathbf{P}_{wx} - \mathbf{t}_{x0})'^2 + (\mathbf{P}_{wz} - \mathbf{t}_{z0})'^2 \geq R^2 \cdot \sin^2(\omega).$$

---

[1] Actually, ω is determined by calibration. Here we assume an ideal camera rotation and show that determination of ω is possible if sensor attitude and i are known by using one pair of corresponding points (i.e., one control point $\mathbf{P}_w$).

Second, we can assume that λ is positive, because of using λ as a scaling factor without changing the viewing direction. We notice that λ_2 is not a valid solution for the case where $\cos(\omega) \geq 0$. λ_1 is a valid solution if the following equation is true:

$$\lambda_1 \cdot f = -R\cos(\omega) + \sqrt{(\mathbf{P}_{wx} - \mathbf{t}_{x_0})'^2 + (\mathbf{P}_{wz} - \mathbf{t}_{z_0})'^2 - R^2\sin^2(\omega)} > 0$$

or, equivalently,

$$R^2 \cdot \cos^2(\omega) < (\mathbf{P}_{wx} - \mathbf{t}_{x_0})'^2 + (\mathbf{P}_{wz} - \mathbf{t}_{z_0})'^2 - R^2 \cdot \sin^2(\omega).$$

Thus, λ_1 is a valid solution if

$$R^2 < (\mathbf{P}_{wx} - \mathbf{t}_{x0})'^2 + (\mathbf{P}_{wz} - \mathbf{t}_{z0})'^2$$

for the case where $\cos(\omega) \geq 0$ (i.e., the outward case where the camera is "looking away from the rotation axis").

For the inward case with $\cos(\omega) < 0$, it follows that λ_1 is a valid solution if the radicand is a non-negative value. λ_2 is a valid solution if the radicand, and thus also λ_2, are both non-negative. This is given if the following is satisfied:

$$R^2\sin^2(\omega) \leq (\mathbf{P}_{wx} - \mathbf{t}_{x_0})'^2 + (\mathbf{P}_{wz} - \mathbf{t}_{z_0})'^2 \leq R^2.$$

Thus, in this case we have two valid solutions in the area of the shaded disc, and this is illustrated in Figure 10.1 (left). Objects which are located in this area will be seen twice (depicted by projection centers C_1 and C_2). Furthermore, the figure shows invisible areas (shaded circles) which depend on ω. The radii of these invisible circles are calculated as $R_{inv} = R \cdot |\sin(\omega)|$ for the case that $\cos(\omega) < 0$, and as $R_{inv} = R$ for the case where $\cos(\omega) \geq 0$.

The required image coordinate j is determined by using equation (10.7) and the (by now) known value of λ. We obtain

$$j = \frac{(\mathbf{P}_{wy} - \mathbf{t}_{y_0})'}{\lambda \cdot \tau}.$$

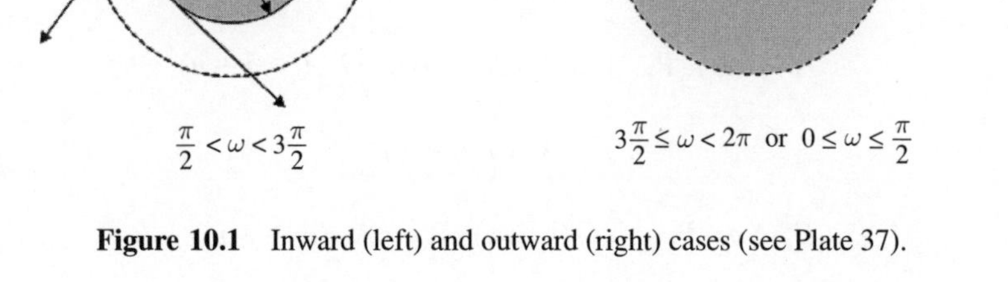

Figure 10.1 Inward (left) and outward (right) cases (see Plate 37).

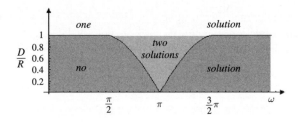

Figure 10.2 Number of solutions depending on ω, R and projected object distance in $\mathbb{R}^2$, with $D = \sqrt{(\mathbf{P}_{wx} - \mathbf{t}_{x_0})^{\prime 2} + (\mathbf{P}_{wz} - \mathbf{t}_{z_0})^{\prime 2}}$.

Finally, the image coordinate i is determined by dividing equation (10.6) by equation (10.8), after factoring out $\cos(i \cdot \Delta\varphi)$, and we obtain

$$\frac{(\mathbf{P}_{wx} - \mathbf{t}_{x_0})' + (\mathbf{P}_{wz} - \mathbf{t}_{z_0})' \tan(i \cdot \Delta\varphi)}{-(\mathbf{P}_{wx} - \mathbf{t}_{x_0})' \tan(i \cdot \Delta\varphi) + (\mathbf{P}_{wz} - \mathbf{t}_{z_0})'} = \frac{-\sin(\omega)\lambda f}{\cos(\omega)\lambda f + R}.$$

By solving this equation for i, the required image coordinate is finally given by

$$i = \arctan\left(\frac{-(\mathbf{P}_{wx} - \mathbf{t}_{x_0})' + k(\mathbf{P}_{wz} - \mathbf{t}_{z_0})'}{k(\mathbf{P}_{wx} - \mathbf{t}_{x_0})' + (\mathbf{P}_{wz} - \mathbf{t}_{z_0})'}\right) \frac{1}{\Delta\varphi} \tag{10.9}$$

with

$$k = \frac{-\sin(\omega)\lambda \cdot f}{\cos(\omega)\lambda f + R}.$$

Figure 10.2 illustrates the cases where we have exactly one, two, or no solutions.

10.1.4 General Case

The general case is defined by solving the general equation (3.16) for the required image coordinates i and j, for known object coordinates $\mathbf{P}_w$. Thus, we are also able to describe multi-projection-center panoramas, and without assuming an ideal case. We consider all parameters to be calculated by calibration. Furthermore, the option of a rotating (CCD or CMOS) matrix sensor is also covered by this general camera model (using Δ_x for describing horizontal distances of the matrix sensor).

As a first step we determine the unknown λ as a function of j. By also applying the exterior attitude $\mathbf{R}^{-1}$ to the difference vector, and taking $\mathbf{R}_{\varphi(i)}$ to the left, we obtain

$$\cos(i \cdot \Delta\varphi) \cdot \tilde{\mathbf{P}}_{wx} + \sin(i \cdot \Delta\varphi) \cdot \tilde{\mathbf{P}}_{wz} = \lambda \cdot (a \cdot j + u), \tag{10.10}$$

$$\tilde{\mathbf{P}}_{wy} = \lambda \cdot (b \cdot j + v), \tag{10.11}$$

$$-\sin(i \cdot \Delta\varphi) \cdot \tilde{\mathbf{P}}_{wz} + \cos(i \cdot \Delta\varphi) \cdot \tilde{\mathbf{P}}_{wz} = \lambda \cdot (c \cdot j + w) + R. \tag{10.12}$$

To eliminate the unknown $i \cdot \Delta\varphi$, the sum of squares of equations (10.10) and (10.12) is computed (written in quadratic normal form) as follows:

$$0 = \lambda^2[(a \cdot j + v)^2 + (c \cdot j + w)^2] + 2 \cdot \lambda \cdot (c \cdot j + w)^2 \cdot R - \tilde{\mathbf{P}}_{wx}^2 - \tilde{\mathbf{P}}_{wz}^2 + R^2,$$

$$\lambda_{1,2} = \frac{-R \cdot W \pm \sqrt{R^2 W^2 - (R^2 - \mathbf{P}_{wx}^2 - \mathbf{P}_{wy}^2)(U^2 + W^2)}}{U^2 + W^2},$$

with $U = u + a \cdot j$ and $W = w + c \cdot j$. Inserting λ into equation (10.11), and solving for j, gives

$$j_{1,2} = M \pm \frac{\sqrt{M^2 - K \cdot L}}{K}$$

with substituted terms K, L and M defined as follows:

$$K = c^2(R^2 - \tilde{\mathbf{P}}_{wx}^2 - \tilde{\mathbf{P}}_{wy}^2) + 2b \cdot c \cdot R \cdot \tilde{\mathbf{P}}_{wz} + \tilde{\mathbf{P}}_{wz}^2(a^2 + b^2),$$

$$L = w^2(R^2 - \tilde{\mathbf{P}}_{wx}^2 - \tilde{\mathbf{P}}_{wy}^2) + 2v \cdot w \cdot R \cdot \tilde{\mathbf{P}}_{wz} + \tilde{\mathbf{P}}_{wz}^2(u^2 + v^2),$$

$$M = c \cdot w(R^2 - \tilde{\mathbf{P}}_{wx}^2 - \tilde{\mathbf{P}}_{wy}^2) + \tilde{\mathbf{P}}_{wz}(c \cdot v \cdot R - b \cdot w \cdot R) - \tilde{\mathbf{P}}_{wz}^2(a \cdot u + b \cdot v).$$

Because j is a function of λ, it follows that, for each solution of λ, there exists one solution for j. Instead of a general theoretical discussion of the solutions of $\lambda_{1,2}$ and $j_{1,2}$, we provide only numerical results. These results show that solutions satisfy the same constrains as discussed in Section 10.1.3.

The image coordinate i can be calculated analogously to its calculation in the ideal case of a multi-projection-center panorama, using equation (10.9), but now with

$$k = \frac{\lambda \cdot (a \cdot j + u)}{\lambda \cdot (c \cdot j + w) + R}.$$

10.2 Texture Mapping

Every 3D point identified by the LRF in the 3D scene, $\mathbf{P}$ (which we call an *LRF point*), has unique coordinates in our reference coordinate system in the Euclidean space $\mathbb{R}^3$. After generating the triangulated 3D surface model, each vertex has to be assigned to a corresponding pixel recorded with the panoramic sensor. The calculation of the corresponding panorama pixel (also called the texture coordinate) was discussed in Section 10.1.

When using different sensor attitudes, any ray (between a camera at one panoramic sensor location and a visible 3D point in the 3D scene, identified by the calculated viewing direction) can be disturbed by obstacles in the scene, and a raytracing routine has to check whether an LRF point can be colored properly. Here it would be helpful if the LRF and camera setup allowed both principal points to be centered in such a way that we could map any camera pixel into the reference coordinate system directly (i.e., no raytracing routine would be necessary in this ideal case). However, such an ideal positioning is very time-consuming.

Thus, for an efficient raytracing routine we may use, for example, the idea of a stencil buffer (as common in computer graphics). This means that the panoramic image (or a part of it)

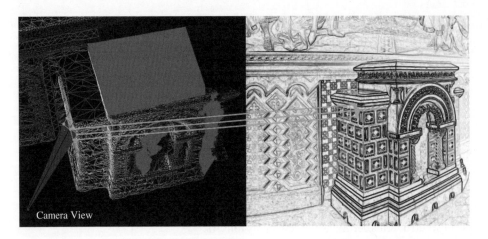

Figure 10.3 Illustration of the raytracing approach using a stencil buffer. Right: Triangles closest to the camera are masked in the stencil buffer. Left: Lines from right to left indicate such a masked triangle in the left image; triangles which are behind this triangle do not obtain valid texture coordinates using the stencil test (see Plate 38).

is loaded into the memory as a mirror of the original texture. One bit per pixel is sufficient for the mirror image (stencil buffer) to save memory. Figure 10.3 illustrates such a raytracing algorithm. For better visualization, the stencil buffer image in this figure is shaded (using 8 bits per pixel).

All LRF points $\mathbf{P}_w$ are sorted by distance to the panoramic sensor location $\mathbf{t}_0$ of the panoramic image. The distance D can be determined as follows:

$$D^2 = (\mathbf{P}_{wx} - \mathbf{t}_{x_0})^2 + (\mathbf{P}_{wy} - \mathbf{t}_{y_0})^2 + (\mathbf{P}_{wz} - \mathbf{t}_{z_0})^2. \qquad (10.13)$$

To save CPU time, the square root is actually not necessary. The raytracing routine starts with the triangle that is nearest to the current panoramic sensor location. After determining the texture coordinates of a triangle, the same triangular area has to be masked in the stencil buffer. If the current triangle, or a part of it, is already masked, the triangle is not visible from this viewpoint of the panoramic sensor. If the triangular area is not (yet) masked, the texture coordinates i and j are valid and placed in a texture map.

This step has to iterate for each panoramic sensor attitude (panoramic image). After all iterations, an algorithm has to decide which panoramic image, or part thereof, should be used for the final texture of a particular triangle. For this decision, the following constraints proved to be useful:

- select attitudes of panoramic sensors such that short distances are guaranteed (for a better image resolution);
- attempt to ensure a small angle between the normal vector of surface triangles and viewing direction $\mathbf{v}_j$ of the rotating camera (a zero angle would be ideal);
- avoid switching between different panoramic images within a reasonably large surface area (to preserve radiometric uniformity).

This part of the rendering procedure is quite sensitive, and the influence of each constraint can be set by weights in the program. Plate 12 illustrates partial texture mapping (just for a subwindow of the given range-scan) and Plate 13 shows a texture mapped 3D model.

Note that we calculated texture coordinates just for the corners of all surface triangles. The 3D hardware accelerator of the graphics adapter maps (fills) the triangulated surface automatically, ignoring (in general) additional perspective projections (i.e., from cylindrical to planar and vice-versa). The related error can be ignored as long as the triangles are small enough. Figure 10.4 illustrates such a mapping error. The three corner points (shown in white) of the triangle are rectified and calculated correctly. However, the mapped surface of the triangle is incorrect, if no further calculations are used.

However, in our applications, the triangular mesh generated was typically small enough that any resulting displacement error is less than a pixel.

Possible mapping errors depend on the size of the triangles generated, and the angle between the normal vector of a triangle and the principal axis of the rotating camera (at the moment when recording the texture values). The error is defined by image coordinates i (an offset angle $i \cdot \Delta \varphi$ to the theoretical principal axis) and j (the vertical pixel offset to the principal axis). Figure 10.5 illustrates the mapping error depending on image offset coordinates. The principal axis, as used in this figure, is defined for the rotating sensor-line camera at $(x, y, z) = (0, 0, -1)$. The vertical offset j is set to the maximum value (e.g., to half of the total length of the sensor line, measured in pixels).

To overcome this mapping problem, a rectified texture has to be calculated; see Section 3.7. The calculation of the central perspective geometry of an individual line or matrix image defines the attitude of the principal axis. There might be a "visually unfavorable" perspective for such a triangle. Actually, an individually rectified texture has to be calculated for each triangle. (The optimum is defined by a viewing direction in the direction of the normal of

Figure 10.4 Mapping error caused by panoramic geometry. The white corner vertices of the triangle are calculated correctly; if the area of the triangle is mapped using a (standard) graphics adapter then inner points of the triangle are textured slightly incorrectly. This is also illustrated by the curved line between both lower vertices, which would be (incorrectly) mapped as a straight line segment.

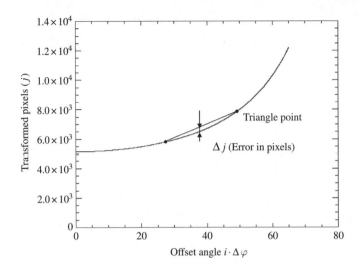

Figure 10.5 Mapping error depending on position and pixel resolution of a triangle as illustrated in Figure 10.4. The error is defined by the maximum difference between an ideal straight edge and the actual curved arc in the panorama, and is calculated in a tangential plane of a cylindrical panoramic image.

a triangle.) The advantage of using an unrectified panoramic image as texture is a dynamically specified principal axis, defined by the relevant projection center during the camera's rotation.

Modern graphics adapters allow to specify the texture mapping routine. This has the advantage of using a hardware-accelerated rectifying micro code. However, for a correct mapping, each pixel inside the triangulated surface has to be calculated. Exact geometric images are, for example, calculated in case of high resolution orthophotos (see next section).

10.3 High Resolution Orthophotos

For computing orthophotos, a DSM is required (see Section 9.2.2). In fact, the DSM itself is the rectified orthophoto because it is already orthogonally projected to a given plane (see Figure 9.5). Instead of depth values, the original gray or color value has to be stored in the image. A common requirement is that high-resolution orthophotos (as the final product) are stored in a common file format independent of the resolution of the viewport of OpenGL. Secondly, each pixel of the orthophoto is calculated. That means that each pixel is shaded (colorized), and cannot be influenced by mapping errors, as described in Section 10.2.

To compute colored orthophotos from data provided by LRF or panoramic sensors with different attitudes, the intersection between camera rays $\mathbf{t}_0 + \mathbf{R}\mathbf{R}_{\varphi(i)}\lambda\mathbf{v}_j$ and DSM, defined by points $\mathbf{P}_w$, has to be determined. A disadvantage is that a fast raytracing routine cannot be used because of working with single pixel. However, raytracing in 2.5D images is easier to perform than volume raytracing. Here, a standard approach (using backward raytracing) can be used

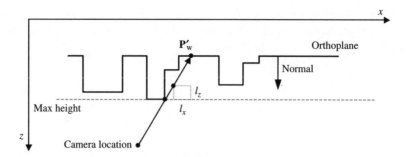

Figure 10.6 Raytracing for 2.5D digitized images. The upward-pointing arrow is the resulting viewing direction towards a transformed point $\mathbf{P}'_w$ of the aligned orthoplane, and the lengths l_x and l_z define the search areas in the DSM.

which is illustrated in Figure 10.6, and briefly described by the following steps (i and j refer to a raster element in the DSM):

- Transform panoramic sensor location $\mathbf{t}_0$ and image vector $\mathbf{v}_j$ into $\mathbf{t}'_0$ and $\mathbf{v}'_j$, related to the attitude of the DSM in a reference xyz coordinate system (by aligning the normal of the orthoplane to the z-axis).
- Determine corresponding image coordinates i' and j' of the panoramic image related to a DSM point $\mathbf{P}_w(i, j)$ (by using the equations of Section 10.1).
- Scan backwards through the DSM raster, along the camera ray, between $\mathbf{P}_w(i, j)$ and the transformed camera location $\mathbf{t}'_0$.
- Determine $l_x(i, j), l_y(i, j)$, and $l_z(i, j)$ (see Figure 10.6) depending on the step size as given by the resolution of the DSM.
- Abort the depth check if $l_z(i, j)$ is greater than the maximum height.
- Map the camera pixel (i', j') to the DSM point $\mathbf{P}(i, j)$ if no depth value (i.e., DSM value) on the determined ray is greater than $l_y(i_{0...n}, j_{0...n})$.

Plate 14 illustrates spatial relations between the panoramic image and the DSM, and depicts a rendered orthophoto. Plate 15 is an anaglyphic panorama of the throne room at Neuschwanstein.

10.4 Fusion of Panoramic Images and Airborne Data

At the beginning of this book we mentioned that sensor-line cameras were originally designed and developed in the context of aerial sensing, from an airplane or a satellite. This section introduces a very recent development – the use of terrestrial panoramic images for enhancing airborne photogrammetry.

Basically this is a fusion of panoramic images (e.g., a side view to a building) with top view images taken from an airplane. The goal is to replace unfavorable bird's-eye perspectives of facades of buildings with orthogonal views, where possible. Furthermore, this can also be extended to a fusion of several sensors, for example merging indoor and outdoor scene models together into one unified model of a virtual 3D world, allowing flights of a virtual camera within buildings and also outdoors.

Actually, data fusion of panoramic images with any other recorded 3D data is straightforward using approaches as already proposed or discussed in this book. As a first step, the attitude of the panorama camera needs to determined in the coordinates of the available 3D data; then the required texture coordinates of each 3D point can be determined, followed by raytracing checks.

10.4.1 Airborne Cameras

In the early days of aerial photography, one option when capturing the image was to use flying equipment which was nearly not moving (such as a balloon or a set of connected kites). This is very similar to today's panoramic scanning approach. George Lawrence, a pioneer of airborne images, used curved capturing surfaces (the "Lawrence camera") when experimenting with balloons or kites (see Figure 1.4 for an example of an image recorded in 1906). A rotating CCD line is basically nothing more than a cylindrical curved plate with respect to the imaging geometry. Another very similar (historic) approach was using a rotating lens camera, and a result is shown in Figure 10.7.

Modern airborne or satellite cameras are fully digital. Many inventions and improvements are responsible for today's photo quality or geometric accuracy.

The scanning principle of an airborne sensor-line camera is similar to that of a flat-bed scanner. Instead of a rotation of the sensor-line camera (as discussed in this book), the image is captured during translative forward motion of an airplane (or satellite). To be precise, the forward motion of an airplane is actually also a rotation around the Earth, as long as it stays at constant altitude above sea level. An undisturbed motion would be very similar to the panoramic off-axis model discussed, with parameters R and ω.

Additionally, the movement of an airplane does not follow an ideal curve; we briefly discuss geometric rectification of aerial images further below. Here we simply note that the exterior attitude **R** of the airborne camera is defined by the attitude of the airplane (in aerial photogrammetry, this is characterized by roll, pitch and yaw angles). The position of the airplane

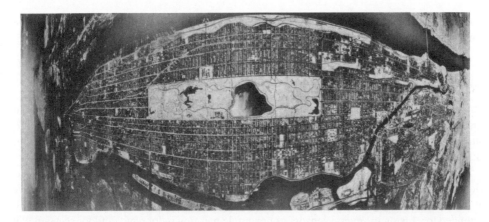

Figure 10.7 View of Manhattan taken in 1949 by a rotating lens camera from an altitude of about 10,000 feet (image from the public domain).

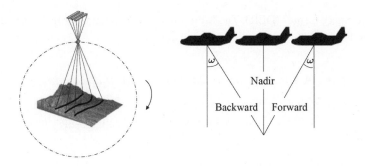

Figure 10.8 Illustration of the push-broom scanning principle, using three sensor lines.

corresponds to vector $\mathbf{t}_0$. All six parameters, of this vector and roll, pitch and yaw, change continuously during a flight, and may be recorded by an *initial measuring unit*.

Instead of using a camera with only one sensor line, typically three or more sensor-lines are used to digitize visible signals at different viewing angles, or at different wavelengths. One main reason is to support stereo capability. Modern airborne cameras use a multi-sensor-line approach. Figure 10.8 shows a standard three-CCD-line push-broom scan principle as already used in airborne photogrammetry for about 20 years.

Of course, matrix cameras are another possibility for use in airborne photography or photogrammetry. Instead of composing the image line by line, a larger flight track may be composed by different images. The advantage of using matrix cameras is the use of more robust bundle block adjustment techniques for geometric rectification, which may help to achieve sufficient accuracy without using expensive measuring units. The disadvantage is mostly the more complex camera design because of the use of multiple sensor matrices (of high resolution). In contrast to the multiple-line concept of line-based cameras, the current design of digital airborne matrix sensors is focused on enlarging the resolution, or the field of view. A stereo concept would then use a bundle of overlapping images. Figure 10.9 shows five airborne cameras: the historic Lawrence camera, three line-based cameras, and the matrix-based UltraCam.

The figure illustrates the (fairly complex) design of the sensor-matrix-based UltraCam, which uses four panchromatic matrix sensors in the middle row, and RGB and near-infrared matrix sensors below and on top, which are stitched with the higher-resolution images recorded by the panchromatic sensors. The three line-based cameras in this figure were all developed the

Figure 10.9 A selection of airborne cameras (from left to right): Lawrence's camera, Leica's ADS40, Vexel's UltraCam, and two DLR cameras, the MFC and HRSC – the latter is currently in use on a Mars mission (see Plate 39).

German Aero Space Center (DLR). The Multi Functional Camera or MFC is the most current one, and also the smallest and most cost-efficient.[2]

10.4.2 Rectifying Airborne Images and Generation of DSMs

Multi-line cameras, especially with symmetric angles ω, or multi-matrix sensor cameras support the generation of 2.5D DSM data from airborne photos. (In our experiments, we processed data captured by the CCD-line-based HRSC and MFC cameras, and the matrix-based UltraCam camera.)

The model of an airborne multi-line camera corresponds to the panoramic camera model as used in this book (of course, without the additional rotation $\mathbf{R}_\varphi$), and is given as follows:

$$\mathbf{P}_w = \mathbf{t}_0(i) + \mathbf{R}(i) \cdot \lambda \cdot \mathbf{v}_j.$$

This models the case of a sensor-line camera, where i is the index of the line, and image vector $\mathbf{v}_j$ is defined by the image coordinate j.

For a matrix camera, where k is the index of a matrix, an image vector $\mathbf{v}_{ij}$ is defined by two matrix coordinates i and j. The equation

$$\mathbf{P}_w = \mathbf{t}_0(k) + \mathbf{R}(k) \cdot \lambda \cdot \mathbf{v}_{ij}$$

models such a multi-sensor-matrix camera. λ is the unknown scale factor.

Image rectification may start with a mapping on an assumed ground plane (obviously, the Earth's surface is not a plane, but the plane assumption allows an initial rectification), which includes the consideration of the exterior attitude at the time when the individual line images were taken. The projection formulas for this mapping are given by

$$x' = \left(\mathbf{t}_{x_0} + \frac{h \cdot \mathbf{v}'_x}{\mathbf{v}'_z}\right)\frac{1}{\tau'}$$

and

$$y' = \left(\mathbf{t}_{y_0} + \frac{h \cdot \mathbf{v}'_y}{\mathbf{v}'_z}\right)\frac{1}{\tau'},$$

where τ' is the targeted resolution on the ground plane, and $h = H - t_{z_0}$, where H is the assumed altitude of the ground plane (typically the mean height of the recorded terrain). In fact, h is the distance from the projection center (of the camera) to the ground plane. By projecting pixels to this ground plane (i.e., basically replacing λ by a general distance value H) we cause perspective "inaccuracies", similar to the central perspective involved when capturing a "normal" sensor-matrix image.

---

[2] A modular MFC concept allows a choice of up to five different CCD lines of different pixel counts or different filters; the standard MFC has five 14K RGB lines which look straight downward (nadir line), two lines looking forward, and two lines looking backward, with $\omega_1 = \pm 10°$ and $\omega_2 = \pm 20°$. All five lines are behind the same optics, with a maximum image size of 80×80 mm. The camera design is very compact such that only one ethernet cable is necessary to control the camera from a laptop. All electronics, including mass memory, are included in the camera's body such that no separate boxes or cables are needed. The compact design allows a small size of $20 \times 30 \times 30$ cm^3 and a maximum weight of 20 kg if all five CCD channels (modules) are contained.

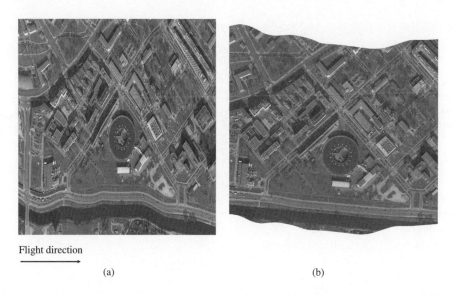

Flight direction

(a)   (b)

Figure 10.10   Rectification of (a) image data recorded by an airborne CCD line camera (showing the circular Synchrotron at Berlin-Adlershof and an approximately straight multi-lane road); the arrow shows the flight direction. (b) The results use data provided by an initial measuring system.

The challenge is that an individual projection center needs to be used for each captured line. This defines "quasi"-orthogonal projection along the flight path, but perspective projection perpendicular to the flight path for this kind of mapping. Figure 10.10(a) shows an unrectified image recorded by using a CCD line scanning approach, and Figure 10.10(b) shows the rectified image. The upper or lower border of the rectified image illustrates the corrected roll movement of the airplane perpendicular to the flight direction.

Second, the unknown scale factor λ may be determined by stereo processing of the data. This may result in a DSM. Plate 16 shows a processed orthophoto based on HRSC data.

To use the HRSC data it is also necessary to filter the data as described in Section 9.3.4. The orthophoto can be used for supporting edge detection (e.g., separation of roofs). Figure 10.11 illustrates the positive impact of using filtered HRSC data, where filtering is based on triangulated surfaces, as described before for "indoor scenes".

10.4.3 Mapping of Terrestrial Side-View Images

There is currently much active research in progress into generating 3D maps, showing (for example) cities not only as 2.5D top views, but also more details at close range, for example by means of fly-through visualization.

The 2.5D representation of a top-view DSM allows each point to be labeled by its elevation above ground level. The 2.5D data set is thus transformed into an initial 3D model. Points which are above the same grid element (i, j) are represented at this step by the highest point (i.e., the largest z-value). A cavity (e.g., the area below a balcony), as sketched in Figure 10.12, is not yet represented in this initial 3D model.

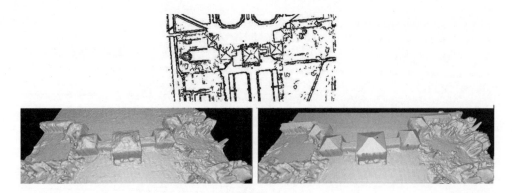

Figure 10.11 Left: visualization of a 3D model (Schloss Nymphenburg, Munich) created from a DSM (calculated from HRSC image data). Right: the same, but now based on filtered HRSC data using planar surface patches, also using an orthophoto (as shown on the top) for semi-automatic surface detection.

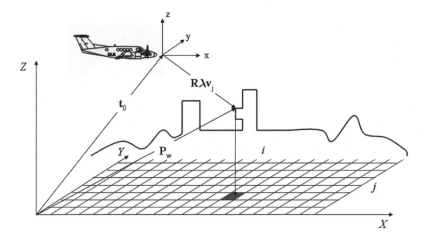

Figure 10.12 Illustration of an airborne camera, with coordinates and mappings, and a sketch of a DSM which models a top view.

For a further improvement of the initial 3D model, terrestrial side-view images, such as panoramic images (see Plate 17), can be used, which allow calculations of a side-view DSM. These side-view DSMs are then integrated into the 3D model of the scene. The geometry generated also supports proper texturing, using the same procedures as described in Section 10.2.

Resolution and accuracy of the initial 3D model depend on the altitude of the airplane, sensor properties, and the quality of the initial measuring unit. Differences in accuracies increase difficulties in merging side-view DSMs into the top-view DSM. It is becoming standard for side-view DSMs to be supported by LRF data.

For a simplified mapping approach, we assume at first that all facades of buildings in one particular section of a road are planar, which defines a ground plane. Note that each object, which actually has a non-zero distance to this plane (e.g., bays, window indentations, or balconies), would be projected incorrectly. Using this simple model, rectangular facades may be addressed by two points (i.e., two corner points of such a facade) in the ground plane (say, bottom left and bottom right), and then extruded by one z-value (with respect to the assumed ground plane of the facades). This specifies a projection plane, which is parallel to the ground plane.

Single textures (say, of one house, or a part of a larger building) are calculated for color mapping such that these are conformal with a perspective projection. Therefore, recorded sideview panoramic images have to be be aligned with captured top-view data (see Section 3.7). Attitudes may be just approximate because landmarks used as control points in the top-view data do not (in general) have the necessary accuracy to obtain accuracy comparable to, for example, a calibration area. The remaining error (e.g., a small tilt) will be removed by a perspective transformation without knowledge of the attitude. A priori information (e.g., windows are expected to form right angles) can be used for this transformation. We used a transformation specified by the equations

$$i' = a_1 \cdot i + b_1 \cdot j + c_1 \cdot i \cdot j + d_1$$

and, analogously, for j',

$$j' = a_2 \cdot i + b_2 \cdot j + c_2 \cdot i \cdot j + d_2.$$

Coefficients in these equations are determined by using at least four control points (e.g., corners on facades forming a rectangle). Overdetermination is also possible using least-squares optimization for improving accuracy. The rectification of a facade is illustrated in Figure 10.13.

10.4.4 Mapping of Perspective Airborne Images

Assuming aerial imaging with multiple sensor-lines, the different viewing angles of those lines actually allow the texturing process of facades to be enhanced.

The nadir, backward, or forward lines not only prove useful for stereo reconstruction, they also provide a particular contribution to color texture as seen from different angles. Forward and backward sensor-lines do have the benefit that they also show side views of buildings. Similar to a matrix sensor, pixels further away from the projection center are better suited for texturing.

An important benefit is also that image data of the airborne sensor do have the same accuracy, resolution and perspective projection as used for generating the top-view geometry model. This enhances the photorealistic 3D impression. Figure 10.14 shows a textured top-view, and

Figure 10.13 Rectification of a facade. Left: before. Right: after.

Figure 10.14 Textured 2.5D top view of the scene shown in Figure 10.10 (see Plate 40).

Figure 10.15 Textured 3D view of the scene shown in Figures 10.10 and 10.14 (see Plate 41).

Figure 10.15 an enhanced 3D view, only using textures recorded by nadir sensor line (and not those recorded by forward or backward lines) during the same imaging campaign (using the MFC in 2007, with a potential ground-sample resolution of less than 10 cm).

Plates 18 and 19 illustrate 3D textured city scenes (of Berlin); a view of the area around the Sony Center, and a close-up on the Reichstag building. In those cases, a matrix sensor was used, and textures recorded close to the border of the sensor were also used for improved texturing (compare with Figure 10.15).

10.5 Exercises

10.1. Verify the model equations for the three different panoramic sensor categories, by calculating (synthetic) object points as a function of $\mathbf{P}_w = F(i, j, \lambda)$ in 3D space, and by then determining the resulting image coordinates as a function of $i/j = F(\mathbf{P}_w)$. Use a λ of your choice when calculating the test object points.

10.2. [possible lab project] This exercise is about pose estimation of a panoramic sensor.

(1) For this purpose, use your screen dumps of Exercise *9.4*(3). Calculate image coordinates with subpixel accuracy for provided control points. Control points (with object coordinates) can be achieved within the VRML model, or by using the map which is also included in the data provided on the website accompanying this book. Calculate the camera's attitude using a standard pinhole model of a central perspective camera model (i.e., a planar capturing surface).

(2) Now use the panoramic images provided, which are ideal cylindrical panoramas (i.e., $R = 0$ and $\omega = 0$), for camera pose estimation.

10.3. [Possible lab project] This exercise is about mapping texture coordinates onto the VRML model. Use the untextured test room from Exercise *9.4*(2).

For each vertex, calculate the corresponding image coordinate, with given camera attitude from Exercise *10.2*(1); save the texture coordinates.

Load the VRML file, now with the saved texture coordinates, as in Exercise *9.4*. When mapping texture coordinates, use multiple camera locations.

10.4. [Possible lab project] Modify Exercise *10.3*, now also including a raytracing check as follows:

(1) Use conventional octree tracing (i.e., each mode has up to eight successors).

(2) Use the stencil buffer test.

Evaluate both raytracing algorithms with respect to quality achieved and computing time.

10.5. [Possible lab project] Use the data available for this exercise on the book's website for rectifying an airborne image which was recorded with a CCD line camera. For each line, the attitude of the camera in world coordinates is given in a separate attitude file. Project those lines to ground planes of different (assumed) heights and compare results on "geometric correctness"; therefore use a priori knowledge, for example the circular building as shown in figures in this chapter. Obviously, different ground plane heights do not only lead to a different scaling of the image. Explain.

10.6 Further Reading

Differences between single- and multi-projection-center panoramas have been analyzed in Huang et al. (2001b).

The HRSC sensor is discussed in Wewel et al. (2000), Scholten et al. (2002) and Scholten and Gwinner (2004).

For stereo processing of aerial images, see Hirschmüller et al. (2005). Their correspondence algorithm uses a semi-global matching strategy. Plate 16 is from Hirschmüller et al. (2005); it shows a processed orthophoto based on HRSC data, and Plates 18 and 19 show processed 3D views based on UltraCam data.

Augmented panoramas (for visualizing 3D scenes) are discussed in Chen et al. (2006); for 3D interaction based on augmented panoramas, see Chan et al. (2005).

The 3D modeling process in general is reviewed in Scheibe et al. (2006), and this paper also contains many references to related work. Route panoramas (see, Shi and Zheng, 2003) are used for texturing facades of buildings.

References

Albertz, J., and W. Kreiling (1989) *Photogrammetric Guide*, 4th edition. Wichmann, Karlsruhe.

Albertz, J. and A. Wiedemann (1996) From analogue to digital close-range photogrammetry. In *Proc. Turkish-German Joint Geodetic Days*, pp. 245–253.

Anandan, P., P. J. Burt, K. Dana, M. Hansen, and G. van der Wal (1994) Real-time scene stabilization and mosaic construction. In *Proc. IEEE Works. Applications Computer Vision*, pp. 54–62.

Ayache, N. J., and B. Faverjon (1987) Efficient registration of stereo images by matching graph descriptions of edge segments. *Int. J. Computer Vision*, **1**:107–132.

Baker, H. (2004) *Computer Graphics with OpenGL*. 3rd edition. Prentice Hall, Upper Saddle River, NJ.

Baker, S., and S. K. Nayar (1999) A theory of single-viewpoint catadioptric image formation. *Int. J. Computer Vision*, **35**:1–22.

Barnard, S. T., and M. A. Fischler (1982) Computational stereo. *ACM Comput. Surv.*, **14**:553–572.

Benosman, R., and S. B. Kang, (eds) (2001) *Panoramic Vision: Sensors, Theory, and Applications*. Springer, New York.

Blinn, J. (1978) Simulation of wrinkled surfaces. In *Proc. SIGGRAPH*, pp. 286–292.

Blundell, B. G. (2007) *Enhanced Visualization. Making Space for 3-D Images*. Wiley-Interscience, Hoboken, NJ.

Bodenmueller, T., and G. Hirzinger (2004) Online surface reconstruction from unorganized 3d-points for the DLR hand-guided scanner system. In *Proc. 3D Data Processing, Visualization, and Transmission*, pp. 285–292.

Boehler, W., M. B. Vincent, and A. Marbs (2003) Investigating laser scanner accuracy. In *Proc. Int. Symp. CIPA*, pp. 696–701.

Börner, A., H. Hirschmüller, K. Scheibe, M. Suppa, and J. Wohlfeil (2008) MFC - A modular line camera for 3D world modelling. In *Proc. Robot Vision*, LNCS 4931, pp. 319–326. Springer, Berlin.

Bourke, P. (2007) web pages on stereographics. *http://astronomy.swin.edu.au/pbourke/* (last accessed May 2007).

Burt, P. J., and E. H. Adelson (1983) A multiresolution spline with applications to images mosaics. *ACM Trans. Graphics*, **2**:217–236.

Burtch, R. (2007) History of photogrammetry. *www.ferris.edu/faculty/burtchr/sure340/notes/History.pdf* (last accessed June 2008).

Chan, L.-W., Y.-Y. Hsu, Y.-P. Hung, and J. Y.-J. Hsu (2005) A panorama-based interface for interacting with the physical environment using orientation-aware handhelds. In *Proc. Int. Conf. Ubiquitous Computing*.

Chen, C.-W., L.-W. Chan, Y.-P. Tsai, and Y.-P. Hung (2006) Augmented stereo panoramas. In *Proc. Asian Conf. Computer Vision*, LNCS 3581, pp. 41–49. Springer, Berlin.

Chen, C.-Y., and R. Klette (1999) Image stitching – comparisons and new techniques. In *Proc. Computer Analysis of Images and Patterns*, LNCS 1689, pp. 615–622. Springer.

Chen, S. E. (1995) QuickTimeVR – an image-based approach to virtual environment navigation. In *Proc. SIGGRAPH*, pp. 29–38.

Cox, I. J. (1994) A maximum likelihood n-camera stereo algorithm. In *Proc. Computer Vision Pattern Recognition*, pp. 733–739.

Coxeter, H. S. M. (2003) *Projective Geometry*, 2nd edition. Springer, New York.

Crow, F. C. (1977) Shadow algorithms for computer graphics, parts 1 and 2. In *Proc. SIGGRAPH*, volume 11-2, pages 242–248 and 442–448.

Daniilides, K., and R. Klette (eds) (2006) *Computer Vision Beyond the Pinhole Camera*. Springer, Amsterdam.

Daniilides, K., and N. Papanikolopoulos (eds) (2004) *Panoramic Robots*. Special issue of *IEEE Robotics Automation Magazine*, Volume 11, December.

Däumlich, F., and R. Steiger, (eds) (2002) *Instrumentenkunde der Vermessungstechnik*. H. Wichmann, Heidelberg.

Davis, A., and P. Stanbury (1985) *The Mechanical Eye in Australia. Photography 1841–1900*. Oxford University Press, Melbourne.

Davis, J.E. (1998) Mosaics of scenes with moving objects. In *Proc. Computer Vision Pattern Recognition*, pp. 354–360.

Doyle, F. (1964) The historical development of analytical photogrammetry. *Photogrammetric Engineering*, **XXX**(2): 259–265.

Evers-Senne, J.-F., J. Woetzel, and R. Koch (2004) Modelling and rendering of complex scenes with a multi-camera rig. In *Proc. European Conf. Visual Media Production*, pp. 11–19.

Faugeras, O., and Q.-T. Luong (2001) *The Geometry of Multiple Images*. MIT Press, London.

Gernsheim, H. (1982) *The Origins of Photography*. Thames and Hudson, New York.

Gibson, J. J. (1950) *The Perception of the Visual World*. Houghton Mifflin, Boston.

Gill, P. E., W. Murray, and M. H. Wright (1981) *Practical Optimization*. Academic Press, London.

Gouraud, H. (1971) Illumination for computer generated pictures. *Comm. ACM*, **18**:311–317.

Gluckman, J., and S. K. Nayar (1999) Planar catadioptric stereo: Geometry and calibration. In *Proc. Computer Vision Pattern Recognition*, Volume I, pp. 22–28.

Greene, N. (1986) Environment mapping and other applications of world projections. *IEEE Trans. Computer Graphics Applications*, **6**:21–29.

Griesbach, D., A. Börner, M. Scheele, K. Scheibe, and S. Sujew (2004) Line scanner in combination with inertial measurement unit. In *Proc. Panoramic Photogrammetry Workshop*, Volume XXXIV, pp. 19–22.

Gruen, A., and T.S. Huang (eds) (2001) *Calibration and Orientation of Cameras in Computer Vision*, *Informations Sciences*, Volume 34. Springer, Berlin.

Gupta, R., and R. Hartley (1997) Linear pushbroom cameras. *IEEE Trans. Pattern Analysis Machine Intelligence*, **19**:963–975.

Hartley, R. and A. Zisserman (2000) *Multiple View Geometry in Computer Vision*. Cambridge Univ. Press, Cambridge.

Hecht, E. (2002) *Optics*. Addison-Wesley, Reading, MA.

Hecht, E., and A. Zajac (1974) *Optics*. Addison-Wesley, New York.

Heinz, I., F. Haertl, and C. Fröhlich (2001) Semi-automatic 3D CAD model generation of as-built conditions of real environments using a visual laser radar. In *Proc. IEEE Int. Workshop Robot-Human Interactive Communication*, pp. 400–406.

Hilbert, D., and S. Cohn-Vossen (1932) *Anschauliche Geometrie*. Springer-Verlag, Berlin.

Hill, F. S., Jr., and S. M. Kelley (2007) *Computer Graphics using OpenGL*. 3rd edition. Prentice Hall, Upper Saddle River, NJ.

Hirschmüller, H., F. Scholten, and G. Hirzinger (2005) Stereo vision based reconstruction of huge urban areas from an airborne pushbroom camera HRSC. In *Proc. Pattern Recognition (DAGM)*, LNCS 3663, pp. 58–66, Springer, Berlin.

Hornberg, A. (ed.) (2006) *Handbook of Machine Vision*. Wiley-VCH, Weinheim.

Howard, I. P., and B. J. Rogers (1995) *Binocular Vision and Stereopsis*. Oxford University Press, New York.

Huang, F., and T. Pajdla (2000) Epipolar geometry in concentric panoramas. Research Report CTU–CMP–2000–07, Center for Machine Perception, Czech Technical University, Prague.

Huang, F., R. Klette, S. K. Wei, A. Börner, R. Reulke, M. Scheele, and K. Scheibe (2001a) Hyper-resolution and polycentric panorama acquisition and experimental data collection. Technical Report, CITR-TR-90, Dept. of Computer Science, University of Auckland.

Huang, F., S. K. Wei, and R. Klette (2001b) Geometrical fundamentals of polycentric panoramas. In *Proc. Int. Conf. Computer Vision*, Volume I, pages 560–565.

Huang, F., S. K. Wei, and R. Klette (2001c) Stereo reconstruction from polycentric panoramas. In *Proc. Robot Vision*, LNCS 1998, pp. 209–218. Springer, Berlin.

Huang, F., S. Wei, and R. Klette (2002) Calibration of line-based panoramic cameras. In *Proc. Image Vision Computing New Zealand*, pp. 107–112.

Huang, H.-C., and Y.-P. Hung (1997) SPISY: the stereo panoramic imaging system. In *Proc. Workshop Real-Time Media Systems*, pp. 71–78, Taipei.

Huang, H.-C., and Y.-P. Hung (1998) Panoramic stereo imaging system with automatic disparity warping and seaming. *Graph. Models Image Processing*, **60**:196–208.

Hung, Y.-P., C.-S. Chen, Y.-P. Tsai, and S.-W. Lin (2002) Augmenting panoramas with object movies by generating novel views with disparity-based view morphing. *J. Visualization Computer Animation*, **13**:237–247.

Irani, M., P. Anandan, and S. Hsu: Mosaic based representations of video sequences and their applications. In *Proc. Int. Conf. Computer Vision*, pp. 605–611.

Ishiguro, H., M. Yamamoto, and S. Tsuji (1992) Omni-directional stereo. *IEEE Trans. Pattern Analysis Machine Intelligence*, **14**:257–262.

Jiang, W., M. Okutomi, and S. Sugimoto (2006) Panoramic 3D reconstruction using rotational stereo camera with simple epipolar constraints. In *Proc. Computer Vision Pattern Recognition*, Volume I, pp. 371–378

Johansson, G., and E. Börjesson (1989) Toward a new theory of vision studies in wide-angle space perception. *Ecological Psychology*, **1**:301–331.

Kanatani, K. (1993) *Geometric Computation for Machine Vision*. Oxford University Press, New York.

Kang, S.-B., and P. K. Desikan (1997) Virtual navigation of complex scenes using clusters of cylindrical panoramic images. Technical Report, CRL 97/5, Digital Equipment Corporation, Cambridge Research Laboratory.

Kang, S.-B., and R. Szeliski (1997) 3-d scene data recovery using omnidirectional multibaseline stereo. *Int. J. Computer Vision*, **25**:167–183.

Kang, S.-B., and R. Weiss (1997) Characterization of errors in compositing panoramic images. In *Proc. Computer Vision Pattern Recognition*, pp. 103–109.

Kangni, F., and R. Laganiere (2006) Epipolar geometry for the rectification of cubic panoramas. In *Proc. Canadian Conference Computer Robot Vision*, pages 70–79.

Keam, R. F. (1998) *Tarawera - The Volcanic Eruption of 10 June 1886*. Author, Auckland.

Kelvin of Largs, Lord (William Thompson) (1889) Lecture on "Electrical Units of Measurement" (delivered 1883). in *Popular Lectures and Addresses*, Volume I, pp. 73–74. Macmillan, London.

Klette, R., and A. Rosenfeld (2004) *Digital Geometry. Geometric Methods for Digital Picture Analysis*. Morgan Kaufmann, San Francisco.

Klette, R., K. Schlüns, and A. Koschan (1998) *Computer Vision: Three-Dimensional Data from Images*. Springer, Singapore.

Koetzle, H.-M. (1995) *Photo Icons. Die Geschichte hinter den Bildern. 1827–1991*. Benedikt Taschen, Cologne.

Kraus, K. (1997) *Photogrammetrie*, 6th edition. Dümmler, Bonn.

Kumar, R., P. Anandan, M. Irani, J. Bergen, and K. Hanna (1995) Representation of scenes from collections of images. In *Proc. IEEE Works. Representation Visual Scenes*, pp. 10–17.

Laveau, S., and O. Faugeras (1994) 3-d scene representation as a collection of images. In *Proc. Int. Conf. Pattern Recognition*, pp. 689–691.

Lee, A. (2000) Building your own subdivision surfaces. *http://gamasutra.com/features/20000908/lee_01.htm* (last accessed May 2008).

Lee, M. C., W. G. Chen, C. L. B. Lin, C. Gu, T. Markoe, S. I. Zabinsky, and R Szcliski (1997) A layered video object coding system using sprite and affine motion model. *IEEE Trans. Circuits Systems Video Technology*, **7**:130–145.

Lee, S. E. and W. Fong (1955) Streams and mountains without end: A Northern Sung handscroll and its significance in the history of early Chinese painting. *Harvard J. Asiatic Studies*, **18**: 494–497.

Lenhardt, K. (2005) Optical measurement techniques with telecentric lenses. *www.schneiderkreuznach.com/pdf/div/* (last accessed June 2008).

Liu, G., R. Klette, and B. Rosenhahn (2005) Structure from motion in the presence of noise. In *Proc. Image Vision Computing New Zealand*, pp. 138–143.

Liu, Y.-C., K.-Y. Lin, and Y.-S. Chen (2008) Bird's eye view vision system for vehicle surrounding monitoring. In *Proc. Robot Vision*, LNCS 4931, pp. 207–218. Springer, Berlin.

Luhmann, T. (2003) *Nahbereichsphotogrammetrie*. Wichmann, Heidelberg.

Luhmann, T., S. Robson, S. A. Kyle, and I. A. Harley (2006) *Close Range Photogrammetry: Principles, Techniques and Applications*. Whittles, Dulwich, South Australia, 2006.

Mann, S., and R. W. Picard (1994) Virtual bellows: Constructing high quality stills from video. In *Proc. IEEE Conf. Image Processing*, Volume I, pp. 363–367.

Marr, D., and T. A. Poggio (1979) A computational theory of human stereo vision. *Proc. R. Soc. Lond. B.*, **204**:301–328.

Mayer, U., M. D. Neumann, W. Kubbat, and K. Landau (2000) Is eye damage caused by stereoscopic displays? In *Proc. SPIE Stereoscopic Displays Virtual Reality Systems*, pp. 4–11.

McMillan, L., and G. Bishop (1995) Plenoptic modeling: an image-based rendering system. In *Proc. SIGGRAPH*, pp. 39–46.

Murray, D.W. (1995) Recovering range using virtual multicamera stereo. *Computer Vision Image Understanding*, **61**:285–291.

Nene, S. A., and S. K. Nayar (1998) Stereo with mirrors. In *Proc. Int. Conf. Computer Vision*, pp. 1087–1094.

Ohta, Y., and T. Kanade (1985) Stereo by intra- and inter-scanline search using dynamic programming. *IEEE Trans. Pattern Analysis Machine Intelligence*, **7**:139–154.

Peleg, S., and M. Ben-Ezra (1999) Stereo panorama with a single camera. In *Proc. Computer Vision Pattern Recognition*, pp. 395–401.

Peleg, S., and J. Herman (1997) Panoramic mosaic by manifold projection. In *Proc. Computer Vision Pattern Recognition*, pp. 338–343.

Peleg, S., Y. Pritch, and M. Ben-Ezra (2000) Cameras for stereo panoramic imaging. In *Proc. Computer Vision Pattern Recognition*, pp. 208–214.

Petty, R., M. Robinson, and J. Evans (1998) 3D measurement using rotating line-scan sensors. *Measurement Science Technology*, **9**:339–346.

Phong, B. (1975) Illumination for computer generated pictures. *Comm. ACM*, **18**:311–317.

Priest, A. (1950) Southern Sung landscapes: The hanging scrolls. *Metropolitan Museum of Art Bulletin*, New Series, **8** (6): 168–175.

Rademacher, R., and G. Bishop (1998) Multiple-center-of-projection images. In *Proc. SIGGRAPH*, pp. 199–206.

Rav-Acha, A., V. Pritch, D. Lischinski, and S. Peleg (2005) Dynamosaics: video mosaics with non-chronological time. In *Proc. Computer Vision Pattern Recognition*, Volume 1, pp. 58–65.

Rayleigh, J. W. Strutt, third Baron (1891) On pin-hole photography. *Philosophical Magazine*, **31**:87–99.

Reulke, R., and M. Scheele (1998) Der Drei-Zeilen CCD-Stereoscanner WAAC: Grundaufbau und Anwendungen in der Photogrammetrie. *Photogrammetrie Fernerkundung Geoinformation*, **3**:157–163.

Reulke, R., A. Wehr, R. Klette, M. Scheele, and K. Scheibe (2003) Panoramic mapping using CCD-line camera and laser scanner with integrated position and orientation system. In *Proc. Image Vision Computing New Zealand*, pp. 72–77.

Roberts, J. W., and O. T. Slattery (2000) Display characteristics and the impact on usability for stereo. In *Proc. SPIE Stereoscopic Displays Virtual Reality Systems*, pp. 128–137.

Sandau, R., and A. Eckardt (1996) The stereo camera family WAOSS/WAAC for spaceborne/airborne applications. *Int. Archives Photogrammetry Remote Sensing*, **XXXI(B1)**:170–175.

Šára, R. (1999) The class of stable matchings for computational stereo. Technical Report CTU-CMP-1999-22, Center for Machine Perception, Czech Technical University, Prague.

Sawhney, H. S., and S. Ayer (1996) Compact representations of videos through dominant and multiple motion estimation. *IEEE Trans. Pattern Analysis Machine Intelligence*, **18**:814–830.

Scheele, M., H. Jahn, and R. Schuster (2004) Geometrische Kalibration von CCD Kameras mittels Beugungsbildern. In *Publikationen der DGPF*, Volume 13, pp. 513–518.

Scheibe, K., H. Korsitzky, R. Reulke, M. Scheele, and M. Solbrig (2001) EYESCAN – a high resolution digital panoramic camera. In *Proc. Robot Vision*, LNCS 1998, pp. 77–83. Springer, Berlin.

Scheibe, K., M. Suppa, H. Hirschmüller, B. Strackenbrock, F. Huang, R. Lui, and G. Hirzinger (2006) Multi-scale 3D-modeling. In *Proc. Advances in Image and Video Technology*, LNCS 4319, pp. 96–107. Springer, Berlin.

Schneider, D., and H.-G. Maas (2004) Development and application of an extended geometrical model for high resolution panoramic cameras. In *Proc. ISPRS Geo-Imagery Bridging Continents*, pp. 366–371.

Schwidefsky, K. (1971) Albrecht Meydenbauer – Initiator der Photogrammetrie in Deutschland. *Bildmessung Luftbildwesen*, **39**:183–189.

Scholten, F., and G. Gwinner (2004) Operational parallel processing in digital photogrammetry – strategy and results using different multi-line cameras. In *Proc. Int. Congr. Photogrammetry Remote Sensing*, pp. 408–413.

Scholten, F., K. Gwinner, and F. Wewel (2002) Angewandte digitale Photogrammetrie mit der HRSC. *Photogrammetrie Fernerkundung Geoinformation*, **5**:317–332.

Schulz, T., and H. Ingensand (2004) Influencing variables, precision and accuracy of terrestrial laser scanners. In *Proc. Ingeo*, Bratislava, p. 8.

Seitz, S.M. (2001) The space of all stereo images. In *Proc. Int. Conf. Computer Vision*, pp. 26–33.

Sgrenzaroli, M. (2005) Cultural heritage 3D reconstruction using high resolution laser scanner: New frontiers data processing. In *Proc. Int. Symp. CIPA*, pp. 544–549.

Shi, M., and J. Y. Zheng (2003) Spatial resolution analysis of route panorama. In *Proc. Int. Conf. Image Processing*, Volume II, pp. 311–314.

Shum, H.-Y., and L.-W. He (1999) Rendering with concentric mosaics. In *Proc. SIGGRAPH*, pp. 299–306.

Shum, H.-Y., and R. Szeliski (1999) Stereo reconstruction from multiperspective panoramas. In *Proc. Int. Conf. Computer Vision*, pp. 14–21.

Shum, H.-Y., M. Han, and R. Szeliski (1998) Interactive construction of 3D models from panoramic mosaics. In *Proc. Computer Vision Pattern Recognition*, pp. 427–433.

Shum, H., A. Kalai, and S. Seitz (1999) Omnivergent stereo. In *Proc. Int. Conf. Computer Vision*, pp. 22–29.

Siegel, M., and S. Nagata (2000) Just enough reality: comfortable 3-D viewing via microstereopsis. *IEEE Trans. Circuits Systems Video Technology*, **10**:387–396.

Siegel, M., Y. Tobinaga, and T. Akiya (1999) Kinder gentler stereo. In *Proc. SPIE Stereoscopic Displays and Applications*, pp. 18–27.

Silicon Graphics (2007) *www.opengl.org* (last accessed May 2007).

Soper, A. C. (1941) Early Chinese landscape painting. *Art Bulletin*, **23** (2): 141–164.

Southwell, D., J. Reyda, M. Fiala, and A. Basu (1996) Panoramic stereo. In *Proc. Int. Conf. Pattern Recognition*, Volume A, pp. 378–382.

Stelmach, L., W. J. Tam, D. Meegan, and A. Vincent (2000) Stereo image quality: effects of mixed spatio-temporal resolution. *IEEE Trans. Circuits Systems Video Technology*, **10**:188–193.

Sugaya, Y., and K. Kanatani (2007) Highest accuracy fundamental matrix computation. In *Proc. Asian Conf. Computer Vision*, Volume II, LNCS 4844, pp. 311–321. Springer, Berlin.

Suto, M. (2008) Stereo panoramas. *www.stereomaker.net/panorama/panoramae.htm* (last accessed June 2008).

Szeliski, R. (2006) *Image Alignment and Stitching – A Tutorial*. World Scientific, Hackensack, NJ.

Valyrus, N. A. (1966) *Stereoscopy*. Focal Press, London.

Viire, E. (1997) Health and safety issues for VR. *Comm. ACM*, **40**:40–41.

Ware, C., C. Gobrecht, and M. Paton (1998) Dynamic adjustment of stereo display parameters. *IEEE Trans. Systems Man Cybernetics*, **28**:56–65.

Watt, A. (2002) *3D-Computergrafik*. Pearson Studium, Munich.

Wei, S. K., F. Huang, and R. Klette (1998) Color anaglyphs for panorama visualizations. Technical Report, CITR-TR-19, Dept. of Computer Science, University of Auckland, Auckland.

Wei, S. K., F. Huang, and R. Klette (2002a) The design of a stereo panorama camera for scenes of dynamic range. In *Proc. Int. Conf. Pattern Recognition*, Volume III, pp. 635–638.

Wei, S. K., F. Huang, and R. Klette (2002b) Determination of geometric parameters for stereoscopic panorama cameras. *Machine Graphics & Vision*, **10**:399–427.

Wewel, F., F. Scholten, and K. Gwinner (2000) High resolution stereo camera (HRSC) – multispectral 3D-data acquisition and photogrammetric data processing. *Canadian J. Remote Sensing*, **26**:466–474.

Williams, L. P. (1978) *Album of Science: Nineteenth Century*. Charles Scribner's Sons, New York.

Willsberger, J. (1977) *The History of Photography. Cameras, Pictures, Photographers*. Doubleday & Company, Garden City, NY.

Wright, R. S., Jr., and B. Lipchak (2005) *OpenGL Superbible*, 3rd edition. SAMS Publishing, Indianapolis, IN.

Xu, G., and Z. Zhang (1996) *Epipolar Geometry in Stereo, Motion and Object Recognition – A Unified Approach*. Kluwer, Amsterdam.

Yamanoue, H., M. Okui, and I. Yuyama (2000) A study on the relationship between shooting conditions and cardboard effect of stereoscopic images. *IEEE Trans. Circuits Systems Video Technology*, **10**:411–416.

Zhang, X., J. Morris, and R. Klette (2005) Volume measurement using a laser scanner. In *Proc. Int. Conf. Image Vision Computing New Zealand*, pp. 177–182.

Zhang, Z., and G. Xu (1997) A general expression of the fundamental matrix for both perspective and affine cameras. In *Proc. Int. Joint Conf. Artificial Intelligence*, pp. 23–29, Nagoya.

Zheng, J.-Y., and S. Tsuji (1992) Panoramic representation for route recognition by a mobile robot. *Int. J. Computer Vision*, **9**:55–76.

Zhu, J., G. Humphreys, D. Koller, S. Steuart, and R. Wang (2007) Fast omnidirectional 3D scene acquisition with an array of stereo cameras. In *Proc. Int. Conf. 3-D Digital Imaging Modeling*, pp. 217–224.

Zhu, Z., L. Zhao, and J. Lei (2005) 3D measurements in cargo inspection with a gamma-ray linear pushbroom stereo system. In *Proc. IEEE Workshop Advanced 3D Imaging Safety Security*, online volume.

Index